Handbook of Antennas and Propagation Technology

Volume I

Handbook of Antennas and Propagation Technology
Volume I

Edited by **Edmond Thor**

NY RESEARCH
PRESS

New York

Published by NY Research Press,
23 West, 55th Street, Suite 816,
New York, NY 10019, USA
www.nyresearchpress.com

Handbook of Antennas and Propagation Technology: Volume I
Edited by Edmond Thor

International Standard Book Number: 978-1-63238-224-5 (Hardback)

Printed in China.

Contents

Preface

Antennas are electrical devices which convert electric power into radio waves and vice versa. Antennas are usually requisite while using any type of radio receiver or transmitter to link itself to the electromagnetic field. They are usually coupled with radio transmitters or receivers. Antennas can be distinguished by a number of execution measures which a user would be involved with in opting for or designing antennas for a particular application in the real world. One can say that antennas are crucial components of all equipments that use radio. They are utilized in applications like two-way radio, radio broadcasting, communications receivers, broadcast television, radar, satellite communications and cell phones as well as other day to day devices such as wireless microphones, bluetooth-enabled devices, RFID tags on merchandise, wireless computer networks and even baby monitors. Radio propagation is the behavior of radio waves when they are broadcast or propagated from one point to another over land or into various parts of the atmosphere and one can say that antennas play a major role in such propagation. As a type of electromagnetic radiation, radio waves are influenced by the phenomena of refraction, reflection, absorption, diffraction, scattering and polarization. Grasping the effects of differing conditions on propagation has many practical applications, such as choosing frequencies for international shortwave broadcasters, radio navigation, designing reliable mobile telephone systems and operation of radar systems. Though there are also various services like emergency locator transmitters, in-flight communication with ocean-crossing aircraft, and television broadcasting that have been moved to communications satellites. Skilled knowledge of antennae and propagation is required in this age of constant connectivity and thus the demand for new innovations is also on the rise.

This book is an attempt to compile and collate all available research on antennas and propagation under one umbrella. I am grateful to those who put their hard work, effort and expertise into these researches as well as those who were supportive in this endeavour. I wish to thank the publishing team for providing excellent technical assistance, as and when required.

Editor

Evaluation of the Inductive Coupling between Equivalent Emission Sources of Components

Moisés Ferber,[1] Sanâa Zangui,[1] Carlos Sartori,[2] Christian Vollaire,[1] Ronan Perrussel,[3] and Laurent Krähenbühl[1]

[1] Laboratoire Ampère, CNRS UMR5005, Ecole Centrale de Lyon, Université de Lyon, 69130 Ecully, France
[2] Laboratorio de Eletromagnetismo Aplicado, Universidade de São Paulo, 05508900 São Paulo, SP, Brazil
[3] LAPLACE, CNRS UMR5213, INP-ENSEEIHT, Université de Toulouse, 31071 Toulouse, France

Correspondence should be addressed to Moisés Ferber, moisesferber@gmail.com

Academic Editor: Dragan Poljak

The electromagnetic interference between electronic systems or between their components influences the overall performance. It is important thus to model these interferences in order to optimize the position of the components of an electronic system. In this paper, a methodology to construct the equivalent model of magnetic field sources is proposed. It is based on the multipole expansion, and it represents the radiated emission of generic structures in a spherical reference frame. Experimental results for different kinds of sources are presented illustrating our method.

1. Introduction

The development of semiconductor technology in the last decades has greatly increased the use of power electronics in various applications, such as computer power supplies, voltage converters, electronic ballasts, and variable-speed drives [1]. Recently, new applications of power electronics have also appeared in the vehicle industry, such as electric cars and airplanes. However, the commutation of the switches (rectifiers, SCRs and triacs, BJTs, MOSFETs, and IGBTs) generates high currents with high *di/dt*, and, thus, a wide bandwidth of unwanted electromagnetic interference (EMI) pollutes the electromagnetic environment [2].

The electromagnetic compatibility (EMC) is an engineering domain responsible to ensure that systems, equipment, and devices can coexist satisfactorily in the same electromagnetic environment [3]. Electric cars, for instance, may encounter malfunction in its electronic systems (ESP, ABS, ALS, etc) if special care is not taken. The EMI between the cables of the power electronics and the cables carrying electronic signals, if they are too close to each other without proper shielding, may prevent the correct operation of certain systems [4].

There are not many reliable methods to predict the EMC of a complex system in the design phase [5], and, thus, in practice, the EMC design is still carried out by trial and error [3] causing high development cost in case of malfunctioning due to EMI, when the prototype is tested.

To ensure the compatibility of cables, equipment, and systems at the design phase, EMC predictive tools must be improved [5]. In order to achieve this requirement, frequency domain simulations can be performed utilizing equivalent models for the EMI sources. For instance, in power electronics, the range of frequency analyzed can be restricted from 10 kHz to 50 MHz, which comprises the common operating range of semiconductor switches utilized in power converters and frequency harmonics produced by them.

The EMI is usually established in different ways, for zinstance, the near-field coupling between filter components [6–9] or the coupling between wires [10]. Each coupling phenomenon is thus best modeled by different mathematical models.

The near-field coupling between filter components can be well modeled by a methodology based on the multipole expansion, which represents the radiation emission in a spherical reference system (r, θ, φ) [6–9], whereas the coupling between wires is usually well modeled by the PEEC method [10].

This paper presents a methodology to determine the first two coefficients of the multipole expansion (Q_{10} and Q_{20}) of a generic magnetic field source, by a numerical or an experimental approach, depending upon the complexity of the source. The numerical approach is rather limited to simple sources, but the experimental approach has no limitations over the geometrical complexity of the source.

The experimental approach utilizes an antenna consisting of four loops around the magnetic field source. The mutual coupling between the loops must be taken into account when modeling the source, in order to avoid a significant error, which can be up to 40%.

Finally, the methodology is validated by comparing the calculated and measured mutual inductance of a modeled power transformer and a well-known loop.

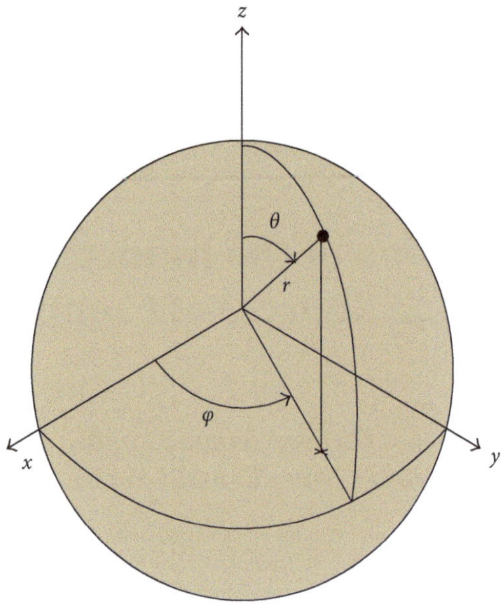

Figure 1: Reference adopted in the field computation.

2. Theory of Multipole Expansion

The multipole expansion can be used to represent electromagnetic fields in 3D, assuming that the field is computed outside a sphere of a given radius that contains the equivalent source. Figure 1 shows the reference sphere considered [11].

In the case of outgoing radiated emission sources, the multipole expansion allows expressing the electric and magnetic fields as [12]

$$E(r,\theta,\varphi) = \sum_{n=1}^{\infty}\sum_{m=-n}^{n} Q_{nm}^{\mathrm{TE}}\mathbf{F}_{1nm}(r,\theta,\varphi) + Q_{nm}^{\mathrm{TM}}\mathbf{F}_{2nm}(r,\theta,\varphi),$$

$$H(r,\theta,\varphi) = \frac{j}{\eta}\sum_{n=1}^{\infty}\sum_{m=-n}^{n} Q_{nm}^{\mathrm{TM}}\mathbf{F}_{1nm}(r,\theta,\varphi) + Q_{nm}^{\mathrm{TE}}\mathbf{F}_{2nm}(r,\theta,\varphi), \tag{1}$$

where

(i) $\eta = \sqrt{\mu/\varepsilon}$ is the intrinsic impedance of the considered environment;

(ii) Q_{nm}^{TE} and Q_{nm}^{TM} are the magnetic and electric coefficients, respectively. The coefficients Q_{nm}^{TE} describe the strength of the transverse-electric (TE) components of the radiated field, while coefficients Q_{nm}^{TM} describe the strength of the transverse-magnetic (TM) components. Each of them corresponds to the equivalent radiated source. Thus, these coefficients are the parameters to be identified that characterize the equivalent model of the radiated field components;

(iii) $\mathbf{F}_{1nm}$ and $\mathbf{F}_{2nm}$ are the vector spherical harmonics which are solutions of Maxwell's equations in free space, excluding the sphere that involves the sources;

(iv) n is the degree, and m is the azimuthal order.

In our study, only the magnetic source in the near-field is considered. That is $Q_{nm}^{\mathrm{TM}} = 0$, and it is assumed that the electric field component is low when compared with the magnetic field. Thus, the computation of the Q_{nm}^{TE}, wrote as Q_{nm} in (2), is carried out by the radial component H_r, in near field [12, 13]:

$$H_r = -\frac{1}{4\pi}\sum_{n=1}^{+\infty}\sum_{m=-n}^{n} Q_{nm}\frac{\partial}{\partial r}\left(\frac{1}{r^{n+1}}\right)Y_{nm}(\theta,\varphi), \tag{2}$$

where Y_{nm} are the normalized spherical harmonics given by the following expression:

$$Y_{nm}(\theta,\varphi) = \sqrt{\frac{(2n+1)(n-m)!}{4\pi(n+m)!}}P_n^m(\cos\theta)e^{jm\varphi}. \tag{3}$$

One of the main properties of the multipole expansion to be emphasized is the decrease of the terms of order n with $r^{-(n+1)}$. This ensures a hierarchy between each order of the decomposition. The larger is the distance to the source, the fewer are the terms required to reconstruct the field. Thus, the accuracy of the mutual inductance computation is related to the choice of the maximum order description, noted $N_{\max}$. It should be observed that there are $(2n + 1)$ components for each n order. For an order source equal to $N_{\max}$, it will correspond to $N_{\max}(N_{\max} + 2)$ components, but due to the previously mentioned property (hierarchy between each order), $N_{\max}$ can be limited to 5, based on the present experience of the authors.

3. Multipole Identification

3.1. Numerical Approach. This approach consists in identifying the source utilizing the software Flux2D based on the finite element method. The software calculates the radial component of the magnetic induction on a measurement sphere S_M, which contains the source, as shown in Figure 2. The computation of the Q_{nm} coefficients is achieved by integrating these components on S_M.

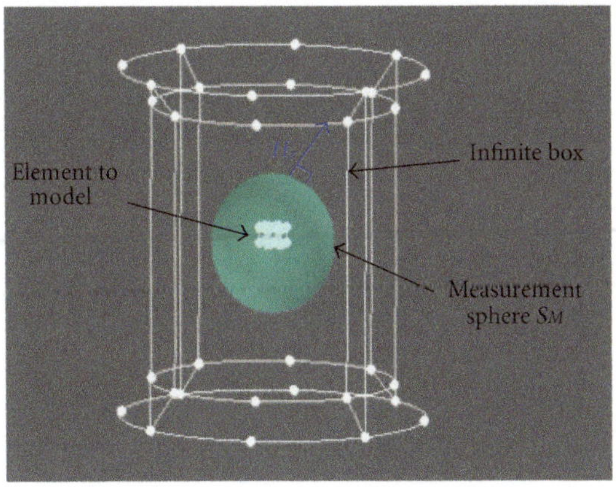

Figure 2: Source modeling Flux2D.

The coefficients of the multipole expansion can be deduced from (2), based on the following expression:

$$Q_{nm} = 4\pi \frac{r_0^n}{(n+1)} \int_0^{2\pi} \int_0^{\pi} H_r(r_0, \theta, \varphi) Y_{nm}(\theta, \varphi) dS, \quad (4)$$

where H_r corresponds to the radial component of the magnetic field on the sphere S_M of a radius of r_0. This result is due to the orthogonal property of Y_{nm} base:

$$\iint_{S_M} Y_{nm} Y_{n'm'} dS = \begin{cases} r_0^2, & \text{if } (n,m) = (n',m'), \\ 0, & \text{otherwise.} \end{cases} \quad (5)$$

The order of the approximation is not limited for this identification method. However, the computational time increases with the order. Moreover, the discretization of the sphere surface must respect the Shannon theorem in order to avoid spatial aliasing. For instance, with $n = 1$, the axes theta and phi require at least two points each, whereas, for $n = 2$, four points are required. For $n = N_{\max}$, $2N_{\max}$ points are necessary for each axis.

The numerical approach can be excessively time consuming or require too much memory, if the modeled source is geometrically complex. The experimental approach is thus an alternative, and it is suitable for practically any source.

3.2. Experimental Approach. This approach consists in identifying the source utilizing an antenna and a measurement equipment. Figure 3 shows the prototype antenna with its loop sensors corresponding to the dipole (2 loops for the dipole component Q_{10}), the quadrupole (2 loops for the quadrupole Q_{20}), and the loop from the standard CISPR16-1. All mentioned loops were built initially only in the z-direction. The complete measurement setup is surrounded by a sphere of radius r_M equal to 0.225 m. The short-circuited loops were proposed as sensors with a flat response within the 9 kHz–30 MHz frequency range. Although the use of short-circuited loops corresponds to high values of currents and thus high sensitivity, the magnetic coupling between them imposes some constraints to the measurement and calibration methodology.

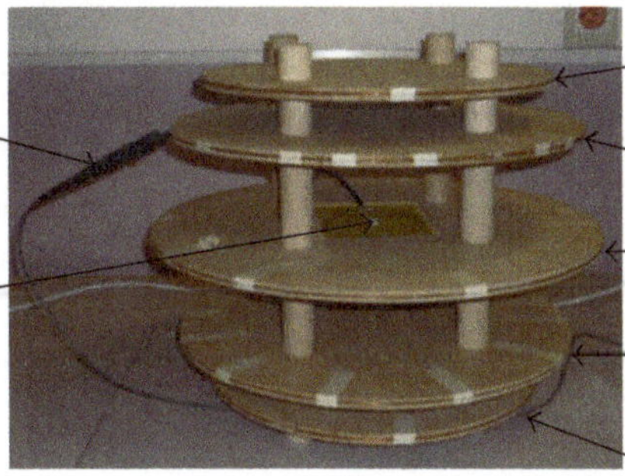

Figure 3: Prototype antenna.

Based on the multipole expansion of the magnetic field, the relationship between the fluxes across the surface delimited by the "sensors set" and the Q_{nm} components of the expansion can be directly obtained [14, 15]. The quasi-static approximation was adopted, and, for the maximum frequency of 30 MHz, it is valid for a $r_M \leq 1.7$ m. In our case, assuming the expansion limited to the second order and in the z-direction ($m = 0$), we have [15]

$$\begin{aligned} Q_{10} &= \frac{10^8 r_M}{32\pi} (\varphi_{10_1} + \varphi_{10_2}), \\ Q_{20} &= \frac{6125 * 10^4 r_M^2}{3\pi\sqrt{21}} (\varphi_{20_1} - \varphi_{20_2}), \end{aligned} \quad (6)$$

where φ_{nm} corresponds to the flux through the loop antenna given by Figure 3. Thus, for the loop configuration given by the same figure, the fluxes through the sensors due to a multipole source will be determined based on the current measured on each loop, after taking into account the magnetic coupling effects of the loops and applying the corresponding antenna factors (AF_{nm}) for $i = 1, 2$:

$$\begin{aligned} \varphi_{10\_i} &= \mathrm{AF}_{10} i_{10\_i}, \\ \varphi_{20\_i} &= \mathrm{AF}_{20} i_{20\_i}. \end{aligned} \quad (7)$$

The correction of the magnetic coupling between the loop sensors, which can be considered as a postprocessing in the identification procedure, is treated as follows: the total concatenated magnetic flux in each loop can be expressed as the sum of the flux produced by the multipole source (desired) and the fluxes produced by all the other antenna loops (undesired). The measured current in loop n is denoted as $i_{\mathrm{MES}}^{(n)}$ and can be obtained by the following expression:

$$i_{\mathrm{MES}}^{(n)} = i_{\mathrm{DUT}}^{(n)} - \sum_{\substack{k=1 \\ k \neq n}}^{5} \frac{j\omega M_{kn} i_{\mathrm{MES}}^{(k)}}{r_n + j\omega L_n}, \quad (8)$$

where r_n is the resistance and L_n is the self-inductance of the loop n, M_{kn} is the mutual inductance between loops k and n, ω is the angular frequency, and $i_{\mathrm{DUT}}^{(n)}$ is the current in loop n due to the multipole source only.

Thus, considering now the measured currents for all the five loops, one can write (8) in matrix form, solved for $i_{DUT}^{(n)}$:

$$[i_{DUT}]_{n,1} = [M]_{n,n}[i_{MES}]_{n,1}. \quad (9)$$

The elements of $[M]_{n,n}$ are unitary in the diagonal and given by $(j\omega M_{kn}/(r_n + j\omega L_n))$ otherwise. The coefficients Q_{10} and Q_{20} can then be determined by (6), and (7) utilizing the set of currents in (9). This procedure was validated numerically and experimentally, and the results are presented in the following section.

4. Multipole Identification Results

A vector network analyzer (VNA) and large bandwidth current probes were utilized to measure the current ratio of each sensor loop relative to the source, in dB. The frequency range of all experiments was from 20 kHz to 10 MHz.

Three different magnetic field sources were studied: a dipole, a quadrupole, and a generic power transformer. The accuracy of the methodology can be easily verified for the first two sources by utilizing the following expressions:

$$\begin{aligned} \text{Dipole: } & Q_{10} = \pi r^2 i; \; Q_{20} = 0, \\ \text{Quadrupole: } & Q_{10} = 0; \; Q_{20} = \pi r^2 h_0 i, \end{aligned} \quad (10)$$

where r is the radius of the loop and h_0 is the distance between the loops in the quadrupole source. Moreover, for these 2 sources, it is only necessary to present the results for the upper (or lower) loop antennas due to the symmetry on the z-axis with respect to the origin.

The accuracy of the coefficients Q_{10} and Q_{20} of the power transformer can be verified indirectly by calculating [16] and measuring its mutual inductance with a known circular loop and then comparing these results. The VNA and the probes are again used for the measurement.

4.1. Dipole. The measured current ratios loop/source for a dipole of radius 5 cm aligned in the z-axis for the loop sensors Q_{10_1} (loop 2) and Q_{20_1} (loop 1) are presented in Figures 4 and 5, respectively. For each figure, there are 6 curves, in which the 3 upper ones correspond to the lower ones after applying the postprocessing described previously.

Supposing a current of 1 A_{rms} in this dipole and the symmetry in the z-axis and utilizing the plots in Figures 4 and 5, we can determine the 4 currents of (7) and finally the components Q_{10} and Q_{20} with (6).

4.2. Quadrupole. The measured current ratios for a quadrupole of parameters r and h_0 both equal to 5 cm are presented in Figures 6 and 7 in a similar fashion done for the dipole.

4.3. Power Transformer. The measured current ratios loop/source for a power transformer rated 220 V—20 A are presented in Figure 8. The experiment was conducted in a similar manner to the previous ones, although there is no longer symmetry in the z-axis.

The components of the multipole expansion of the dipole, the quadrupole, and the transformer and the errors of the first two sources relatively to (10) are presented in Table 1. The frequency considered for these results was 200 kHz, located on the flat part of the measured curves.

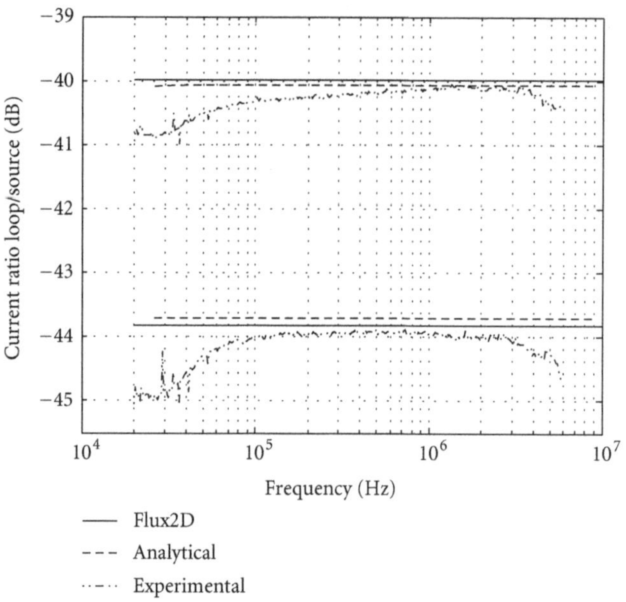

FIGURE 4: Dipole dB current ratio, sensor Q_{10_1} (loop 2).

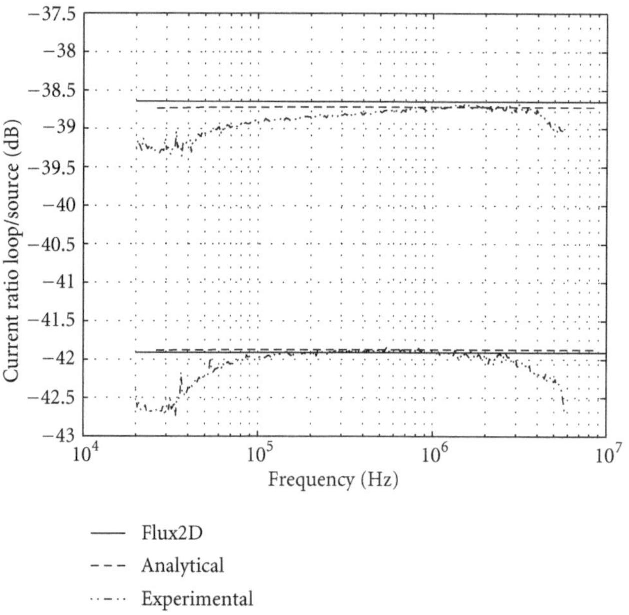

FIGURE 5: Dipole dB current ratio, sensor Q_{20_1} (loop 1).

TABLE 1: Components of the multipole expansion.

Component	Q_{10} (m·Am^2)	Q_{20} (m·Am^3)	Error (%)
Dipole	7.3	0	7.6 (Q_{10})
Quadrupole	0	0.63	20 (Q_{20})
Transformer	66.5	0.82	—

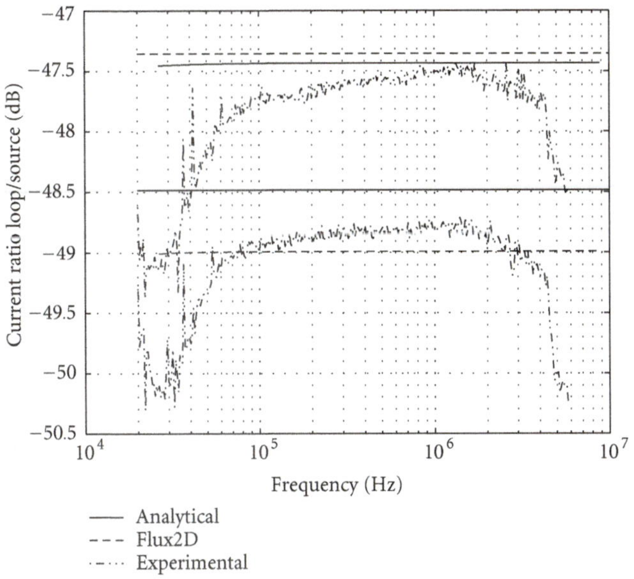

FIGURE 6: Quadrupole dB current ratio, sensor Q_{10_1} (loop 2).

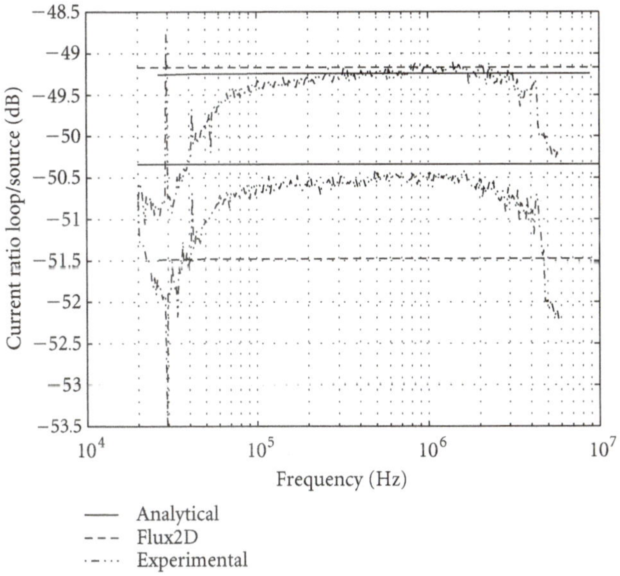

FIGURE 7: Quadrupole dB current ratio, sensor Q_{20_1} (loop 1).

It should be noted that there is a −4 dB and −3.5 dB difference between the current ratios in Figure 4 and in Figure 5, respectively. Therefore, the effect of the mutual inductances in the antenna would correspond to a difference of 37% and 33% in the calculation of Q_{10} and Q_{20}, respectively, for the dipole.

The mutual inductance between this transformer and a loop can be determined by the following expression:

$$M_{12} i_1 = L_2 i_2, \tag{11}$$

where M_{12} is the mutual inductance, i_1 the current in the transformer, L_2 the self-inductance of the loop, and i_2 the current in the loop.

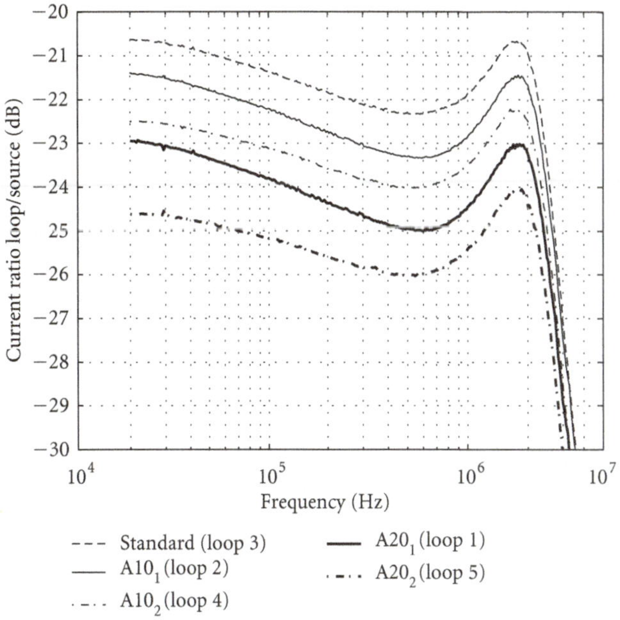

FIGURE 8: Transformer dB current ratio for all sensor loops.

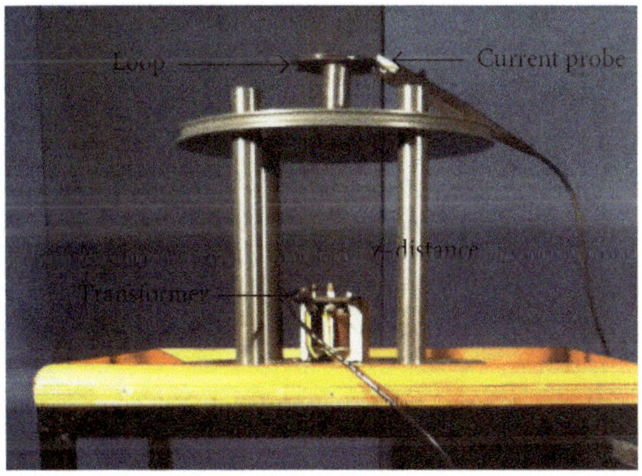

FIGURE 9: Mutual inductance measurement.

This was done by measuring the ratio of currents using the VNA and applying an analytical formula for the self-inductance of a loop [16] as shown in (11). Figure 9 shows the measurement setup for this case.

5. Computing the Mutual Inductance

Using the equivalent radiated field source model, we can determine the coupling between two equivalent sources through the computation of the mutual inductance. Figure 10 illustrates the configurations regarding the representation of two radiating sources (models 1 and 2).

The computation of the mutual impedance between source 1 and source 2 can be expressed in terms of the electrical field **E** and magnetic field **H** for each source [12]:

$$Z_{12} = -\frac{1}{i_1 i_2} \oiint_{\Sigma_1} (\mathbf{E}_1 \times \mathbf{H}_2 - \mathbf{E}_2 \times \mathbf{H}_1). \tag{12}$$

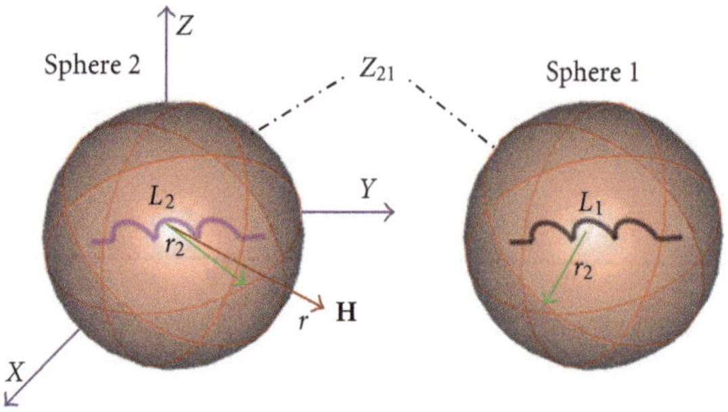

FIGURE 10: Representation of two radiating sources.

When the spheres that contain the sources do not intersect each other, the mutual impedance can be expressed according to the coefficients of the multipole expansion:

$$Z_{12} = \frac{1}{i_1 i_2}\frac{1}{k^2}\sqrt{\frac{\varepsilon_0}{\mu_0}}\sum_{n=1}^{N_{\max}}\sum_{m=-n}^{n}(-1)^m\left(Q_{1n,-m} * Q_{2n,m}\right). \tag{13}$$

The expression of the mutual inductance is

$$M_{12} = \frac{1}{j\omega i_1 i_2}\frac{1}{k^2}\sqrt{\frac{\varepsilon_0}{\mu_0}}\sum_{n=1}^{N_{\max}}\sum_{m=-n}^{n}(-1)^m\left(Q_{1n,-m} * Q_{2n,m}\right), \tag{14}$$

where i_1 and i_2 are, respectively, the current that flows in sources 1 and 2, and k is the phase constant.

The coefficients associated to the magnetic transverse modes of the multipole expansion of sources 1 and 2 must be expressed in the same reference: a translation is required, for example, the coefficients of the source 2 can be expressed in the reference of the source 1.

The rotation of the coefficients Q_{nm} is expressed by using Euler angles. It should be mentioned that only two angles are necessary because of the spherical symmetry. The details of the methodology for determining the rotation matrices for complex or real coefficients Q_{nm}, are presented in [17, 18]. The translation is based on the "Addition Theorem for Vector Spherical Harmonics" [18].

The addition theorem links the harmonics evaluated on r to those evaluated on r', where r is measured from the origin of the second spherical basis, whose axes are parallel to the first as shown in Figure 11. The origin of the second spherical basis is located in the first by r''. These 3 vectors are connected by the relation $r = r' + r''$.

The expression of the translation coefficients Q_{snm} are:

$$\begin{aligned} Q_{n'm'}^{\mathrm{TE}} &= \sum_{n=1}^{\infty}\sum_{m=-n}^{n} Q_{mn}^{\mathrm{TE}} A_{n',m',n,m} + Q_{nm}^{\mathrm{TM}} B_{n',m',n,m}, \\ Q_{n'm'}^{\mathrm{TM}} &= \sum_{n=1}^{\infty}\sum_{m=-n}^{n} Q_{nm}^{\mathrm{TM}} A_{n',m',n,m} + Q_{mn}^{\mathrm{TE}} B_{n',m',n,m}. \end{aligned} \tag{15}$$

The coefficients $A_{n',m',n,m}$ and $B_{n',m',n,m}$ involve computing of the Wigner $3j$ symbol according to quantum mechanics [19].

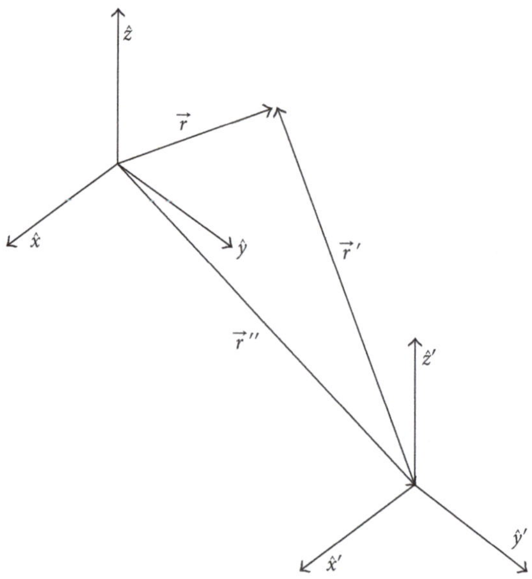

FIGURE 11: Translation of a spherical basis.

TABLE 2: Comparison between mutual inductances.

Height (cm)	Measured (nH)	Estimated (nH)	Error (%)
29.3	3.65	3.17	13
35.8	1.98	1.92	3

Utilizing the Q_{10} and Q_{20} of the transformer, it is also possible to estimate its mutual inductance with the loop in a simpler manner. This is achieved by using an analytical expression for the mutual inductance between 2 loops [16], by considering that Q_{10} is represented by a loop and Q_{20} is represented by 2 loops, both in the z axis. These results are presented in Table 2.

6. Conclusion

The presented methodology enables the evaluation of coupling parameters of components by using equivalent emission sources. This method is composed by two steps. At first, the equivalent sources which represent the radiated field component using the multipole expansion representation are identified. It can be obtained by a numerical or an experimental approach. Both of them were discussed in the paper. Secondly, the equivalent sources will be used to compute the coupling between them, which was represented by a mutual inductance as a function of the distance that separates them.

Other kind of multipole expansions like the cylindrical one can be more suitable for modeling components such as tracks or cables, and it will also be considered. For example, in the case of the coupling between a track and a component, the spherical harmonics method is not very adequate and other harmonics method should be used.

The method proposed could be helpful when used together with other circuit simulator methods in the evaluation of equivalent circuit of power electronics devices (R-L-M-C). This will give us a considerable gain of memory space concerning the full model configuration used in EMC filter numerical simulations.

References

[1] R. Redl, "Electromagnetic environmental impact of power electronics equipment," *Proceedings of the IEEE*, vol. 89, no. 6, pp. 926–938, 2001.

[2] R. Redl, "Power electronics and electromagnetic compatibility," in *Proceedings of the 27th Annual IEEE Power Electronics Specialists Conference (PESC '96)*, vol. 1, pp. 15–21, January 1996.

[3] E. Hoene, "EMC in Power Electronics," in *Proceedings of the 5th International Conference on Integrated Power Systems (CIPS '08)*, pp. 1–5, March 2008.

[4] A. R. Ruddle, X. Ferrières, J. P. Parmantier, and D. D. Ward, "Experimental validation of time-domain electromagnetic models for field coupling into the interior of a vehicle from a nearby broadband antenna," vol. 151, no. 6, pp. 430–433.

[5] D. S. Dixon, M. Obara, and N. Schade, "Finite-element analysis (FEA) as an EMC prediction tool," *IEEE Transactions on Electromagnetic Compatibility*, vol. 35, no. 2, pp. 241–248, 1993.

[6] S. Zangui, K. Berger, C. Vollaire, E. Clavel, R. Perrussel, and B. Vincent, "Modeling the near-field coupling of EMC filter components," in *Proceedings of the IEEE International Symposium on Electromagnetic Compatibility*, pp. 825–830, July 2010.

[7] S. Zangui, K. Berger, B. Vincent et al., *Evaluation of Coupling Parameters of EMC Filter Components by Using Equivalent Emission Sources*, MOMAG, Brazil, 2010.

[8] S. Zangui, K. Berger, R. Perrussel, E. Clavel, C. Sartori, and C. Vollaire, *Using Equivalent Emission Sources to Evaluate the Coupling between Components*, Compumag, Sydney, Australia, 2011.

[9] S. Zangui, B. Vincent, K. Berger et al., "Near-field coupling between EMC filter components," in *Proceedings of the 14th Biennial IEEE Conference on Electromagnetic Field Computation (CEFC '10)*, vol. 1, pp. 9–12, May 2010.

[10] A. E. Ruehli, "Equivalent circuit models for three-dimensional multiconductor systems," *IEEE Transactions on Microwave Theory and Techniques*, vol. 22, no. 3, pp. 216–221, 1974.

[11] S. Zangui, B. Vincent, K. Berger, E. Clavel, R. Perrussel, and C. Vollaire, *Equivalent Source Corresponding to Radiated Field of EMC Filter Components*, EMC Europe, Wroclaw, Poland, 2010.

[12] B. C. Brock, "Using vector spherical harmonics to compute antenna mutual impedance from measured or computed fields," SANDIA Report, 2001.

[13] A. Gati, Y. Adane, M.-F. Wong, J. Wiart, and V. Fouad-Hanna, "Inverse characterization of antennas by equivalent sources using spherical harmonics," *Comptes Rendus Physique*, vol. 6, no. 6, pp. 640–646, 2005.

[14] B. Vincent, O. Chadebec, J. L. Schanen, and K. Berger, "Loop antennas for near-field multipolar-expansion identification: first experimental validations," *IEEE Transactions on Instrumentation and Measurement*, vol. 59, no. 12, pp. 3086–3092, 2010.

[15] B. Vincent, O. Chadebec, J.-L. Schanen et al., "New robust coil sensors for near field characterization," *Journal of Microwaves and Optoelectronics*, vol. 8, no. 1, pp. 64S–77S, 2009.

[16] E. Durand, *Magnétostatique*, Masson et Cie Editeurs, 1968.

[17] M. A. Blanco, M. Flórez, and M. Bermejo, "Evaluation of the rotation matrices in the basis of real spherical harmonics," *Journal of Molecular Structure: THEOCHEM*, vol. 419, no. 1–3, pp. 19–27, 1997.

[18] C. A. Balanis, *Advanced Engineering Electromagnetics*, John Wiley & Sons, New York, NY, USA, 1989, chapter 10.

[19] A. Messiah, "Clebsch-Gordan (C.-G.) Coefficients and "3j" Symbols," in *Appendix C.I in Quantum Mechanics*, vol. 2, pp. 1054–1060, North-Holland, Amsterdam, The Netherlands, 1962.

2

Overview on Multipattern and Multipolarization Antennas for Aerospace and Terrestrial Applications

Aixin Chen,[1] Weiwei Jiang,[1] Zhizhang Chen,[2] and Jiaheng Wang[1]

[1] *School of Electronic and Information Engineering, Beihang University, Beijing 100191, China*
[2] *Department of Electrical and Computer Engineering, Dalhousie University, Halifax, NS, Canada B3J 2X4*

Correspondence should be addressed to Aixin Chen; axchen@buaa.edu.cn

Academic Editor: Haiwen Liu

In recent years, reconfigurable antennas, with the ability to radiate wave in more than one pattern and polarization, play a great role in modern telecommunication systems. Compared with conventional antennas, multipattern and multipolarization antennas have more advantages and better prospects. They can be used to improve system gain and security, satisfy system requirements, avoid noisy environment, and adapt to the environment flexibly. This paper discusses different patterns and polarizations of reconfigurable antennas according to current research work in this area. In the opinion of this paper, the radiation pattern states of antennas include beam direction, shape, and gain. The polarization states of antennas include horizontal/vertical linear, ±slant 45° linear, left-hand or right-hand circular polarized. Different multipattern and multipolarization antennas with various structures and working mechanisms are compared and discussed. Multipattern and multipolarization antennas have been well applied for aerospace and terrestrial applications, such as dynamic scenarios, adaptive beam scanning, and multiple-input-multiple-output (MIMO) systems.

1. Introduction

Generally, as we know, theconventional antennas only can work in one radiation pattern and polarization. When the environmental conditions or system requirements change, we need to reconfigure an antenna. It is inconvenient and costs too much. Reconfigurable antennas are a solution to solve this problem. They can change their radiation pattern and polarization timely to address complex system requirements by modifying their geometry and electrical behavior. Reconfigurable antennas can either increase the capabilities of wireless integrated information systems, expand their functionality, orwiden their bandwidths, with efficient spectrum and power utilization.

Compared to today's radio technology which depends on incompatible communication systems with inflexible hardware, the technology of reconfigurability will significantly reduce the hardware complexity, the number of components, and the cost. As described above, reconfigurable antennas promise to bring a host of benefits to future generation of wireless systems. Next-generation communication systems will rely upon reconfigurable antennas, such as wireless systems in dynamic scenarios (e.g., satellite/terminal tracking), adaptive beam scanning (e.g., radar/remote sensing), and MIMO systems (e.g., active management of channel correlation/diversity/interference). Among diversity of schemes at antenna level, multipattern and multipolarization antennas are important ones. They can be used to improve system gain and security, satisfy system requirement, avoid noisy environment, and adapt to the environment flexibly.

The concept of reconfigurable antenna was first presented in 1983 [1] and further investigation and realization were carried out and results were reported in [2]. In the recent years, reconfigurable antennas attracted more and more attention due to their capability of providing multipattern and multipolarization. One of the important applications is in aerospace and terrestrial wireless communication systems.

In the following sections, the classification of the multipattern and multipolarization antennas, the comparison between the different antennas with various structures and working mechanisms, the field of their application, and the vision and goal are to be discussed.

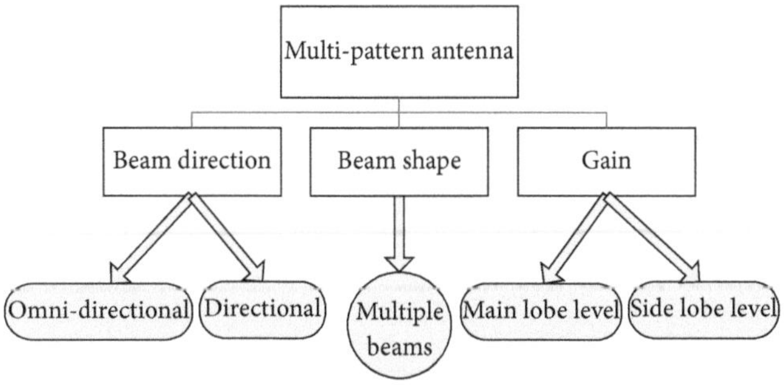

FIGURE 1: Classification of multipattern antennas.

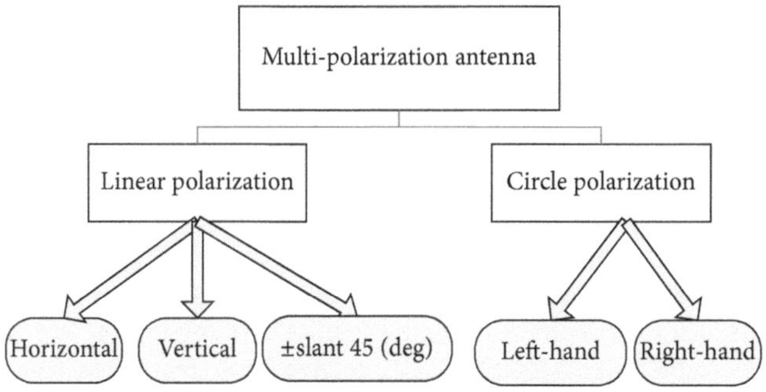

FIGURE 2: Classification of multipolarization antennas.

2. Classification and Technologies of Reconfigurable Antennas

In this paper, the reconfigurable antennas (mainly multipattern and multipolarization antennas here) can be classified into three categories: multipattern antennas, multipolarization antennas, and multipattern-multipolarization mixed reconfigurable antennas.

The multipattern antennas are divided into three major types achieved by changing the beam direction, beam shape, and gain, as indicated in Figure 1. The beam direction includes the omnidirectional and directional direction. Antennas which alter between omni-directional and directional radiation patterns [3–8] or vary the direction from one side to another side [9–17] are called the beam direction reconfigurable antennas. Beam shape reconfigurable antennas can be achieved by changing the number of beams (multiple beams) [18–21]. Finally, the gain reconfigurable antennas are able to be realized by changing the shape of beam's main lobe and side lobe [22–24].

The multipolarization antennas include linearly polarized antennas and circularly polarized antennas, as shown in Figure 2. A radiating structure that is able to change its polarization (horizontal, vertical, and radiations, and ±slant 45°) is called linear polarization reconfigurable antennas [25–30]. Circular polarization reconfigurable antennas are achieved by varying polarization between left-hand circular and right-hand circular [31–39]. The antennas switching between the linear polarization and circular polarization are mixed polarization reconfigurable antennas [40–51].

The multipattern-multipolarization mixed reconfigurable antennas are the antennas which combine multipattern with multipolarization [52–61].

In order to realize various reconfigurable antennas, five major technologies are introduced.

(i) Electronically tuned reactance: alter the capacitors or inductors by varying the bias voltage. The common elements are varactor diodes and microelectromechanical-system- (MEM-S) tunable capacitors/inductors.

(ii) Radio frequency (RF) electrical switches: by controlling the voltage, achieve the switches to be "on" state or the "off" state. PIN diodes, field effect transistor (FET) switches, and MEMS are popularly used.

(iii) Photoconductive switches: activate the semiconductor material (silicon, gallium arsenide) by controlling the laser light to realize the "on" state or the "off" state. It results in exciting electrons from the valence to the conduction band and thus creates a conductive connection.

(iv) Exotic materials: a static applied electric/magnetic field or voltage can change the relative material permittivity/permeability. The smart materials which are used in the substrate of the reconfigurable antennas include nematic liquid crystals, nonlinear materials, and ferroelectric films.

(v) Structural alteration: by altering the height, acreage, and shape of antenna, the physical structural is able to be modified. The altering structure is the antenna radiating parts.

The multipattern and multipolarization antennas are designed by implying the five major technologies to change surface current distribution, and the feeding network, the physical structure, the effective radiating aperture of antennas. The different patterns or polarizations correspond to different states caused by the technologies.

3. Multipattern Antennas

3.1. Beam Direction Reconfigurable Antennas. The conventional antennas can only work in the omni-directional or directional beam direction. The beam of omni-directional antenna homogeneously directs all the direction. The beam of directional antenna only directs in some range, but its radiated effect is better than that the omni-directional antenn. Consequently, we can choose the suitable type to meet the need of communication system. For example, in the case that the wireless access points distribute in different places and are close to the antenna, we should choose the omni-directional antenna. On the contrary that the wireless access points are put together in one direction and long away from the antenna, the directional antenna is our choice. So we can see that the reconfigurable antenna with either omni-directional or directional radiation pattern is very useful in wireless system. In addition, when the target is unique and always moves, for example, in the radar and terminal

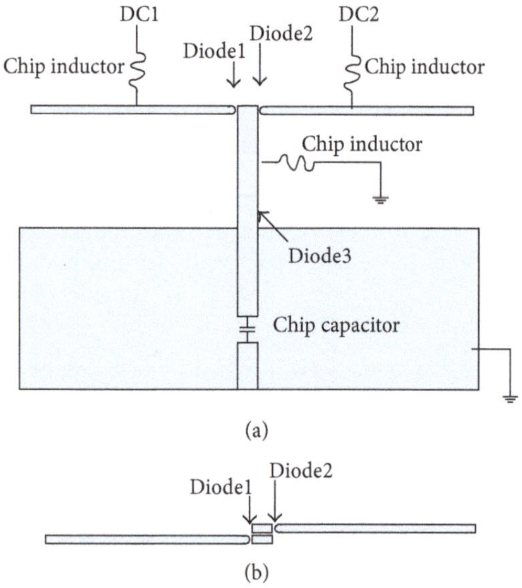

FIGURE 3: Radiation pattern reconfigurable antenna. (a) Side view. (b) Top view (redrawn after).

tracking system, the beam direction of antenna should change at the same time. The directional reconfigurable antenna fits it well. Next, three different examples of beam direction of reconfigurable antennas are described.

3.1.1. Beam Direction Reconfigurable Antennas with Omnidirectional and Directional Characteristics. This type of antennas can realize omni-directional and directional radiation. The antenna shown in Figure 3 is a simple radiation pattern reconfigurable antenna. The antenna is a combination of a monopole antenna and a dipole antenna with reflector [3]. Three switches are inserted to control the radiation pattern of antenna. The reconfigurable antenna can be operated as either a monopole antenna with an omnidirectional radiation pattern at the switches "on" state or a dipole antenna with reflector, which has directional radiation pattern at the switches "off" state. The antenna is compact and easy to control.

3.1.2. Beam Direction Reconfigurable Antennas with Varied Directional Characteristics. This is a type of complex radiation pattern reconfigurable antennas. A reconfigurable leaky-wave antenna capable of steering its radiation beam over a wide range is discussed in [9]. The antenna system is a microstrip composite right/left-hand (CRLH) leaky-wave antenna composed by 25 cascaded metamaterial unit cells, as shown in Figure 4. To achieve the CRLH behavior, insert an artificial series capacitance and a shunt inductance into a conventional microstrip line by means of an interdigital capacitor and a shorted stub, respectively. By tuning the bias voltage of varactor diodes D_S and D_{SH} separately, we can steer the radiation beam from -55° to 50°.

The fabricated antenna shown in Figure 5(a) is a planar 6×6 fully reconfigurable array operating at 5.7 GHz, capable of functioning both as a reconfigurable array lens and a reconfigurable reflectarray [10]. The array element consists of two varactor diode-loaded patches coupled by a varactor diode-loaded slot, as shown in Figure 5(b). In the lens mode, the antenna has a broadside directivity of 20.8 dBi and a beam-scanning range of 50° by 50°. The array as a reflectarray can achieve a directivity of 19.4 dBi at broadside and a beam-scanning range of 60° by 30°. This array is able to demonstrate full 2D beamforming with low cost and is easy to fabricate.

3.2. Beam Shape Reconfigurable Antennas. The antennas which radiate multiple beams at the same time are called multibeam antennas. Multibeam antennas are able to both transmit signal in multiple directions at the same time with high gain and achieve the low side lobe by controlling the amplitude and phase of the feeds. To meet the different demand of communication system, we can change the number and shape of beams. Beam shape reconfigurable antennas are important and have been studied by many scientists.

A multifunctional reconfigurable antenna (MRA) capable of steering its radiation beam in the azimuth plane into one of 4 and 8 directions with variable beamwidth is presented in [20] (see Figure 6). The antenna is composed of a planar array of electrically small metallic square-shaped pixels which are interconnected by MEMS switches. By activating or deactivating the interconnecting MEMS switches, the geometry of the MRA dipole is changed. The antenna with optimum performance is able to apply for a wireless communication system.

3.3. Gain Reconfigurable Antennas. The beam makes up by main lobe and side lobe. In general, to improve the gain of antenna, we should increase the mainlobe level and lower the side lobe level as far as possible.

A flexible microstrip patch antenna incorporates a novel multilayer construction consisting of a liquid metal (eutectic gallium indium) encased in an elastomer [24]. The antenna is flexible and durable as its combined properties of the fluid and the elastomeric substrate, as shown in Figure 7(a). Injecting the metal into micro-fluidic channels provides a simple way to define the shape of the liquid, which is stabilized mechanically by a thin oxide skin that forms spontaneously on its surface. In Figure 7(b), a multilayer patch antenna is fabricated using specially designed serpentine channels that take advantage of the unique rheological property of the liquid metal alloy. And in Figure 7(c), by controlling the shape of antenna static, curving a low dielectric mandrel with a radius of 12.7 mm, and curving around a similar mandrel with a radius of 25.4 cm, separately, the antenna can realize three different values of gain. The flexibility antenna is well suited for conformal antenna applications.

4. Multipolarization Antennas

Comparing with classical antennas with single polarization, the multipolarization antennas offer advantages of reduction in installation space, low coupling effect, and low installation cost. Their intrinsic polarization diversity advantage is very useful in mitigating the multipath fading and increasing the

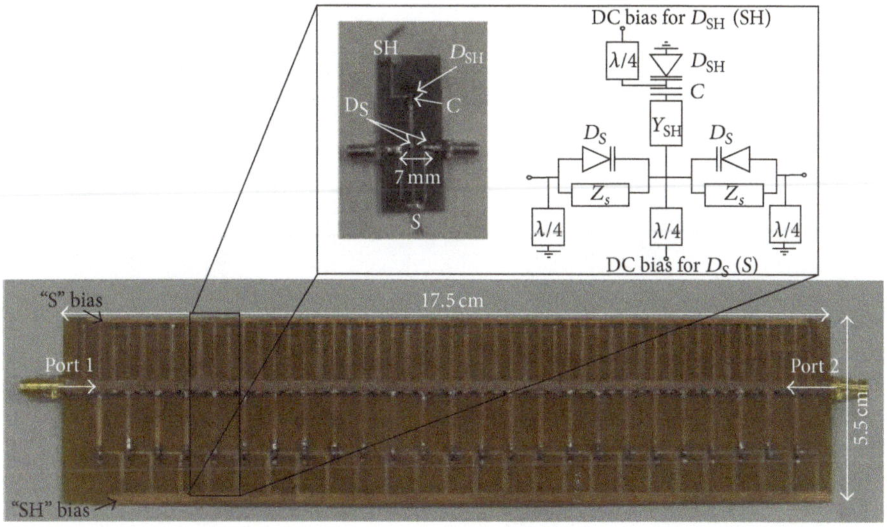

FIGURE 4: Two-port reconfigurable leaky-wave antenna and its tunable unit cell design (with permission from IEEE).

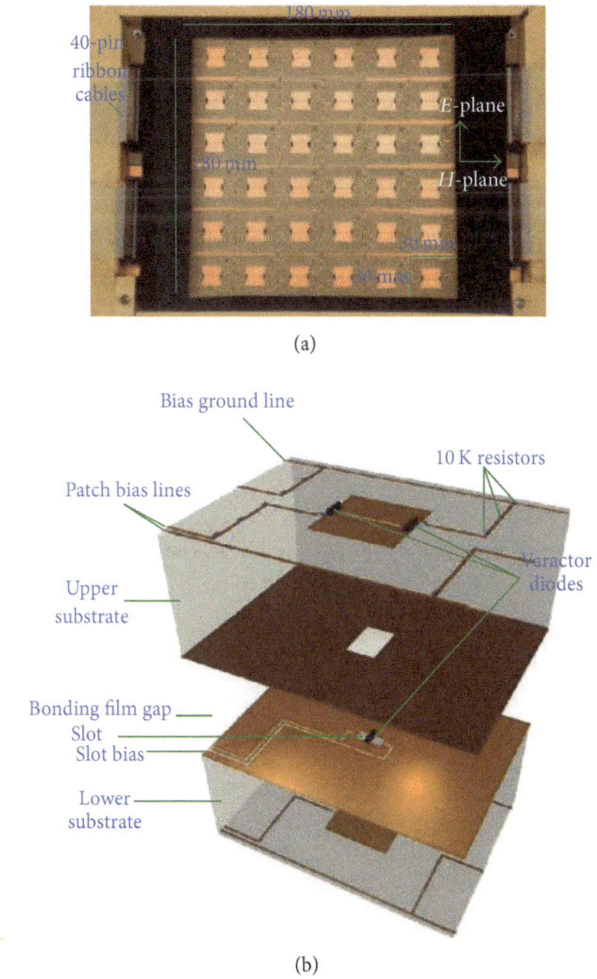

FIGURE 5: (a) Fabricated 6 × 6 planar reconfigurable array. (b) Array element and bias network design (vertically exaggerated) (with permission from IEEE).

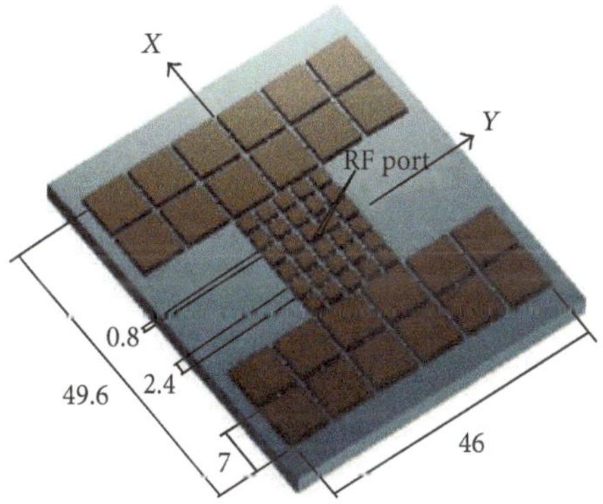

FIGURE 6: Multifunctional reconfigurable antenna (with permission from IEEE).

channel capacity, especially in polarization-varied environment. Such antennas can switch between linear polarization (LP) and circle polarization (CP), depending on the system requirements. The kinds of linear polarization include horizontal, vertical, and ±slant 45° polarization. The types of circular polarization are right-hand circular polarization (RHCP) and left-hand circular polarization (LHCP). In this section, three different polarization reconfigurable antennas to adjust the different propagation environment are presented as examples.

4.1. Linear Polarization Reconfigurable Antennas. A polarization reconfigurable slot antenna with compact feed is presented in [27]. Fed by dual modes of coplanar waveguide, dual orthogonal linear polarizations are excited in the slot and controlled by two PIN diodes. The antenna is composed by a rectangular slot, a coplanar waveguide (CPW), a capacitively coupled strip, and two PIN diodes. The rectangular slot

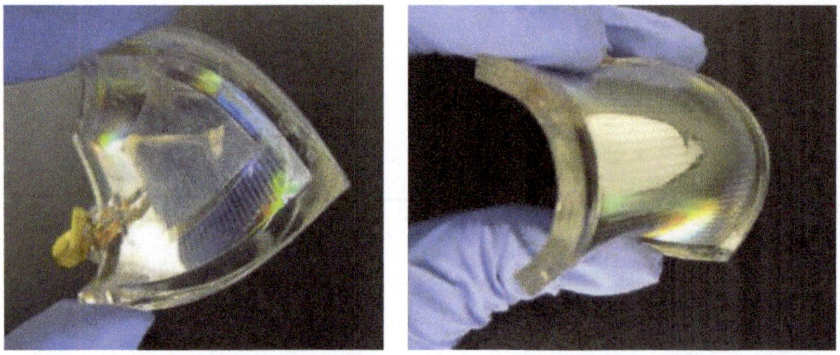

(a)

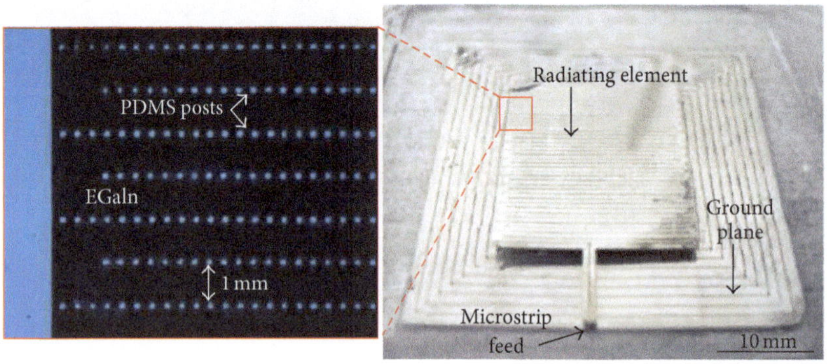

(b)

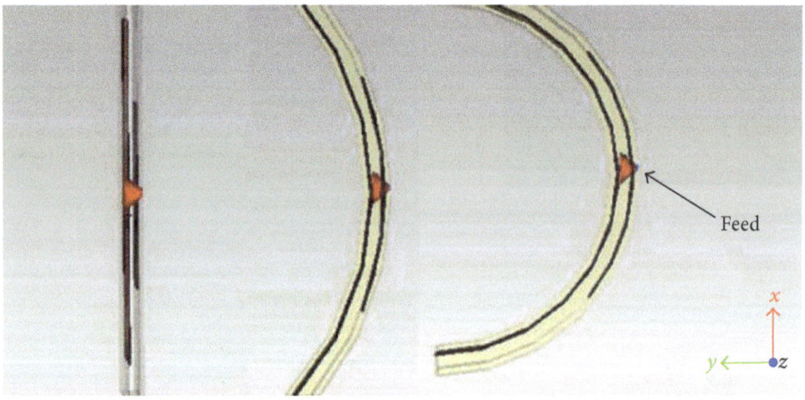

(c)

Figure 7: (a) Reconfigurable liquid metal alloy microstrip patch antenna. (b) The patch antenna and ground plane. (c) Cross-sectional view of the conformal configurations: (right) static, (middle) curved to a 25.4 mm radius, and (left) curved to a 12.7 mm radius (with permission from IEEE).

with the dimensions of $53 \times 50\,\text{mm}^2$ serves as the radiation aperture, etched in the front side of FR4 ($\varepsilon r = 4.4$, $\tan\delta = 0.01$). A switchable feed, including a C-shaped capacitively coupled strip at the back side and a CPW at the front side through, is controlled by two PIN diodes. The front side of fabricated antenna topology is shown in Figure 8(a). The back side of fabricated antenna structure is shown in Figure 8(b). The bias circuit of the feed is shown in Figure 8(c). And the geometry of the slot antenna is shown in Figure 8(d). When PIN1 is "on" and PIN2 is "off," the antenna is fed through the C-shaped strip, which is capacitively coupled to the CPW. As a result, the horizontal polarization of the slot is excited. When PIN1 is "off" and PIN2 is "on," the vertical polarization is excited by typical CPW feed structure. Dual orthogonal polarizations are achieved by dual orthogonal modes of CPW independently, making the feed structure compact.

4.2. Circular Polarization Reconfigurable Antennas. A circularly polarized reconfigurable microstrip array antenna with conical-beam radiation is shown in Figure 9 [39]. The array antenna structure consists of four L-shaped patch elements

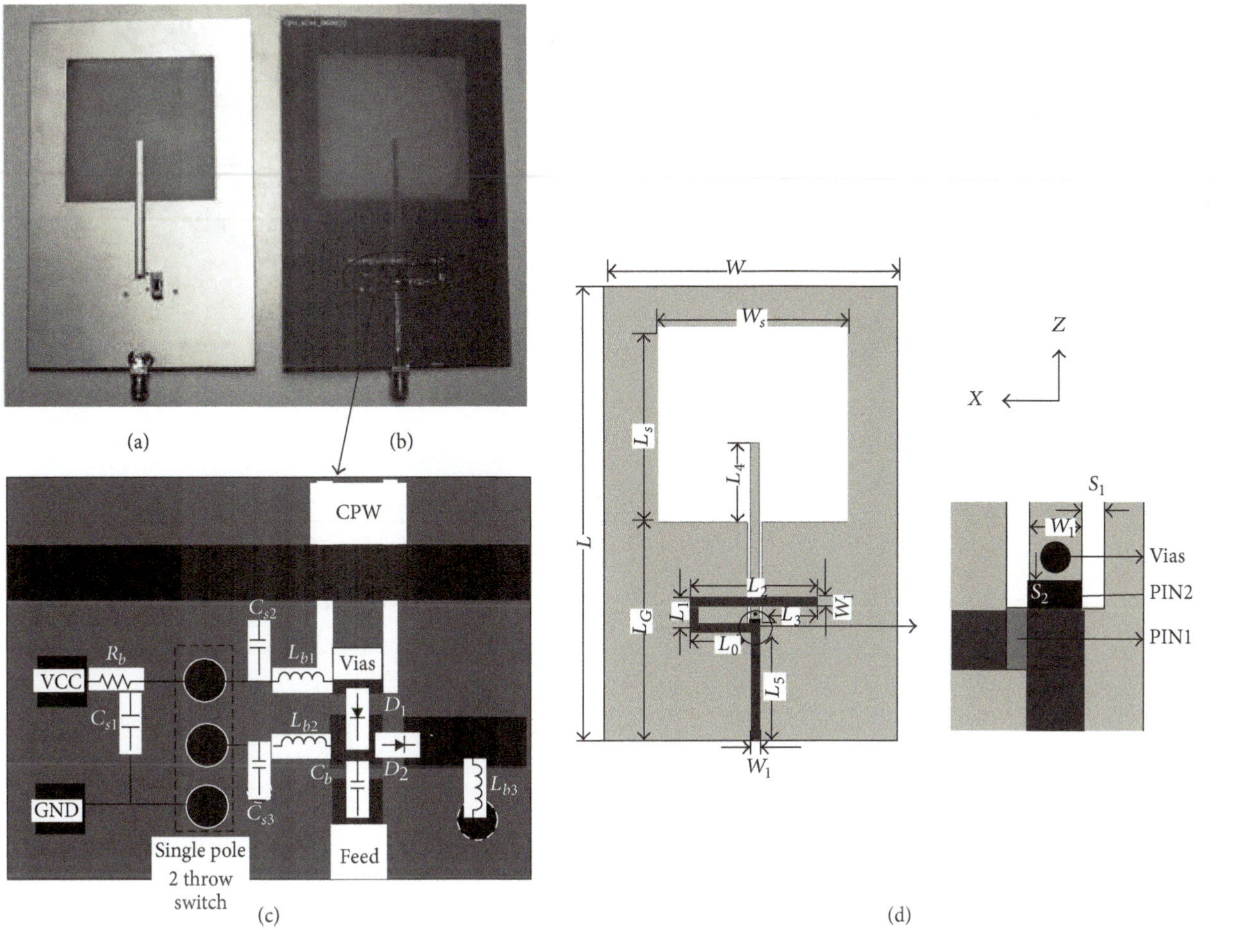

FIGURE 8: (a) Front side of fabricated polarization reconfigurable antenna. (b) Back side. (c) Bias circuit of the feed. (d) Geometric structure (with permission from John Wiley and Sons).

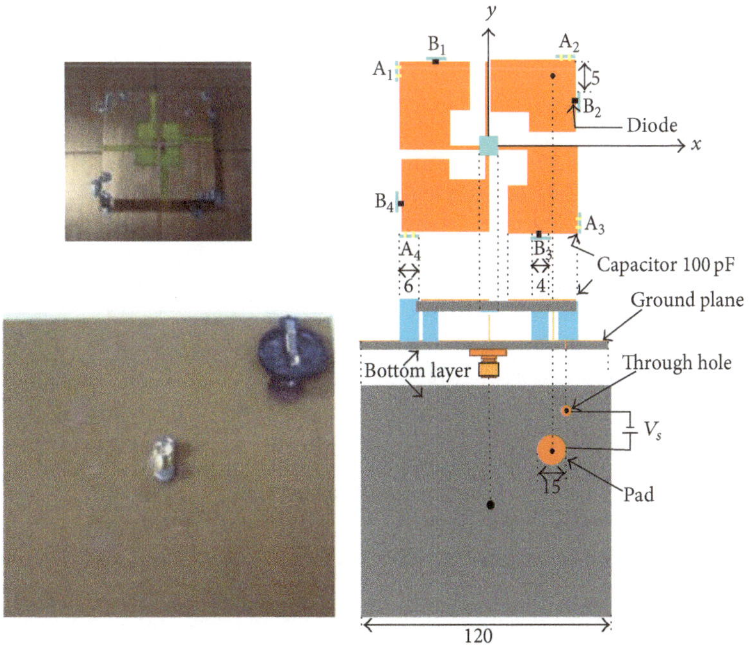

FIGURE 9: Layout of a reconfigurable CP microstrip array and the finished prototype (with permission from IEEE).

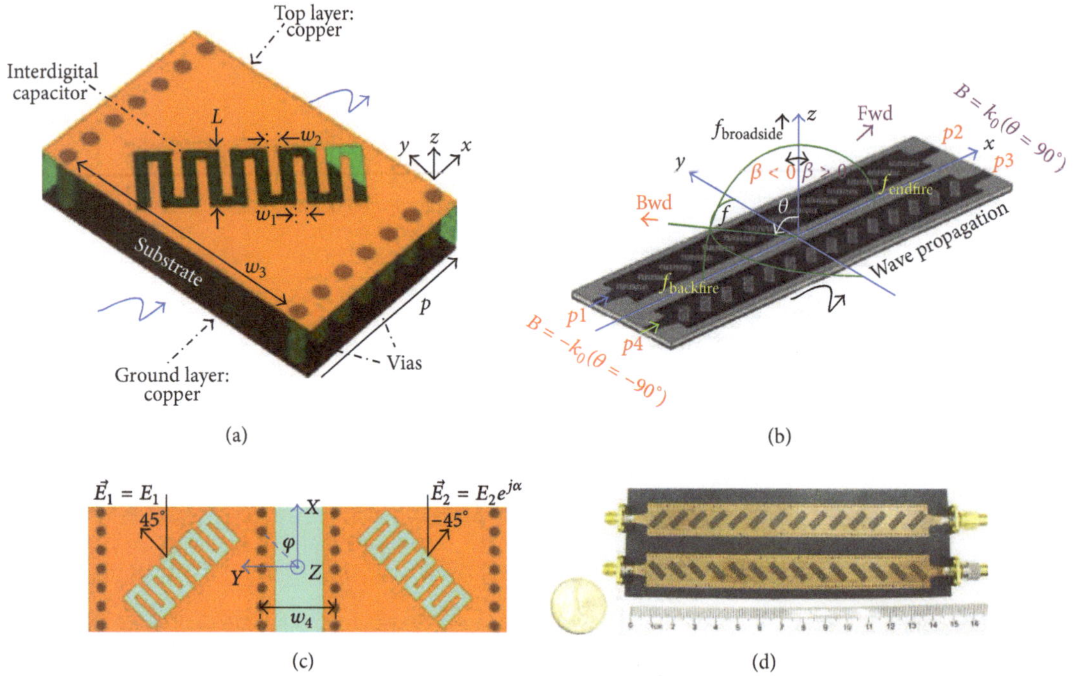

FIGURE 10: Polarization reconfigurable CRLH-SIW antenna. (a) Single radiating element. (b) Two-element unit cell of the whole structure. (c) Overall LW antenna prototype. (d) fabricated antenna (with permission from IEEE).

arranged in a square-ring formation. Each patch is shorted to the ground plane through conducting walls. With a top-loaded monopole feed, two orthogonal resonant modes, loop mode and monopole mode, can be excited simultaneously. When the monopole mode and the loop mode are properly coupled, the microstrip array can generate a CP radiation. By controlling the frequency, the antenna can switch between LHCP and RHCP. In addition the array has the characteristic of omnidirectional radiation and conical beam.

4.3. Mixed Polarization Reconfigurable Antennas. A composite right/left-handed (CRLH) leaky-wave (LW) antenna with flexible polarization is presented in [41]. The leaky transmission line (TL) is a planar passive circuit built using the substrate integrated waveguide technology. It consists of two symmetrical waveguide lines loaded with series interdigital capacitors which radiate orthogonal 45 linearly polarized waves, as shown in Figure 10. The configuration of single radiating element is shown in Figure 10(a). Two-element unit cell of the whole structure is shown in Figure 10(b). The overall LW antenna prototype is shown in Figure 10(c). Photograph of the fabricated LW antennas is shown in Figure 10(d). The antenna is able to generate arbitrary different polarization states by changing the way of excitation, including LP and CP. This leaky TL is fabricated by the standard printed-circuit board process. Two broadband couplers are also designed and fabricated for the specified excitation purpose. Six different polarization states, including four LP cases (horizontal, vertical, and ±slant 45°) and two CP ones (left hand and right hand), can be realized. The LW structure shows some desirable merits, such as the simplicity in design, low-cost fabrication, and beam-steering and polarization-flexible capabilities, providing a high degree of flexibility for the real application.

5. Multipattern-Multipolarization Mixed Reconfigurable Antennas

This type of antennas realizes multipattern and multipolarization characteristics by using one reconfigurable antenna. A novel reconfigurable microstrip antenna with radiation pattern selectivity and polarization diversity is discussed in [57]. As shown in Figure 11(a), A four-way power divider made by three Wilkinson power dividers and interconnected with switches (switch A and switch B) is designed to feed-four-rectangular-radiating-patcheantenna. By controlling the state of the switches, the antenna characteristics can be changed into two modes. When switch A is closed and switch B is open, the four rectangular radiating patches are excited by four sources with equal amplitude and phase. A metamaterial antenna with conical beam and linear polarization is achieved. When switch A is open and switch B is closed, four sources with equal amplitude but 90-phase difference for each adjacent output lead to a wideband antenna with broadside beam and circular polarization. The photographs of measurement setup and fabricated antenna are shown in Figure 11(b). This single-port antenna is useful for terrestrial land mobile or other wireless applications.

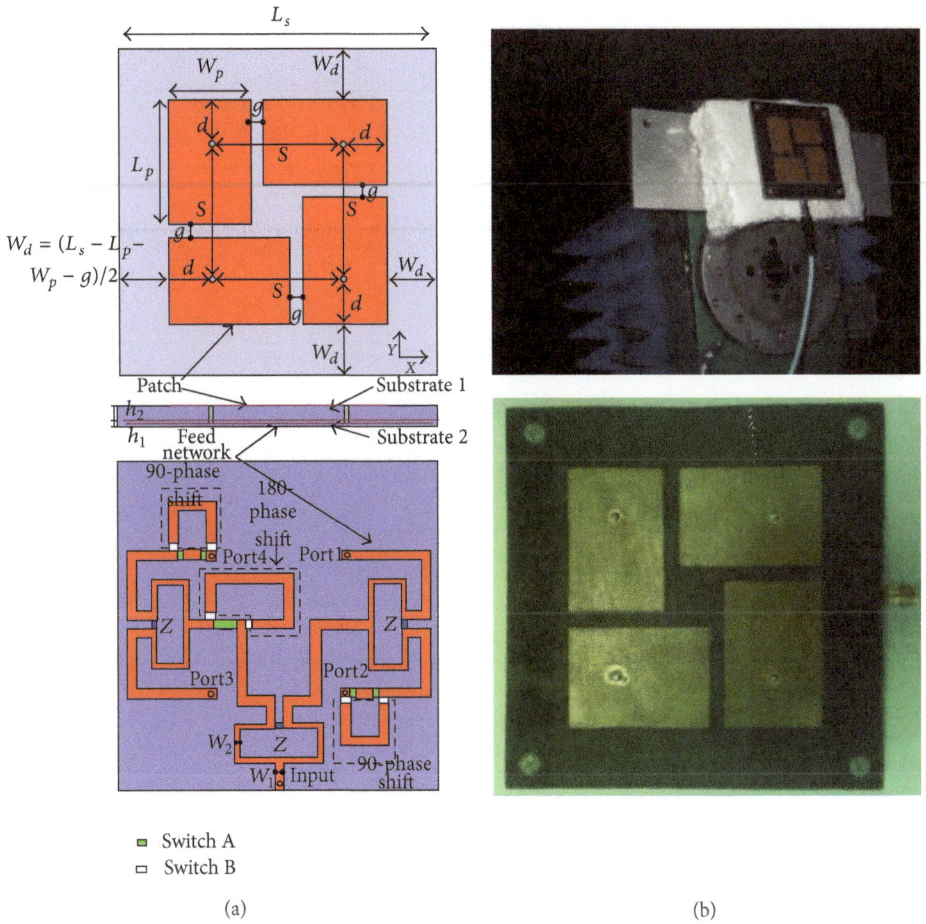

FIGURE 11: (a) The pattern and polarization reconfigurable microstrip antenna. (top) Top view. (middle) Side view. (bottom) Bottom view. (b) Measurement setup photograph and the fabricated antenna (with permission from IEEE).

6. Comparison between Multipattern and Multipolarization Antennas

As shown in Figure 12, multipattern and multipolarization antennas described above can be divided into two types according to the method of feed: connected feed or spatial feed. The classification criteria are whether the feed of the antenna is connected to the radiator or not. The reconfigurable reflectarray antennas and lens antennas, which are classified as spatial feed, generally feed by horn antennas or others, as described in [10, 13, 15, 25, 26, 29, 61]. Most antennas presented in this paper have a connected feed. Compared with the connected-feed antennas, the spatial-feed reconfigurable reflectarray antennas and lens antennas have the advantages of the gain of a reflector antenna changeable with the flexibility of an array, feed losses and layout problems eliminated through the use of a spatial feed, and system amenable to closed-loop control. However, they have limitations of scanning range, feed blockage, and profile.

The reconfigurable antennas with a connected feed include planer antennas and nonplanar antennas. The nonplanar reconfigurable antennas in the form of waveguide antenna and cubic antenna are discussed in [19, 36], respectively. They are able to avoid outside interference and radiation loss due to their simple closed structure. Nevertheless, paralleling with planar reconfigurable antenna, they are more difficult to integrate with the circuit because of nonplanar structure and are less popular in reconfigurable area due to inflexible structure.

The planar reconfigurable antennas fed by coaxial or coplanar waveguide are mainly made up of microstrip antennas [7–9, 33–35, 37–39, 42–46, 48–52, 57–60], substrate integrated waveguide (SIW) antennas [11, 12, 40, 41], and monopole or dipole antennas [3–5, 16, 17, 20, 53, 55]. The microstrip antennas are widely used in reconfigurable antennas with the advantages of small size, light weight, low profile, and ease of integration. The SIW antennas have both the microstrip antenna's advantages of low profile and being easy to integrate and the waveguide antenna's merits of low radiation loss and high efficiency. But, compared to the microstrip-reconfigurable antenna, they are more difficult to process and have lower flexibility. The monopole or dipole antennas have flexible structure, but their ordinary is gain lower than SIW antennas.

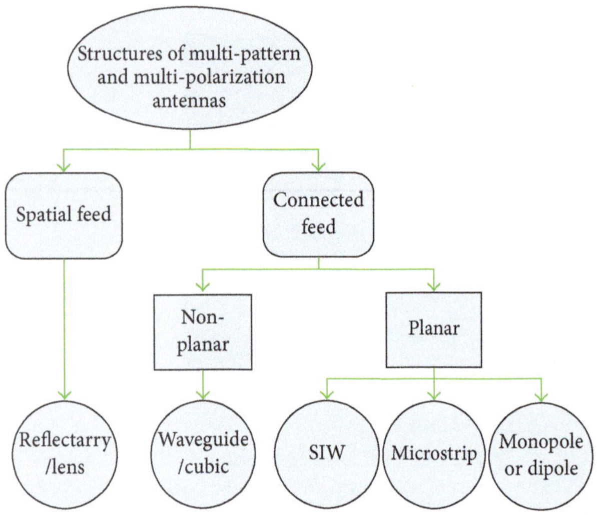

FIGURE 12: The different structures of multipattern and multipolarization antennas.

7. Applications

The reconfigurable antennas with multipattern and multipolarization have been used in both aerospace and terrestrial wireless communication to improve spectrum efficiency, such as wireless systems in dynamic scenarios, adaptive beam scanning, and MIMO systems. In dynamic scenarios, it is necessary to reconfigure the antenna radiation pattern to maintain high data rate, serve a new coverage zone, and limit fading in rainy areas at all possible frequency bands of operation. The dynamic scenarios including satellite and terminal tracking are discussed in [26, 50, 62–64]. In the adaptive beam scanning system, the antenna should automatically change its radiation beam direction corresponding to the various propagation environments or moving target, such as in the system of radar and remote sensing [65–68]. A MIMO system employs multiple antennas at both the transmitter and receiver front ends. Pattern and polarization reconfigurable antennas are widely studied and adopted in MIMO systems for their intrinsic pattern and polarization diversity advantage in mitigating the multipath fading and increasing the channel capacity, especially in pattern and polarization varied environments. The MIMO system with reconfigurable antennas is able to actively manage channel correlation, diversity, and interference [9, 17, 18, 27, 31, 59, 69, 70].

8. Vision and Goal

Even though the research on the multipattern and multipolarization antennas has got some achievements, we still have a long way to go. Until now, the reconfigurable antennas which have been studied and researched are not smart enough that they are not completely multifunctional and software controlled with machine. In addition, the antennas still only play a role of the transducers. A very interesting goal is that we can create field programmable antenna system which enjoys the same flexibility as other field-programmable devices. In the field-programmable antenna system, the reconfigurable antennas can detect changes in their RF environment and react accordingly. If such system is realized, the antennas are evolved from simple transducers to advanced wave processors, along with the advantages of the utilization of radiation pattern selectivity and polarization diversity to transmit over already "busy" frequencies. And the reconfigurable antennas will play more a important role in aerospace and terrestrial applications.

9. Conclusion

This paper presents an overview on the multipattern and multipolarization antennas. The concepts, characteristics, and typical antennas of radiation patterns and polarizations are introduced. A comparison between different multipattern and multipolarization antennas with different structures and working mechanisms are also discussed. The reconfigurable antennas can be well applied to dynamic scenarios, adaptive beam scanning, and MIMO systems. The goal in the future work is to realize the field programmable antenna system to evolve antennas from simple transducers to advanced wave processors.

Acknowledgments

This work was supported in part by the National Science Foundation of China and NSAF under their joint Grant no. 11076031.

References

[1] D. H. Schaubet, F. G. Farrar, S. T. Hayes et al., "Frequency-agile polarization diverse microstrip antennas and frequency scanned arrays," U.S. Patent 4,367,474, Jan. 1983.

[2] C. G. Christodoulou, Y. Tawk, S. A. Lane, and S. R. Erwin, "Reconfigurable antennas for wireless and space applications," *Proceedings of the IEEE*, vol. 100, no. 7, pp. 2250–2261, 2012.

[3] W. S. Kang, J. A. Park, and Y. J. Yoon, "Simple reconfigurable antenna with radiation pattern," *Electronics Letters*, vol. 44, no. 3, pp. 182–183, 2008.

[4] M. I. Lai, T. Y. Wu, J. C. Hsieh, C. H. Wang, and S. K. Jeng, "Design of reconfigurable antennas based on an L-shaped slot and PIN diodes for compact wireless devices," *IET Microwaves, Antennas and Propagation*, vol. 3, no. 1, pp. 47–54, 2009.

[5] L. G. Maloratsky, "Switched directional/omnidirectional antenna module for amplitude monopulse systems," *IEEE Antennas and Propagation Magazine*, vol. 51, no. 5, pp. 90–102, 2009.

[6] F. Peruani and A. Maiti, "Modeling broadcasting using omnidirectional and directional antenna in delay tolerant networks as an epidemic dynamics," *IEEE Journal on Selected Areas in Communications*, vol. 28, no. 4, pp. 524–531, 2010.

[7] M. Ding, R. Jin, J. Geng, X. Guo, and J. Chen, "A high-gain dual-band directional/omnidirectional reconfigurable antenna for WLAN systems," *International Journal of RF and Microwave Computer-Aided Engineering*, vol. 18, no. 3, pp. 225–232, 2008.

[8] M. Donelli, R. Azaro, L. Fimognari, and A. Massa, "A planar electronically reconfigurable wi-fi band antenna based on a

parasitic microstrip structure," *IEEE Antennas and Wireless Propagation Letters*, vol. 6, pp. 623–626, 2007.

[9] D. Piazza, M. D'Amico, and K. R. Dandekar, "Performance improvement of a wideband MIMO system by using two-port RLWA," *IEEE Antennas and Wireless Propagation Letters*, vol. 8, pp. 830–834, 2009.

[10] J. Y. Lau and S. V. Hum, "A planar reconfigurable aperture with lens and reflectarray modes of operation," *IEEE Transactions on Microwave Theory and Techniques*, vol. 58, no. 12, pp. 3547–3555, 2010.

[11] A. Suntives and S. V. Hum, "A fixed-frequency beam-steerable half-mode substrate integrated waveguide leaky-wave antenna," *IEEE Transactions on Antennas and Propagation*, vol. 60, no. 5, pp. 2540–2544, 2012.

[12] A. Suntives and S. V. Hum, "An electronically tunable half-mode substrate integrated waveguide leaky-wave antenna," in *Proceedings of the 5th European Conference on Antennas and Propagation (EUCAP '11)*, pp. 3670–3674, April 2011.

[13] C. Liu and S. V. Hum, "An electronically tunable single-layer reflectarray antenna element with improved bandwidth," *IEEE Antennas and Wireless Propagation Letters*, vol. 9, pp. 1241–1244, 2010.

[14] L. Boccia, I. Russo, G. Amendola, and G. Di Massa, "Multi-layer antenna-filter antenna for beam-steering transmit-array applications," *IEEE Transactions on Microwave Theory and Techniques*, vol. 60, no. 7, pp. 2287–2300, 2012.

[15] W. Hu, M. Y. Ismail, R. Cahill et al., "Liquid-crystal-based reflectarray antenna with electronically switchable monopulse patterns," *Electronics Letters*, vol. 43, no. 14, pp. 744–745, 2007.

[16] C. J. Panagamuwa, A. Chauraya, and J. C. Vardaxoglou, "Frequency and beam reconfigurable antenna using photoconducting switches," *IEEE Transactions on Antennas and Propagation*, vol. 54, no. 2, pp. 449–454, 2006.

[17] D. Piazza, N. J. Kirsch, A. Forenza, R. W. Heath, and K. R. Dandekar, "Design and evaluation of a reconfigurable antenna array for MIMO systems," *IEEE Transactions on Antennas and Propagation*, vol. 56, no. 3, pp. 869–881, 2008.

[18] Z. Li, Z. Du, and K. Gong, "Compact reconfigurable antenna array for adaptive MIMO systems," *IEEE Antennas and Wireless Propagation Letters*, vol. 8, pp. 1317–1320, 2009.

[19] J. Sarrazin, Y. Mahé, S. Avrillon, and S. Toutain, "Pattern reconfigurable cubic antenna," *IEEE Transactions on Antennas and Propagation*, vol. 57, no. 2, pp. 310–317, 2009.

[20] H. Eslami, C. P. Sukumar, D. Rodrigo et al., "Reduced overhead training for multi reconfigurable antennas with beam-tilting capability," *IEEE Transactions on Wireless Communications*, vol. 9, no. 12, pp. 3810–3821, 2010.

[21] Y. Yang, X. Zhao, and T. Wang, "Design of arbitrarily controlled multi-beam antennas via optical transformation," *Journal of Infrared, Millimeter, and Terahertz Waves*, vol. 30, no. 4, pp. 337–348, 2009.

[22] A.-H. Tsai, L.-C. Wang, J.-H. Huang, and R.-B. Hwang, "Stable subchannel allocation for OFDMA femtocells with switched multi-beam directional antennas," in *Proceedings of the IEEE Global Telecommunications Conference (GLOBECOM '11)*, pp. 1–6, December 2011.

[23] G. Krieger, N. Gebert, and A. Moreira, "Multidimensional waveform encoding: a new digital beamforming technique for synthetic aperture radar remote sensing," *IEEE Transactions on Geoscience and Remote Sensing*, vol. 46, no. 1, pp. 31–46, 2008.

[24] G. J. Hayes, J.-H. So, A. Qusba, M. D. Dickey, and G. Lazzi, "Flexible liquid metal alloy (EGaIn) microstrip patch antenna," *IEEE Transactions on Antennas and Propagation*, vol. 60, no. 5, pp. 2151–2156, 2012.

[25] M. F. Jamlos, T. A. Rahman, and M. R. Kamarudin, "The effects of air stacked on reconfigurable quadratic slots antenna for WiMAX and 4G applications," *Microwave and Optical Technology Letters*, vol. 54, no. 9, pp. 2139–2144, 2012.

[26] T. Takano, K. Miura, M. Natori et al., "Deployable antenna with 10-m maximum diameter for space use," *IEEE Transactions on Antennas and Propagation*, vol. 52, no. 1, pp. 2–11, 2004.

[27] Y. Li, Z. Zhang, J. Zheng, and Z. Feng, "Channel capacity study of polarization reconfigurable slot antenna for indoor MIMO system," *Microwave and Optical Technology Letters*, vol. 53, no. 6, pp. 1209–1213, 2011.

[28] B. Q. Wu and K. M. Luk, "The design of wideband reconfigurable antenna with four polarizations," in *Proceedings of the 39th European Microwave Conference (EuMC '09)*, pp. 759–762, September 2009.

[29] A. Clemente, L. Dussopt, R. Sauleau et al., "1-bit reconfigurable unit cell based on PIN diodes for transmit-array applications in X-band," *IEEE Transactions on Antennas and Propagation*, vol. 60, no. 5, pp. 2260–2269, 2012.

[30] G. Wang, R. Bairavasubramanian, B. Pan, and J. Papapolymerou, "Radiofrequency MEMS-enabled polarisation-reconfigurable antenna arrays on multilayer liquid crystal polymer," *IET Microwave Antenna and Propagation*, vol. 5, no. 13, pp. 1594–1599, 2011.

[31] A. Grau, J. Romeu, M. J. Lee, S. Blanch, L. Jofre, and F. de Flaviis, "A dual-Linearly-polarized MEMS-reconfigurable antenna for narrowband MIMO communication systems," *IEEE Transactions on Antennas and Propagation*, vol. 58, no. 1, pp. 4–17, 2010.

[32] P. Seongmin and S. Imseob, "A miniaturized switchable circularly polarized patch antenna controlled by a single diode operation," *Microwave and Optical Technology Letters*, vol. 54, no. 10, pp. 2375–2378, 2012.

[33] K. F. Tong and J. Huang, "New proximity coupled feeding method for reconfigurable circularly polarized microstrip ring antennas," *IEEE Transactions on Antennas and Propagation*, vol. 56, no. 7, pp. 1860–1866, 2008.

[34] W. S. Yoon, J. W. Baik, H. S. Lee, S. Pyo, S. M. Han, and Y. S. Kim, "A reconfigurable circularly polarized microstrip antenna with a slotted ground plane," *IEEE Antennas and Wireless Propagation Letters*, vol. 9, pp. 1161–1164, 2010.

[35] P. Seongmin and S. Youngje, "A circular polarized microstrip antenna with an arrow-shaped slotted ground," *Microwave and Optical Technology Letters*, vol. 54, no. 1, pp. 271–273, 2012.

[36] J. A. Ruiz-Cruz, M. M. Fahmi, S. A. Fouladi, and R. R. Mansour, "Waveguide antenna feeders with integrated reconfigurable dual circular polarization," *IEEE Transactions on Microwave Theory and Techniques*, vol. 59, no. 12, pp. 3365–3374, 2011.

[37] S. H. Hsu and K. Chang, "A novel reconfigurable microstrip antenna with switchable circular polarization," *IEEE Antennas and Wireless Propagation Letters*, vol. 6, pp. 160–162, 2007.

[38] A. Khaleghi and M. Kamyab, "Reconfigurable single port antenna with circular polarization diversity," *IEEE Transactions on Antennas and Propagation*, vol. 57, no. 2, pp. 555–559, 2009.

[39] J. S. Row and M. C. Chan, "Reconfigurable circularly-polarized patch antenna with conical beam," *IEEE Transactions on Antennas and Propagation*, vol. 58, no. 8, pp. 2753–2757, 2010.

[40] Y. J. Cheng, W. Hong, and K. Wu, "Millimeter-wave half mode substrate integrated waveguide frequency scanning antenna with quadri-polarization," *IEEE Transactions on Antennas and Propagation*, vol. 58, no. 6, pp. 1848–1855, 2010.

[41] Y. Dong and T. Itoh, "Millimeter-Wave half mode substrate integrated waveguide frequency scanning antenna with quadri-polarization," *IEEE Transactions on Antennas and Propagation*, vol. 60, no. 2, pp. 760–771, 2012.

[42] X.-X. Yang, B.-C. Shao, F. Yang et al., "A polarization reconfigurable patch antenna with loop slots on the ground plane," *IEEE Antennas and Wireless Propagation Letters*, vol. 11, pp. 69–72, 2012.

[43] Y. F. Wu, C. H. Wu, D. Y. Lai, and F. C. Chen, "A reconfigurable quadri-polarization diversity aperture-coupled patch antenna," *IEEE Transactions on Antennas and Propagation*, vol. 55, no. 3, pp. 1009–1012, 2007.

[44] R. H. Chen and J. S. Row, "Single-fed microstrip patch antenna with switchable polarization," *IEEE Transactions on Antennas and Propagation*, vol. 56, no. 4, pp. 922–926, 2008.

[45] O. H. Karabey, S. Bildik, and S. Strunck, "Continuously polarisation reconfigurable antenna element by using liquid crystal based tunable coupled line," *Electronics Letters*, vol. 48, no. 3, pp. 141–143, 2012.

[46] B. Kim, B. Pan, S. Nikolaou, Y. S. Kim, J. Papapolymerou, and M. M. Tentzeris, "A novel single-feed circular microstrip antenna with reconfigurable polarization capability," *IEEE Transactions on Antennas and Propagation*, vol. 56, no. 3, pp. 630–638, 2008.

[47] H. Emami, N. Sarkhosh, and E. R. L. Lara, "Reconfigurable photonic feed for sinuous antenna," *Journal of Light Wave Technology*, vol. 30, no. 16, pp. 2725–2732, 2012.

[48] P. Y. Qin, A. R. Weily, Y. J. Guo, and C. H. Liang, "Polarization reconfigurable U-slot patch antenna," *IEEE Transactions on Antennas and Propagation*, vol. 58, no. 10, pp. 3383–3388, 2010.

[49] M. S. Nishamol, V. P. Sarin, D. Tony et al., "An electronically reconfigurable microstrip antenna with switchable slots for polarization diversity," *IEEE Transactions on Antennas and Propagation*, vol. 59, no. 9, pp. 3424–3427, 2011.

[50] J. H. Lim, G. T. Back, and T. Y. Yun, "Polarization-diversity cross-shaped patch antenna for satellite-DMB systems," *ETRI Journal*, vol. 32, no. 2, pp. 312–318, 2010.

[51] X. Yuan, Z. Li, D. Rodrigo et al., "A parasitic layer-based reconfigurable antenna design by multi-objective optimization," *IEEE Transactions on Antennas and Propagation*, vol. 60, no. 6, pp. 2690–2701, 2012.

[52] X.-S. Yang, B.-Z. Wang, S.-H. Yeung et al., "Circularly polarized reconfigurable crossed-vagi patch antenna," *IEEE Antennas and Propagation Magazine*, vol. 53, no. 5, pp. 65–80, 2011.

[53] Y. Dong and T. Itoh, "Planar ultra-wideband antennas in ku- and k-band for pattern or polarization diversity applications," *IEEE Transactions on Antennas and Propagation*, vol. 60, no. 6, pp. 2886–2895, 2012.

[54] B. Poussot, J. M. Laheurte, L. Cirio, and O. Picon, "Diversity gain measurements of a reconfigurable antenna with switchable polarization," *Microwave and Optical Technology Letters*, vol. 49, no. 12, pp. 3154–3158, 2007.

[55] B. Poussot, J. M. Laheurte, L. Cirio, O. Picon, D. Delcroix, and L. Dussopt, "Diversity measurements of a reconfigurable antenna with switched polarizations and patterns," *IEEE Transactions on Antennas and Propagation*, vol. 56, no. 1, pp. 31–38, 2008.

[56] P. Daniele and D. Michele, "Pattern and polarization reconfigurable CRLH leaky wave antenna," in *Proceedings of the 4th European Conference on Antennas and Propagation (EuCAP '10)*, pp. 1–5, April 2010.

[57] W. Cao, B. Zhang, A. Liu et al., "A reconfigurable microstrip antenna with radiation pattern selectivity and polarization diversity," *IEEE Antennas and Wireless Propagation Letters*, vol. 11, pp. 453–456, 2012.

[58] W. Chen, J. Sun, and Z. Feng, "A novel compact reconfigurable polarization and pattern antenna," *Microwave and Optical Technology Letters*, vol. 49, no. 11, pp. 2802–2805, 2007.

[59] D. Piazza, P. Mookiah, M. D'Amico, and K. R. Dandekar, "Experimental analysis of pattern and polarization reconfigurable circular patch antennas for MIMO systems," *IEEE Transactions on Vehicular Technology*, vol. 59, no. 5, pp. 2352–2362, 2010.

[60] W. L. Liu, T. R. Chen, S. H. Chen, and J. S. Row, "Reconfigurable microstrip antenna with pattern and polarisation diversities," *Electronics Letters*, vol. 43, no. 2, pp. 77–78, 2007.

[61] A. G. Besoli and F. D. Flaviis, "A multifunctional reconfigurable pixeled antenna using mems technology on printed circuit board," *IEEE Transactions on Antennas and Propagation*, vol. 59, no. 12, pp. 4413–4424, 2011.

[62] M. Ali, A. T. M. Sayem, and V. K. Kunda, "A reconfigurable stacked microstrip patch antenna for satellite and terrestrial links," *IEEE Transactions on Vehicular Technology*, vol. 56, no. 2, pp. 426–435, 2007.

[63] M. D. van de Burgwal, K. C. Rovers, KC. H. Blom et al., "Mobile satellite reception with a virtual satellite dish based on a reconfigurable multi-processor architecture," *Microprocessors and Microsystems*, vol. 35, no. 8, pp. 716–728, 2011.

[64] J. A. Lee, S. B. Byun, J. H. Lim, and T. Y. Yun, "Reconfigurable antenna for wideband code division multiple access and korean satellite digital multimedia broadcasting controlled by PIN-diodes," *Microwave and Optical Technology Letters*, vol. 49, no. 6, pp. 1334–1337, 2007.

[65] M. Alshershby and J. Q. Lin, "Reconfigurable plasma antenna produced in air by laser-induced filaments: passive radar application," in *Proceedings of the International Conference on Optoelectronics and Microelectronics (ICOM '12)*, pp. 364–371, August 2012.

[66] N. Romano, G. Prisco, and F. Soldovieri, "Design of a reconfigurable antenna for ground penetrating radar applications," *Progress in Electromagnetics Research*, vol. 94, pp. 1–18, 2009.

[67] F. Soldovieri and N. Romano, "The mutual interaction between the reconfigurable transmitting and receiving antennas in ground penetrating radar surveys," *Journal of Electromagnetic Waves and Applications*, vol. 23, no. 14-15, pp. 1919–1928, 2009.

[68] M. F. Jamlos, O. A. Aziz, T. A. Rahman et al., "A reconfigurable radial line slot array (RLSA) antenna for beam shape and broadside application," *Journal of Electromagnetic Waves and Applications*, vol. 24, no. 8-9, pp. 1171–1182, 2010.

[69] P. A. Martin, P. J. Smith, and R. Murch, "Improving space-time code performance in slow fading channels using reconfigurable antennas," *IEEE Communications Letters*, vol. 16, no. 4, pp. 494–497, 2012.

[70] F. Fazel, A. Grau, H. Jafarkhani, and F. de Flaviis, "Space-time-state block coded MIMO communication systems using reconfigurable antennas," *IEEE Transactions on Wireless Communications*, vol. 8, no. 12, pp. 6019–6029, 2009.

3

Interference-Based Decode and Forward Scheme Using Relay Nodes in Heterogeneous Networks

Kentaro Nishimori,[1] Koshiro Kitao,[2] and Tetsuro Imai[2]

[1] *Faculty of Engineering, Niigata University, Niigata-shi, Niigata 950-2181, Japan*
[2] *NTT DOCOMO, INC., 3-5 Hikari-no-oka, Yokosuka-shi, Kanagawa 239-8536, Japan*

Correspondence should be addressed to Kentaro Nishimori, nishimori@ie.niigata-u.ac.jp

Academic Editor: Ding-Bing Lin

This paper proposes interference-based decode and forward scheme that utilizes relay stations (RSs). In Long-Term-Evolution (LTE-) Advanced, heterogeneous networks in which femto- and picocells are overlaid onto macrocells are extensively discussed. However, interference between macro- and pico(femto)cells arises due to their different transmit power levels. Unlike conventional cooperative transmission schemes, the RS decodes interference in the first transmit timing period and forwards it to the user equipment (UE) in the second period. Moreover, cooperative transmission can be achieved without stopping the transmission from the base station (BS) to UE when forwarding the interference from the RS to the UE by utilizing the fact that signal-to-noise power ratio (SNR) between the RS and UE is much greater than that between the BS and UE. The basic performance of the proposed method is shown based on computer simulation. Moreover, the interference temperature and shadowing effect are measured when considering the coexistence between macro- and femtocells, and the performance of the proposed method is verified using measured shadowing effect.

1. Introduction

Due to the immense popularity of mobile phones and wireless LAN systems, increasing the data rate within a limited spectrum is one of the most important goals for wireless system design. Macrocells, the service areas which are one to several kilometers, were introduced in conventional cellular systems. On the other hand, femto- and picocells are currently the focus because small cells can enhance the frequency utilization and can be established using a low-power-consuming base station (BS) that requires a small installation space. In Long-Term-Evolution (LTE-) Advanced, heterogeneous networks are extensively discussed in addition to traditional well-planned macrocell deployment to further improve the frequency utilization [1–3]. In heterogeneous network deployment, low power nodes such as femto-, pico-, and relay nodes are placed throughout a macrocell layout, and they are placed generally in an unplanned manner. Hence, interference between macro- and pico(femto)cells occurs because the transmission power of pico(femto)cells is different from that of macrocell and the service coverage areas are different between these cells.

In this paper, we consider a macrocell system consisting of a BS and user terminals (UEs) and propose an interference-based decode and forward technique using a relay station (RS) to cancel the interference from pico- or femtocell. In recent years, cooperative transmission using RSs has been primarily used to improve the overall system capacity or to obtain a diversity gain [4, 5]. On the other hand, RSs can also be utilized for interference cancellation [6, 7]. The RS forwards information regarding the interference that is used by a receiver to decode the desired signal. Ideas on forwarding interference were considered in [6, 7] for the case when a destination node receives the desired signal and interference with small power, and instead of forwarding the desired signal, the RS forwards information regarding the interference. Moreover, interference cancellation techniques with the aid of the RS for cognitive radio are proposed [8, 9]. However, the treatment in [6–9] is information-theoretic and did not explain the concrete procedure of how to use the RS in detail. In this work, we focus on a more practical setting with concrete system/modulation parameters, as well as a complete procedure for channel estimation, which is a prerequisite for actualizing distributed interference cancellation.

Interference-based decode and forward scheme using an RS for cognitive radio was proposed in [10]. In this method, the RS decodes and forwards the interference to a UE in a secondary system from a BS in the primary system using periodic training signals in the primary system. However, the time for interference transfer from the RS to UE is overhead in this method. We proposed a method that addresses this problem by utilizing the fact that the signal-to-noise ratio (SNR) between the RS and BS is greater than that between the BS and UE [11]. The optimal position of the relay stations by the proposed method is clarified based on computer simulation [11]. This paper brings significantly broadened treatment of the proposed method.

(i) The basic performance of the proposed method is evaluated in more detail. In particular, the proposed method is effective compared to the conventional adaptive array, regardless of the condition of SIR, SNRs of BS/RS and RS/UE.

(ii) Not only zero forcing (ZF) but also minimum mean square error (MMSE) algorithms with 2-element adaptive array at the UE are evaluated as the reference schemes, because MMSE algorithm is evaluated for the interference rejection between multicells for LTE-Advanced [12].

(iii) The performance of the proposed method is evaluated using measured data regarding a shadowing effect when considering coexistence between marco and femtocells.

When considering the RS as a virtual antenna which helps the UE from the interference, the configuration using the proposed method can be regarded as one of configurations regarding distributed antenna system.

The rest of the paper is organized as follows. Section 2 describes the target scenario and defines the problem to be addressed. Section 3 presents the proposed technique for interference cancellation using an RS. Section 4 provides quantitative performance analysis of the proposed scheme with theoretical propagation models. The affect on shadowing with the measured data is evaluated in Section 5. Section 6 concludes the paper.

2. Target Scenario and Problem Definition

Figure 1 shows a configuration for heterogeneous networks in LTE-Advanced [1]. In such heterogeneous networks, picocells, femtocells, and relays are deployed inside the service area of a macrocell in order to enhance the frequency utilization and guarantee the quality of service in high traffic areas. The picocell service area is several tens of meters. Femtocells have drawn much attention recently because their transmission power level is very low and thus they only require a low-power-consuming BS and small installation

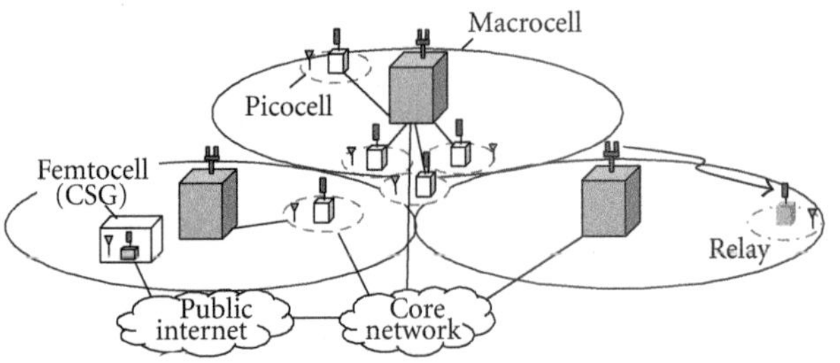

Figure 1: Configuration of heterogeneous network.

space. We focus on the coexistence between macro- and pico(femto)cells that use the same frequency band. However, the interference between macro- and pico(femto)cell occurs because the transmission power of pico(femto)cells is different from that of macrocells and the service coverage areas are different between macro- and pico(femto)cells. In this paper, we focus on the interference from a pico(femto) BS (P(F)-BS) to a set of macro-user equipment (M-UE), and we propose interference-based decode and forward scheme using RSs.

Figure 2 shows the signal format of the BS in macro- and pico(femto)cells. Frequency division duplex (FDD) is generally used in macrocell BS (M-BS). On the other hand, pico- or femto-BS (P-BS or F-BS) employing time division duplex (TDD) is assumed in this paper. Both FDD and TDD are included as the standard of LTE-Advanced [13]. In such a scenario, P-BS transmits a signal using time slot 1 and receives a signal using time slot 2 as in Figure 2. In this paper, we utilize this feature in the proposed method.

3. Proposed Method

The proposed method cancels the interference at the M-UE from the P-BS by using the aid of as RS in the macrocell (M-RS). Let us assume a downlink scenario. The RS with decode and forward generally receives the desired signal from the BS in time slot 1 and this signal is forwarded to the UE via the RS in the next time slot (time slot 2) [4, 5]. Figure 3 shows the concept of the proposed method. Although three macrocells are considered as *general* system configuration in Figure 1, for the sake of ease, a macrocell is considered in Figure 3. In the proposed method, the M-RS decodes the interference from the P-BS (F-BS) at time slot 1 and the decoded interference is forwarded to the M-UE during time slot 2 when the interference from the P-BS (F-BS) arrives at the M-UE. We assume that M-BS has multiple antennas but each RS has only one antenna in this paper. For simplification, P-BS is treated as the interference hereafter but F-BS is also treated as interference in this paper. The transmit timing of the preamble signals for CSI estimation is shown in Figure 4. The proposed scheme can be explained in the following six steps.

(1) All CSI (M-BS/M-UE, M-BS/M-RE, M-RS/M-UE, P-BS/M-UE, and P-BS/M-RS) estimations are carried out using the preamble in Figure 4.

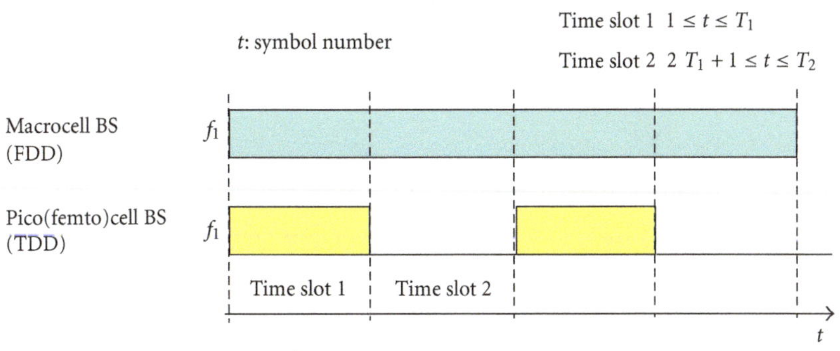

FIGURE 2: Signal format of base station in macro- and pico(femto)cells.

(2) The M-BS controls the transmit weight in order not to arrive the signal at the M-RS by using the CSI in Step 1 during the data period of time slot 1.

(3) The interference incurred during the data period of time slot 1 is decoded using the received signal at the M-RS and the CSI from Step 1.

(4) The M-RS transmits the interference obtained in Step 3 to the M-UE during the next period of time slot 2.

(5) The interference incurred during the data period for time slot 2 is estimated from the received signal at the M-UE and the CSI from Step 1.

(6) The desired signals, $S_1(t)$ and $S_2(t)$, for the M-UE are estimated using the interference decoded in Step 5, the received signal in Steps 2 and 4, and the CSI from Step 1.

Steps 1 to 3 are employed during time slot 1 and Steps 4 to 6 are employed during time slot 2. In Step 1, we divide the period for the training sequence into three parts as shown in Figure 4 in order to estimate all the required CSIs. During Period 1, the M-BS and M-RS stop their transmissions to the M-UE such that only the CSI of the interference from the P-BS to M-UE and M-RS is acquired. The received signals, $X_{P1,U}(t)$ and $X_{P1,R}(t)$, at the M-UE and M-RS during Period 1 ($1 \leq t \leq T_{p1}, T_{p1} = T_p/3$ in Figure 4) are given below:

$$X_{P1,U}(t) = h_{PU} I_{P1}(t) + N_U(t), \tag{1}$$

$$X_{P1,R}(t) = h_{PR} I_{P1}(t) + N_R(t), \tag{2}$$

where h_{PU} and h_{PR} denote the channel responses of the interfering signals for M-UE and M-RS, respectively. T_p denotes the length of the preamble signal for P-BS. $I_{P1}(t)$ represents the known preamble signals from the P-BS in Period 1. $N_U(t)$ and $N_R(t)$ represent the thermal noise for the M-UE and M-RS from the P-BS, respectively.

During Periods 2 and 3, signals are transmitted by the M-BS and M-RS, respectively. The M-BS (M-RS) stops its transmission in Period 2 (3). The received signals, $X_{P2,U}(t)$ and $X_{P3,U}(t)$, at the M-UE for Period 2 ($T_{p1} + 1 \leq t \leq T_{p2}, T_{p2} = 2T_{p1}$) and Period 3 ($T_{p2} + 1 \leq t \leq T_p$) are given as

$$X_{P2,U}(t) = h_{BU} S_P(t) + h_{PU} I_{P2}(t) + N_U(t), \tag{3}$$

$$X_{P3,U}(t) = h_{RU} S_P(t) + h_{PU} I_{P3}(t) + N_U(t), \tag{4}$$

where h_{BU} and h_{PU} denote the channel responses from the M-BS and M-RS to the M-UE, respectively. $I_{P2}(t)$ and $I_{P3}(t)$ represent the known preamble signals from P-BS in Periods 2 and 3, respectively. $S_P(t)$ denotes the known preamble signals from the M-BS and M-RS in Periods 2 and 3, respectively. During Period 2, M-RS receives the signal form M-BS and interference from P-BS. The received signals, $X_{P2,R}(t)$, at the M-RS for Period 2 are given as

$$X_{P2,R}(t) = h_{BR} S_P(t) + h_{PR} I_{P2}(t) + N_R(t), \tag{5}$$

where h_{BR} denotes the channel response from the M-BS to M-RS.

All CSIs are obtained using (1) to (5). The least square (LS) criteria is adopted for CSI estimation. The estimated CSIs of P-BS/M-UE and P-BS/M-RS are obtained using (1) and (2), respectively, as given below:

$$\widetilde{h}_{PU} = \frac{1}{T_{p1}} \sum_{t=1}^{Tp1} \frac{X_{P1,U}(t)}{I_{P1}(t)}, \tag{6}$$

$$\widetilde{h}_{PR} = \frac{1}{T_{p1}} \sum_{t=1}^{Tp1} \frac{X_{P1,R}(t)}{I_{P1}(t)}. \tag{7}$$

The estimated CSIs of M-BS/M-UE, M-BS/M-RS and M-RS/M-UE are obtained by using (3) and (7), respectively, as follows:

$$\widetilde{h}_{BU} = \frac{1}{T_{p1}} \sum_{t=Tp1+1}^{Tp2} \frac{X_{P2,U}(t) - \widetilde{h}_{PU} I_{P2}(t)}{S_P(t)}, \tag{8}$$

$$\widetilde{h}_{BR} = \frac{1}{T_{p1}} \sum_{t=Tp1+1}^{Tp2} \frac{X_{P2,R}(t) - \widetilde{h}_{PR} I_{P2}(t)}{S_P(t)}, \tag{9}$$

$$\widetilde{h}_{RU} = \frac{1}{T_{p1}} \sum_{t=Tp2+1}^{Tp3} \frac{X_{P3,U}(t) - \widetilde{h}_{PU} I_{P3}(t)}{S_P(t)}. \tag{10}$$

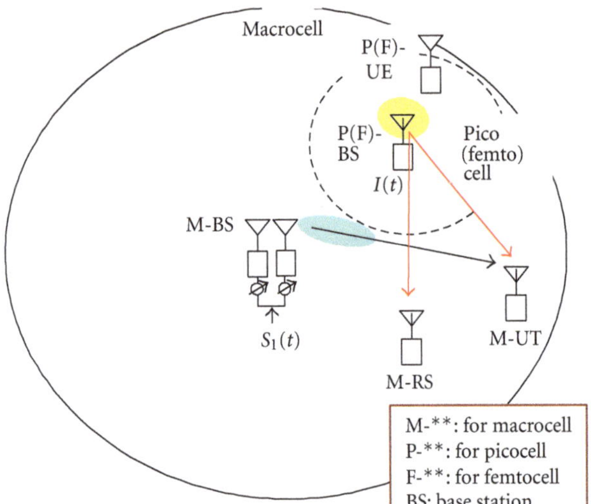

(a) Behavior at $t = t_1$

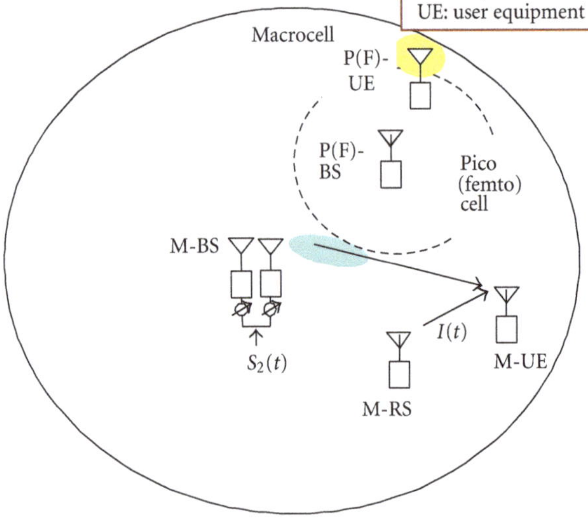

(b) Behavior at $t = t_2$

FIGURE 3: Concept of proposed method.

As shown in (6) to (10), T_{p1} is required as the number of symbols for the channel estimation. Throughout the evaluation, T_{p1} is set to be five.

The propagation channel between M-BS and M-RS is not basically changed because these stations do no move, unlike M-UE. Hence, we can consider that the propagation channel is not changed between Steps 1 and 2. When considering the channel between M-BS and M-UE, the adaptive weight may be changed due to the Doppler shift. On the other hand, transmission speed for one time slot can be assumed to be much shorter than the Doppler frequency when considering recent broadband wireless communication. However, when the Doppler frequency is very high, the adaptive weight might be changed. We must consider this issue as one of future works.

In the second step (Step 2), the M-BS transmits a signal to the M-UE with an antenna pattern that creates a null toward the M-RS during the data period for time slot 1 in Figure 3. The received signals at the M-UE and M-RS are expressed as

$$X_{t_1,U}(t) = h_{BU}S_1(t) + h_{PU}I(t) + N_U(t), \tag{11}$$

$$X_{t_1,R}(t) = h_{PR}I(t) + N_R(t), \tag{12}$$

where $S_1(t)$ and $I(t)$ represent *unknown* data signals for the M-UE from the M-BS and P-BS, respectively. Hence, the M-UE must decode and cancel interference $I(t)$ without any information pertaining to the interference.

$S_1(t)$ cannot be decoded using the procedure in Steps 1 and 2. On the other hand, interference $I(t)$ can be estimated at the M-RS using (12) and $\widetilde{h}_{PR}$ obtained in Step 1. In Step 3, the M-RS decodes the interference that occurs during the data period of time slot 1. The decoded interference, $\widetilde{I}(t)$, at the M-RS is obtained using (7) and (12) as

$$\begin{aligned} \widetilde{I}(t) &= \frac{X_{t_1,R}(t)}{\widetilde{h}_{PR}} \\ &= \frac{h_{PR}}{\widetilde{h}_{PR}}I(t) + \frac{N_R(t)}{\widetilde{h}_{PR}} \\ &\approx I(t) + \frac{N_R(t)}{\widetilde{h}_{PR}}. \end{aligned} \tag{13}$$

Here, interference-to-noise power ratio (INR) at the M-RS is $E[|h_{PR}|^2/|N_R(t)|^2]$. When the INR is large, the second term on the right side of (13) is negligible. In Step 4, the M-RS transmits interference $\widetilde{I}(t)$ estimated in Step 3 to the M-UE using the next data period for time slot 2 in Figure 3. The M-BS transmits a signal to the M-UE using this period. The received signal at the M-UE is denoted by

$$X_{t_2,U}(t) = h_{BU}S_2(t) + h_{RU}\widetilde{I}(t) + N_U(t), \tag{14}$$

where $S_2(t)$ represents *unknown* data signals for the M-UE from the M-BS at time slot 2.

In Step 5, after the transmission of the interference, the interference from the P-BS during the data period for time slot 1, $\widetilde{I}(t)$, is estimated by using the received signal in (14). First, (14) is changed using (10) as:

$$\frac{X_{t_2,U}(t)}{\widetilde{h}_{RU}} = \frac{h_{BU}}{\widetilde{h}_{RU}}S_2(t) + \frac{h_{RU}}{\widetilde{h}_{RU}}\widetilde{I}(t) + \frac{N_U(t)}{\widetilde{h}_{RU}}. \tag{15}$$

When $h_{RU} \approx \widetilde{h}_{RU}$, signal-to-interference power ratio (SIR) and INR at M-UE are denoted as $E[|h_{BU}|^2/|h_{RU}|^2]$ and $E[|h_{RU}|^2/|N_U(t)|^2]$, respectively. It is reasonable to assume that $E[|h_{RU}|^2]$ is much larger than $E[|h_{BU}|^2]$ and $E[|N_U(t)|^2]$, because the M-RS should be located near the M-UE. In other word, the SIR is very small and the INR is very large. Hence, the first and third terms on the right side of (15) can be approximately assumed to be much smaller than the second term on the right side of (15). Note that SNR, $E[|h_{BU}|^2/|N_U(t)|^2]$ is not small and the first term on the right side of (15) is not zero. As a result, when we ignore the first and third terms on the right side of (15), (15) can be approximated as

$$\frac{X_{t_2,U}(t)}{\widetilde{h}_{RU}} \approx \widetilde{I}(t) \equiv \hat{I}(t). \tag{16}$$

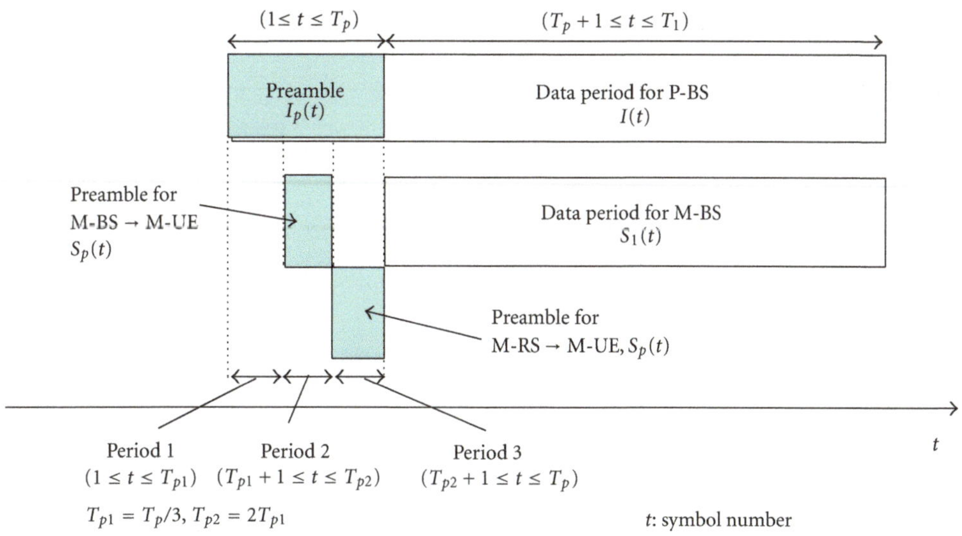

FIGURE 4: Transmit timing of preamble signals for CSI estimation.

In Step 6, the desired signals, $S_1(t)$ and $S_2(t)$, can be estimated using interference $\hat{I}(t)$ estimated in Step 5, received signals $X_{t_1,U}(t)$ and $X_{t_2,U}(t)$ in Steps 2 and 4, and CSIs $\widetilde{h}_{BU}$, $\widetilde{h}_{PU}$, and $\widetilde{h}_{RU}$ obtained in Step 1. The estimated desired signals, $\widetilde{S}_1(t)$ and $\widetilde{S}_2(t)$, can be denoted as.

$$\widetilde{S}_1(t) = \frac{X_{t_1,U}(t) - \widetilde{h}_{PU}\hat{I}(t)}{\widetilde{h}_{BU}},$$
$$\widetilde{S}_2(t) = \frac{X_{t_2,U}(t) - \widetilde{h}_{RU}\hat{I}(t)}{\widetilde{h}_{BU}}. \tag{17}$$

4. Basic Characteristics of the Proposed Method

In this section, we evaluate the performance of the proposed method by computer simulation. Figure 5 represents the configurations of the proposed and reference scheme. We compare the proposed interference cancellation method (Method (A)) to a reference scheme (Method (B)), an adaptive array using ZF/MMSE. For a fair comparison, the total number of antennas at the transmitter and receiver sites are identical for Methods (A) and (B). The numbers of antennas at the M-BS and the M-UE are two, respectively, in Method (B). The number of antennas at the M-BS is two and that at the M-RS plus M-UE is also two in Method (A). In Method (B), the M-UE uses an adaptive array with ZF/MMSE [12]. When using the ZF algorithm, the decoded signal is denoted as

$$\begin{pmatrix} \widetilde{S}(t) \\ \widetilde{I}(t) \end{pmatrix} = \begin{pmatrix} h_{S_1} & h_{I_1} \\ h_{S_2} & h_{I_2} \end{pmatrix}^{-1} \begin{pmatrix} X_{U_1} \\ X_{U_2} \end{pmatrix}$$
$$= \begin{pmatrix} \mathbf{h}_S & \mathbf{h}_I \end{pmatrix}^{-1} \mathbf{X}_U$$
$$= \mathbf{H}^{-1}\mathbf{X}_U, \tag{18}$$

where $\mathbf{h}_S = [h_{S_1}, h_{S_2}]^T$ and $\mathbf{h}_I = [h_{I_1}, h_{I_2}]^T$ denote the channel responses for the M-UE from the M-BS and P-BS, respectively. $\mathbf{X}_U = [X_{U_1}, X_{U_2}]^T$ is the received signal vector at the M-UE. When using the MMSE algorithm, the decoded signal is denoted as

$$\begin{pmatrix} \widetilde{S}(t) \\ \widetilde{I}(t) \end{pmatrix} = \left(\mathbf{H}^H\mathbf{H} + \sigma^2\mathbf{U}\right)^{-1}\mathbf{H}^H\mathbf{X}_U, \tag{19}$$

where $\mathbf{U}$ is 2×2 identity matrix and σ^2 is the noise power. A total transmit power by M-BS and M-RS in the proposed method is the same with that by M-BS in the conventional method.

Transmissions from the M-BS to M-UE in the macrocell use adaptive modulation. Table 1 gives the required SNRs for different modulation levels. The SNR for each modulation scheme is given in Table 1 for the bit error rate (BER) of 5×10^{-3} [14]. For all the methods, the modulation level is decided based on the instantaneous SNR. The modulation scheme for the interference is QPSK.

The total transmission rate, R, is defined as

$$R = \frac{\sum_{k=1}^{N_F} R(k)}{N_F},$$
$$R(k) = M_a[1 - \mathrm{BER}(k)], \tag{20}$$

where N_F represents the number of frames. M_a denotes the number of bits per symbol when applying the adaptive modulation. $R(k)$ and $\mathrm{BER}(k)$ are the transmission rate and BER for the kth frame, respectively. In the simulation, we set the length per frame to 600 and $N_F = 20000$. A Rayleigh fading environment is assumed among all nodes. In the evaluation, we did not consider channel codes in order to evaluate the basic ability of the proposed method.

First, the transmission rate is evaluated versus the SIR in Figure 6, in order to confirm the effectiveness of the approximation in (16). Here, we assume that the desired

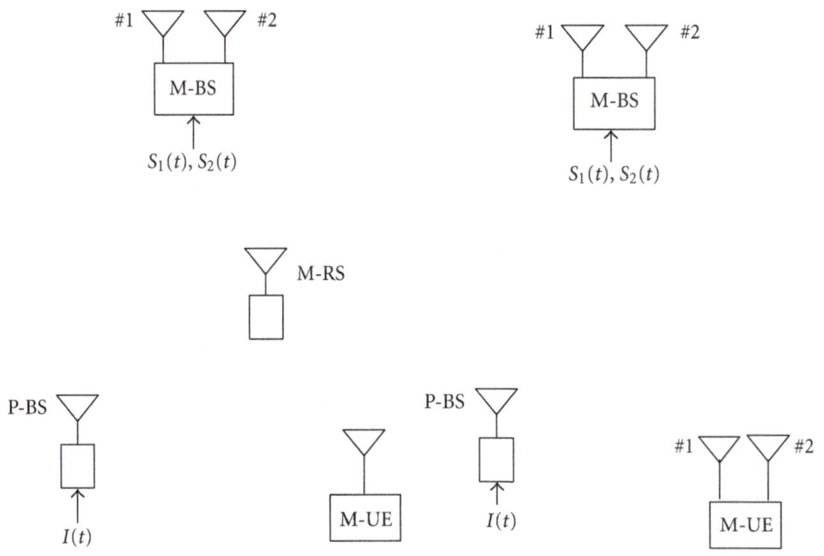

(a) Configuration for Method (A) Method (A): Proposed method

(b) Configuration for Method (B) Method (B): Adaptive array using ZF/MMSE algorithms

FIGURE 5: Proposed and reference schemes.

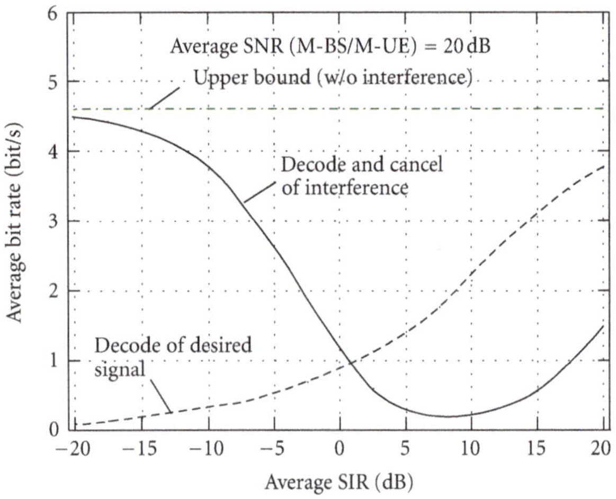

FIGURE 6: Transmission rate versus SIR.

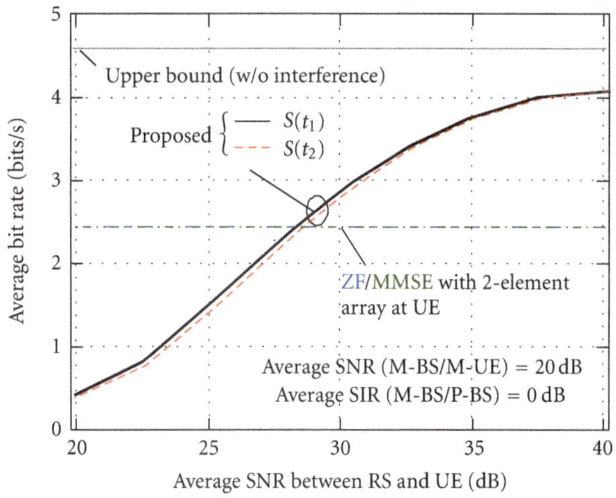

FIGURE 7: Transmission rate versus SNR between M-RS and M-UE.

signal (S) and the interference (I) are the signals from the M-BS and M-RS to the M-UE at time slot 2 in Figure 3, respectively. For comparison, the results when only the desired signal is decoded are shown. Figure 6 shows that the transmission rate is improved when the SIR is decreased. When the SIR is less than −10 dB, the degradation compared to the transmission rate with the upper bound is less than 1 bit/s. Hence, the approximation in (16) is effective when considering the SNR between the M-RS and M-UE to be greater than that between the M-BS and M-UE.

The SNR between the M-RS and M-UE is a very important parameter that determines the ability for decoding the interference. The transmission rates of the proposed method versus the SNR between the M-RS and M-UE are plotted in Figure 7. The average SNR between the M-BS and M-UE and the SIR between the M-BS and P-BS for the M-UE are set to 20 and 0 dB, respectively. The average transmission rate for time slots 1 and 2 is shown in Figure 7. As shown in Figure 7, the transmission rates for time slots 1 and 2 are improved according to the SNR between the M-RS and M-UE and both transmission rates are almost the same. Moreover, the proposed method obtains a higher transmission rate than that for 2-element adaptive array using ZF/MMSE algorithm when the SNR between the M-RS and M-UE is greater than 28 dB. Note that the average bit rate by the MMSE algorithm is almost same with that by the ZF algorithm. The proposed method is effective compared to the adaptive array with ZF/MMSE when the SNR between the M-RS and M-UE is 8 dB higher than that between the M-BS and M-UE.

Figure 8 plots the transmission rates of the proposed method and ZF/MMSE adaptive array (Method (B)) versus

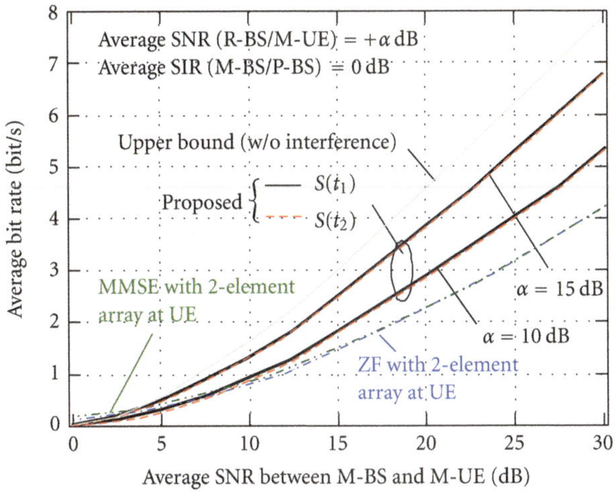

FIGURE 8: Transmission rate versus SNR between M-RS and M-UE.

the SNR between the M-BS and M-UE. Note that the average bit rate by the MMSE algorithm is almost the same with that by the ZF algorithm except very small SNR. The results are plotted when the SNR between the M-RS and M-UE is 10/15 dB (α in Figure 8), greater than that between the M-BS and M-UE. As shown in the figure, by applying the proposed method, the transmission rate is improved compared to that for Method (B), when the SNR between the B-RS and M-UE is greater than 10 and 7 dB, respectively, and when α is 10 and 15 dB.

5. Performance Evaluation Considering Coexistence between Macro- and Femtocells Using Measured Shadowing Effect

In this section, the performance of the proposed method is evaluated when considering the coexistence between macro and femtocells using measured shadowing effect. Since there are UEs that do not belong to a closed subscriber group (CSG) BS in a femtocell, such UEs might give/receive interference to/from the femtocell [1]. Moreover, the femtocell is generally used in indoor scenarios, and the propagation characteristics between the femto- and macrocells are different from those inside a macrocell. We performed measurements to obtain the shadowing characteristics between femto- and macrocells and inside a macrocell and reflected the results in a computer simulation.

Figure 9 shows the considered measurement environment. The measurements are taken in the Faculty of Engineering building of Niigata University in Japan. The M-UE is located in a parking area and moved around Courses 1 to 10 in Figure 9. The M-BS and F-BS are located at the top of the building and the 4th floor in Figure 9, respectively. The antenna heights of the M-BS and F-BS are 22.5 and 16.5 m, respectively. The height of the M-UE is 1 m. The F-BS is located in the place in Figure 10, because the influence of the shadowing might be changed by the difference of the place in Figure 10. The radio frequency is 2.2 GHz and a continuous wave (10 W) is transmitted from the M-BS or F-BS to

TABLE 1: Required SNR for different modulation levels.

SNR	Modulation
3.63 dB	BPSK
8.40 dB	QPSK
12.08 dB	8QAM
15.39 dB	16QAM
18.55 dB	32QAM
21.63 dB	64QAM
24.67 dB	128QAM
27.70 dB	256QAM
30.72 dB	512QAM
33.73 dB	1024QAM

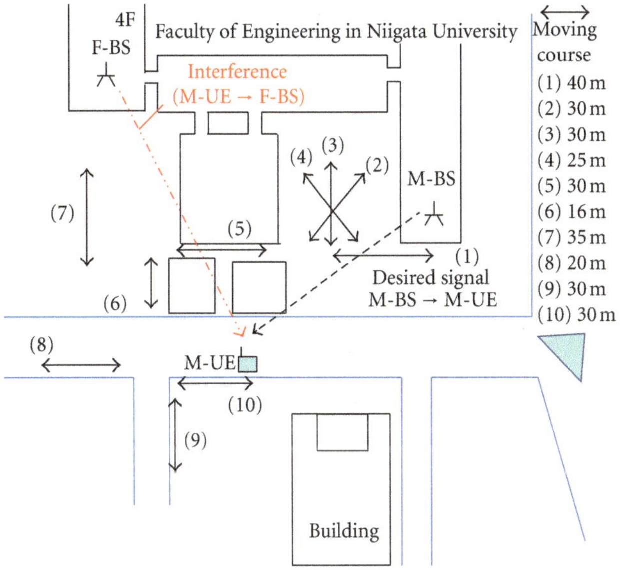

FIGURE 9: Measurement environment.

the M-UE. The received power and shadowing effect are measured. The shadowing effect is obtained by calculating the standard deviation of median values in the received power. The median values are obtained with the distance of 1 m. We assume the scenario where the interference from the femtocell arrives at the M-UE when the M-BS communicates with the M-UE.

Figure 11 denotes the received power for the desired signal (M-BS/M-UE, outdoor) and interference (F-BS/M-UE, Cases 1, 2, and 3) when considering Courses 1 to 10. Table 2 show the standard deviation of the shadowing effect for the M-BS/M-UE and F-BS/M-UE, respectively. As shown in Figure 11, the power of desired signal is 3 dB higher than that of interference (Case 1) when considering CDF = 50%. Due to the influence of penetration loss in indoor, the received powers by Case 2 and 3 are lower than that by Case 1. Table 2 shows that the standard deviation of the shadowing is for the desired signal 5.1 and from 4.8 to 6.6 for the interference, respectively. Since outdoor-to-indoor propagation is considered when considering the interference from/to the femtocell to/from the M-UE, the range of the

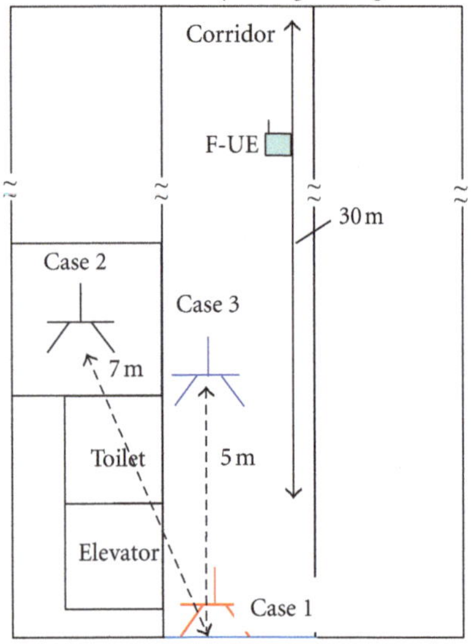

FIGURE 10: Location of femto-BS.

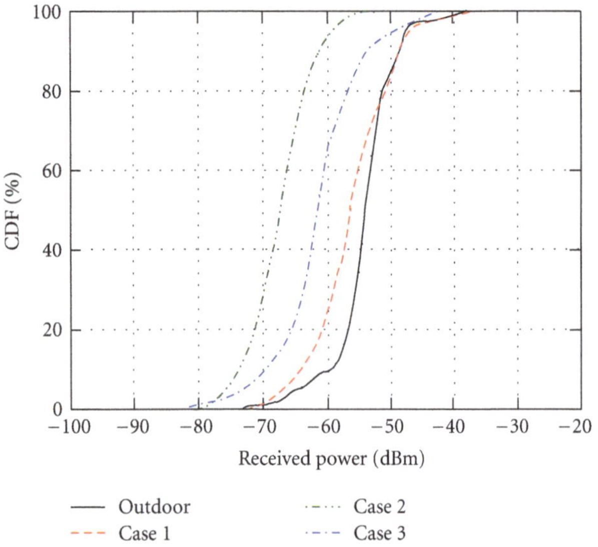

FIGURE 11: Received powers for desired signal and interference. Outdoor: desired signal, interference: Cases 1 to 3.

TABLE 2: Standard deviation of shadowing.

Type	Standard deviation [dB]
Outdoor (M-BS/M-UE)	5.1
Case 1 (F-BS/M-UE)	6.1
Case 2 (F-BS/M-UE)	4.8
Case 3 (F-BS/M-UE)	6.6

standard deviation for the F-BS/M-UE is larger than that for the M-BS/M-UE.

The transmission bit rate is evaluated using the measured standard deviation of the shadowing. Table 3 gives the simulation parameters. Considering the value on the standard deviation of shadowing, the results of outdoor and Case 2 in Table 2 are used for M-BS/M-UE and F-BS/M-UE transmissions, respectively. Figure 12 denotes the locations of M-BS, M-UE, F-BS, and M-RS. Three femtocells are assumed to be located inside a macrocell. F-BSs are randomly located inside the service area in M-BS. M-UE is also located inside the service area in M-BS. We employed the simulation where the location on F-BSs and M-UE is changed. The trial number is 10000. Regarding the basic parameter in Table 3, we refer to the parameters which are used in [2]. Regarding the cell radius of M-BS, when the cell radius is 289 m [2], we assume the cell size of 8 times, because the effectiveness of the relay station should be evaluated.

TABLE 3: Simulation parameters.

Parameter	Macro-BS	Pico-BS	Relay
Cell radius	2312 m	40 m	—
BS Tx power	46 dBm	30 dBm	30 dBm
UE Tx power	27 dBm	27 dBm	—
BS ant. gain	14 dBi	5 dBi	5 dBi
UE ant. gain	0 dBi	0 dBi	—
Noise power	−174 dBm/Hz	—	−174 dBm/Hz
Antenna height	22.5 m	16.5 m	15 m
Path loss (→ UE)	C2(NLOS)	B1(LOS)	B1(LOS)

Since the evaluation using heterogeneous path loss conditions is essential when considering heterogeneous networks [15, 16], the propagation pathloss model proposed in [17] is used in this paper. Model C2, which is used for typical macrocell non-line-of-sight (NLOS) scenarios, is adopted for the path loss between the M-BS and M-UE. Since we consider the interference from the F-BS to M-UE, Model B1, which is used for microcell line-of-sight (LOS) scenarios, is adopted for the path loss between the P-BS and M-UE. Model B1 is also used for the pathloss between the M-RS and M-UE. The path loss between the M-BS and M-RS is Model B5f (NLOS), which is used for relay transmission between the M-BS and M-RS. The relay stations are located at the position of $d/2$ in Figure 12. Four relay stations are assumed in Figure 12. When considering the propagation of M-BS/M-RS, M-RS/M-UE, and F-BS/M-RS, the values on the standard deviation of shadowing are used in those in [17]. The number of relays, N_R, is changed from 4 to 16 in a circular arrangement. The other simulation conditions are the same as those described in Section 4. We assume that the M-UE and M-RS can exchange the information with a certain interval, and that the M-UE can ideally select M-RS from which SNR to M-UE is the highest. We evaluate average transmission rate which means average cell throughput of M-BS. In the evaluation, we did not consider channel codes in order to evaluate the basic ability of the proposed method.

In the simulation, we assume that the M-RS obtains the information regarding the adaptive modulation scheme by using the control channel of F-BS, because the M-RS can receive only interference by the F-BS. Moreover, we assume that the M-RS correctly informs the information regarding adaptive modulation scheme by the F-BS to the M-UE.

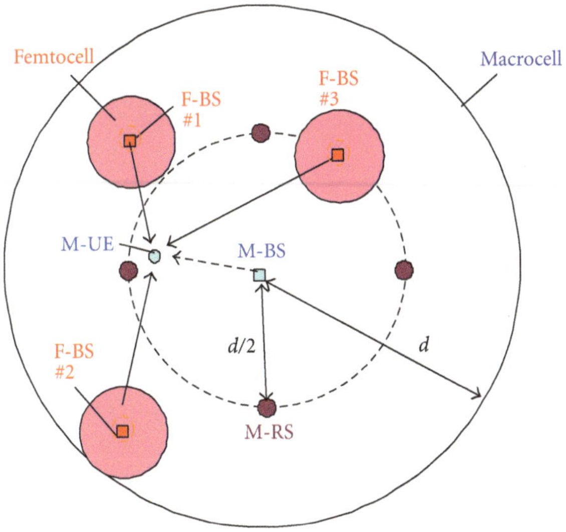

FIGURE 12: Location of M-BS, M-UE, F-BS, and M-RS.

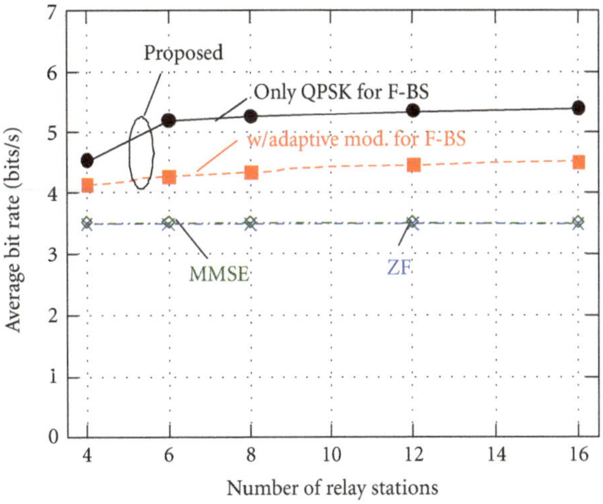

FIGURE 13: Transmission rate versus transmit power of F-BS when considering shadowing effect.

Figure 13 shows the relationship between the transmission rate versus the number of relays, N_R. Not only QPSK modulation but also adaptive modulation for F-BS is assumed in Figure 13. The figure shows that the transmission rate is improved when N_R is increased, but the performance becomes gradually saturated when $N_R \geq 6$ when considering only QPSK modulation for F-BS. Although the transmission rate is degraded compared to the case with only QPSK modulation for F-BS when assuming adaptive modulation for F-BS, we confirm that the transmission rate is gradually increased by the aid of increasing number of relays. The reason why the average bit rate (of M-UE) decreases when F-BS is allowed to use adaptive modulation is that the performance by the proposed method depends on the interference decoding in (13) at the M-RS. When the higher modulation scheme in F-BS is employed, the higher interference-to-noise power Ratio (INR) is required at the M-RS for the interference decoding. Moreover, it is shown that the proposed method outperforms the ZF/MMSE schemes even when $N_R = 4$. The average bit rate by the MMSE algorithm is almost the same with that by the ZF algorithm.

6. Conclusion

This paper proposed interference-based decode and forward method using the RSs in heterogeneous networks. The RS in a macrocell system decodes and forwards the interference from the pico(femto)cell to the UE, and the UE can decode the signal without stopping the signal reception from the BS by utilizing the higher SNR between the RS and UE compared to that between BS and UE in the proposed method. We clarified the effectiveness of the proposed method compared to using the conventional adaptive array with ZF and MMSE algorithms when the SNR between the RS and UE is 8 dB greater than that between the BS and UE. Moreover, we confirmed the effectiveness of the proposed method based on evaluation using the measured data for shadowing effects when considering the coexistence between macro- and femtocells.

As for the future work, we evaluate the influence when changing the service area of macrocell/picocell, because the interference power is largely affected by the service area of macrocell/picocell. In this paper, our aim was to propose the basic technique when the relay nodes can be applied and we focused on the basic ability of the interference cancellation by using the proposed method. On the other hand, hybrid ARQ is adopted as one of the important techniques in LTE-Advanced and the evaluation by using the proposed method including the channel codes is essential for the future work.

References

[1] 3GPP, TR 36.814 (V9.0.0), "Further advancements for E-UTRA physical layer aspects," 2010.

[2] M. Tanno, A. Morimoto, T. Abe, Y. Kishiyama, and T. Nakamura, "Heterogeneous Network in LTE-Advanced," IEICE Technical Report RCS2009-317, 2010.

[3] A. Khandekar, N. Bhushan, J. Tingfang, and V. Vanghi, "LTE-advanced: heterogeneous networks," in *European Wireless Conference*, pp. 978–982, April 2010.

[4] R. U. Nabar, H. Bölcskei, and F. W. Kneubühler, "Fading relay channels: performance limits and space-time signal design," *IEEE Journal on Selected Areas in Communications*, vol. 22, no. 6, pp. 1099–1109, 2004.

[5] K. Yamamoto, H. Maruyama, T. Shimizu, H. Murata, and S. Yoshida, "Spectral efficiency of fundamental cooperative relaying in interference-limited environments," *IEICE Transactions on Communications*, vol. 91, no. 8, pp. 2674–2682, 2008.

[6] R. Dabora, I. Maric, and A. Goldsmith, "Interference forwarding in multiuser networks," in *Proceedings of the IEEE Global Telecommunications Conference*, December 2008.

[7] I. Marié, R. Dabora, and A. Goldsmith, "On the capacity of the interference channel with a relay," in *Proceedings of the IEEE International Symposium on Information Theory (ISIT '08)*, July 2008.

[8] A. Jovičić and P. Viswanath, "Cognitive radio: an information-theoretic perspective," *IEEE Transactions on Information Theory*, vol. 55, no. 9, pp. 3945–3958, 2009.

[9] P.-H. Lin, S.-C. Lin, C.-P. Lee, and H.-J. Su, "Cognitive radio with partial channel state information at the transmitter," *IEEE Transactions on Wireless Communications*, vol. 9, no. 11, pp. 3402–3413, 2010.

[10] K. Nishimori, H. Yomo, P. Popovski, Y. Takatori, R. Prasad, and S. Kubota, "Distributed interference cancellation for dynamic spectrum sharing," in *Proceedings of the IEEE International Conference on Communications (ICC '08)*, pp. 3044–3049, May 2008.

[11] K. Nishimori, K. Kitao, and T. Imai, "Interference cancellation using relay station in heterogeneous networksin," in *Proceedings of the IEEE Vehicular Technology Conference (VTC Fall '11)*, September 2011.

[12] Y. Ohwatari, N. Miki, T. Asai, T. Abe, and H. Taoka, "Performance of advanced receiver employing interference rejection combining to suppress inter-cell interference in LTE-advanced downlink," in *Proceedings of the IEEE Vehicular Technology Conference (VTC Fall '11)*, September 2011.

[13] 3GPP, TS 36.201 (V10.0.0), Evolved universal terrestrial radio access (E-UTRA), LTE physical layer, General description (Release 10), 2010.

[14] S. T. Chung and A. J. Goldsmith, "Degrees of freedom in adaptive modulation: a unified view," *IEEE Transactions on Communications*, vol. 49, no. 9, pp. 1561–1571, 2001.

[15] K. Nishimori, R. D. Taranto, H. Yomo et al., "Spatial opportunity for cognitive radio systems with heterogeneous path loss conditions," in *Proceedings of the IEEE 65th Vehicular Technology Conference (VTC '07)*, Dublin, Ireland, April 2007.

[16] K. Nishimori, R. Di Taranto, H. Yomo, and P. Popovski, "Cognitive radio operation under directional primary interference and practical path loss models," *IEICE Transactions on Communications*, vol. 94, no. 5, pp. 1243–1253, 2011.

[17] WINNER II channel models, http://www.ist-winner.org/deliverables.html.

4

Direction-of-Arrival Estimation of Closely Spaced Emitters Using Compact Arrays

Hoi-Shun Lui[1,2] and Hon Tat Hui[3]

[1] *Department of Signals and Systems, Chalmers University of Technology, 412 96 Gothenburg, Sweden*
[2] *School of Information Technology and Electrical Engineering, The University of Queensland, Brisbane, QLD 4072, Australia*
[3] *Department of Electrical and Computer Engineering, National University of Singapore, Singapore 117576*

Correspondence should be addressed to Hoi-Shun Lui; h.lui@uq.edu.au

Academic Editor: Charles Bunting

Performance evaluation of direction-of-arrival (DOA) estimation algorithms has continuously drawn significant attention in the past years. Most previous studies were conducted under the situation that antenna element separation is about half wavelength in order to avoid the appearance of grating lobes. On the other hand, recent developments in wireless communications have favoured the use of portable devices that utilize compact arrays with antenna element separations of less than half wavelength. Performance evaluation of DOA estimation algorithms employing compact arrays is an important and fundamental issue, but it has not been fully studied. In this paper, the performance of the matrix pencil method (MPM) that applies to DOA estimations is investigated through Monte Carlo simulations. The results show that closely spaced emitters can be accurately resolved using linear compact array with an array aperture as small as around half wavelength.

1. Introduction

Direction-of-arrival (DOA) estimation is one of the most important applications in array signal processing. In the context of array signal processing, these algorithms can be grouped into three categories: conventional methods, subspace methods, and maximum likelihood techniques [1–4]. Throughout the last two decades, subspace methods have achieved great success due to their low complexity and reasonable performance. Out of all the subspace methods, Pisarenko's method [4] was one of the first methods that exploited the structure of the data model. Schmidt [5] then studied the signal properties of a sensor array and introduced the Multiple Signal Classification (MUSIC) algorithm. The MUSIC algorithm exploits the geometrical structure of the signal. Performance evaluations of the MUSIC algorithm have been well studied in the 1980s to mid-1990s, for example [6, 7]. The main advantage of subspace methods over the conventional beamforming method is that the resolution is not limited by the aperture of the entire array, and subspace methods are therefore also known as super resolution techniques [8]. Another well-known subspace method is Estimation of Signal Parameters via Rotational Invariance Techniques (ESPRIT) [9, 10], which exploits the array structure such that the specific knowledge of the array manifold is not required. Compared to MUSIC, it is computationally more efficient without the need of searching peaks throughout the entire range of angles. A review about subspace methods from the statistical prospective can be found in [4]. In addition, there is another subspace method, known as the matrix pencil method (MPM), which was first applied to damped exponential extractions from transient electromagnetic signals [11–13] and later to DOA estimations [14–20]. One significant advantage of MPM over MUSIC and ESPRIT is that DOA estimation can be done using a single snapshot instead of the covariance matrix that requires a large number of snapshots [15, 16]. The incoming directions can be determined directly without the need for searching among all the directions as in the MUSIC. Even with coherent signals, spatial smoothing is not required.

The DOA estimation problem is about estimation of undamped exponentials based on the measured samples of the signal from an array. Performance evaluations of subspace methods and the corresponding Cramer-Rao Lower Bound (CRLB), which gives the lower bound on the variance of any unbiased estimator, have been well studied for the estimation of a single damped/undamped exponential (e.g., [21–28]). Attempts have also been made for the case with two exponentials embedded in the signal. For instance, Stoica and Nehorai [26] have studied the DOA estimation performance of MUSIC for the case of two equal powered signals, and the results showed that the variance of MUSIC algorithm is smaller when (i) the number of sensors (the length of data) increases, (ii) the signals are not highly correlated, and (iii) the two signals are not closely spaced. Steedly and Moses [27] derived the complete expressions for the CRLB of the parameters of an exponential model with one set of damped/undamped exponentials and multiple sets of amplitude coefficients. For the case of two exponentials, numerical examples have shown that the angle CRLB is lower as the data length is large enough, and the signals are not closely spaced. In these studies, multiple snapshots of the signals or a single snapshot with long data length [27] are considered, and the separation between the sensors is usually fixed to the half wavelength. This is equivalent to the case where the array aperture is large (more than a wavelength if more than 3 elements). Such arrays are sometimes known as "standard arrays." The half wavelength separation is essentially the upper limit of Nyquist sampling rate [4] that avoids the appearance of grating lobes in the visible region of an array. At the same time, it provides the largest possible array aperture potentially resulting in the highest resolution, for instance, in conventional beamformers [1]. In addition, another practical consideration is the mutual coupling effect between antenna elements. In most array signal processing algorithms, the algorithms are developed under the assumption that interactions between array elements can be ignored. Mutual coupling is usually treated as a source of noise. Calibrations and preprocessing procedures are required to compensate such an undesirable effect [29–31]. Recent work has demonstrated that the undesirable mutual coupling effect can be effectively compensated for arrays down to $\approx 0.1\lambda$, both numerically [32, 33] and experimentally [34].

In the last two decades, there has been significant interest in reducing the physical dimensions of existing electronic devices with the rapid development of the electronics industry. In line with this development, recent applications in antenna arrays have been focused on compact array design [35–37] with less than a half wavelength element separation. To our knowledge, the performance of subspace methods under the situations where (i) the array aperture is significantly reduced to less than a wavelength and (ii) the increment of the number of sensors is within such a small array aperture has not been well exploited. Perturbation analysis for subspace methods and the corresponding CRLB are well studied for the single exponential process. When there is more than one mode embedded in the signals, however, the variance [21] and the CRLB [27] of the estimation problem are given in the form of matrix expressions. The results of how array aperture and sampling rates change can affect the performance of subspace methods, however, cannot be easily observed.

In view of this, this paper aims to characterize the effect of (i) changes in the array aperture with a fixed number of elements and (ii) changes in the number of elements in a fixed aperture on the accuracies DOA estimations with signals corrupted by Additive Gaussian White Noise (AGWN) using a Monte Carlo approach. In particular, the MPM will be considered. The focus of this study is on the effect of variations of the array aperture on the MPM algorithm under the ideal sensor condition. A brief review of MPM is first given followed by the Monte Carlo simulations and discussions towards the end of the paper.

2. Matrix Pencil Method (MPM) for DOA Estimation

Consider a space where there are $P+1$ signals coming from different directions and a uniform linear array (ULA) of $N+1$ identical sensors. The sensors are located along the x-axis with element separation between adjacent sensors denoted by Δ. Then, the signal $x(n)$ received by the sensors can be written as [11, 14–22]

$$x(n) = \sum_{p=0}^{P} A_p \exp\left(j\gamma_p + \frac{j2\pi\Delta n \cos\phi_p}{\lambda}\right), \quad 0 \le n \le N, \tag{1}$$

where A_p, γ_p, and ϕ_p are, respectively, the amplitude, phase, and DOA of each of the $P+1$ plane wave sources incident on the ULA. λ is the wavelength of the operation frequency. Equation (1) can also be written as

$$x(n) = \sum_{p=0}^{P} c_p y_p^n, \tag{2a}$$

where

$$c_p = A_p \exp\left(j\gamma_p\right), \tag{2b}$$

$$y_p = \exp\left[\frac{\left(j2\pi\Delta\cos\phi_p\right)}{\lambda}\right]. \tag{2c}$$

In matrix form, this can be written as

$$\begin{bmatrix} x(0) \\ x(1) \\ \vdots \\ x(N) \end{bmatrix} = \begin{bmatrix} 1 & 1 & \cdots & 1 \\ y_0 & y_1 & \cdots & y_P \\ \vdots & \vdots & \ddots & \vdots \\ y_0^N & y_1^N & \cdots & y_P^N \end{bmatrix} \begin{bmatrix} c_0 \\ c_1 \\ \vdots \\ c_P \end{bmatrix} \tag{2d}$$

or

$$\mathbf{x} = \mathbf{Y}\mathbf{c}. \tag{2e}$$

Next, we define the $L \times (N - L + 2)$ matrix $\mathbf{X}'$ using $x(k)$, the elements of $\mathbf{X}$,

$$\mathbf{X}' = \begin{bmatrix} x(0) & x(1) & \cdots & x(N-L+1) \\ x(1) & x(2) & \cdots & x(N-L+2) \\ \vdots & \vdots & \ddots & \vdots \\ x(L-1) & x(L) & \cdots & x(N) \end{bmatrix}, \quad (3)$$

where L is the pencil parameter which satisfies the criteria of $(N + 1 - P) \geq L \geq (P + 2)$. Applying a singular value decomposition (SVD) to $\mathbf{X}'$, we have

$$\mathbf{X}' = \mathbf{U}\mathbf{\Sigma}\mathbf{V}^H. \quad (4)$$

The superscript "H" denotes the complex conjugate transpose of a matrix. If the data is not contaminated by noise, the first $P + 1$ singular values are nonzero, and, hence, P can be determined. If the data is noise contaminated, the parameter P is estimated by observing the ratio of the various singular values to the largest one as defined by [11]

$$\frac{\sigma_c}{\sigma_{\max}} \approx 10^{-w}, \quad (5)$$

where w is the number of accurate significant decimal digits of the data $x(n)$. The singular values for which the ratio in (5) is below 10^{-w} are essentially noise singular values, and they should be discarded. It should be noted that when the incoming signals are spatially well resolved, these results in larger singular values are easier to be solved. For instance, if we have two cases with incoming signals from the directions of (i) 5° and 6° and (ii) 5° and 46°, the second singular value in case (ii) is larger than that in case (i). As a result, only the cases of closely spaced emitters are considered in this work. The proposition here is that if we could find out how MPM performs under various array configurations for closely spaced emitters, it should also be able to work for spatially well-resolved signals.

Once P is determined based on the condition given in (5), the submatrix of $\mathbf{U}$ with the first $P + 1$ columns can be formed and denoted as $\mathbf{U}_s$. To estimate the DOAs, the following matrix pencil problem is solved:

$$\mathbf{U}_2 - \lambda' \mathbf{U}_1 = 0, \quad (6a)$$

or equivalently,

$$\mathbf{U}_1^{+} \mathbf{U}_2 - \lambda' \mathbf{I} = 0, \quad (6b)$$

where $\mathbf{U}_1$ is $\mathbf{U}_s$ with the last row deleted, $\mathbf{U}_2$ is $\mathbf{U}_s$ with the first column deleted, λ' is the eigenvalue of the matrix pencil problem, $\mathbf{I}$ is the $(P + 1) \times (P + 1)$ identity matrix, and the superscript "+" denotes the Moore-Penrose pseudoinverse of a matrix [21] defined as

$$\mathbf{U}_1^{+} = \left\{\mathbf{U}_1^{H} \mathbf{U}_1\right\}^{-1} \mathbf{U}_1^{H}. \quad (6c)$$

The superscript "−1" denotes the inverse of a matrix. Once (6a), (6b), and (6c) are solved, the DOAs can be obtained by the following formula [16, 18]:

$$\phi_p = \cos^{-1}\left[\frac{\lambda \ln(\lambda_p')}{j2\pi\Delta}\right], \quad (7)$$

where $p = 1, 2, \ldots, P + 1$ ($P + 1$ singular values from (6a), (6b) and (6c)). It is well known that if the argument of the function $\cos^{-1}$ in (7) is greater than 1 or a complex number, the resultant angle will become complex which is considered as a fail estimate for MPM. The resultant complex estimation angle can be used as a sanity check which indicates that MPM fails to resolve all the signals in that particular snapshot.

3. Monte Carlo Simulations

This section presents Monte Carlo simulations of DOA estimations using ULAs with different array apertures and different number of elements. Three cases are presented to illustrate our findings. First, a four-element ULA with different apertures under the illumination of two closely spaced emitters is studied. Then, ULAs with fixed apertures but different numbers of elements are considered. Lastly, DOA estimations of three closely spaced incoming signals using arrays with a fixed aperture but different numbers of elements are studied.

Case 1 (Four-Element ULAs with Different Apertures (2 Incoming Signals)). A ULA with four ideal isotropic sensors is studied. The objective is to investigate how the accuracy of DOA estimation is affected by the reduction of the array aperture. The separation of the array elements is changed from $\Delta = 0.5\lambda$ down to 0.05λ, with a step size of 0.05λ. This corresponds to a reduction of the array aperture (defined as the physical length of the array) from 1.5λ to 0.15λ. The case with $\Delta = 0.5\lambda$ is the usual "standard" interelement separation used in those conventional arrays.

Two equal-power and equal-phase coherent sources come from the azimuth angles of $\phi_1 = 5°$ and $\phi_2 = 6°$, which are close to the endfire direction ($\phi = 90°$ corresponds to the broadside direction). The signals (both magnitude and phase) are contaminated with AGWN at a signal-to-noise ratio (SNR) of 5 dB. To evaluate the performance of the DOA estimations, the average bias, $\bar{b}$, is introduced and used as a metric throughout this paper. It is given by

$$\bar{b} = \sum_{i=1}^{Q} \frac{b_i}{Q}, \quad (8)$$

where $b_i = |\hat{\phi}_i - \phi_i|$ is the absolute value of the bias of the estimated direction ($\hat{\phi}_i$ being the mean of the estimated directions; $\hat{\phi}_i$ is determined from 10,000 independent Monte Carlo simulations), Q is the number of incoming signals which is 2 in this case. For a successful DOA estimation, all the signals should be accurately estimated. A small value of average bias is obtained only if all the incoming signals are accurately estimated. At each value of interelement separation (Δ) shown in the figures in this paper, the average bias is obtained by 10,000 independent Monte Carlo simulations.

The results are shown in Figure 1. It is shown that the average biases decrease as the interelement separation increases. This indicates that the estimated directions are more accurate when the interelement separation is large. However, when $\Delta = 0.05\lambda$, the average bias is 3° when the SNR levels of the two incoming signals are both 5 dB. This means that MPM

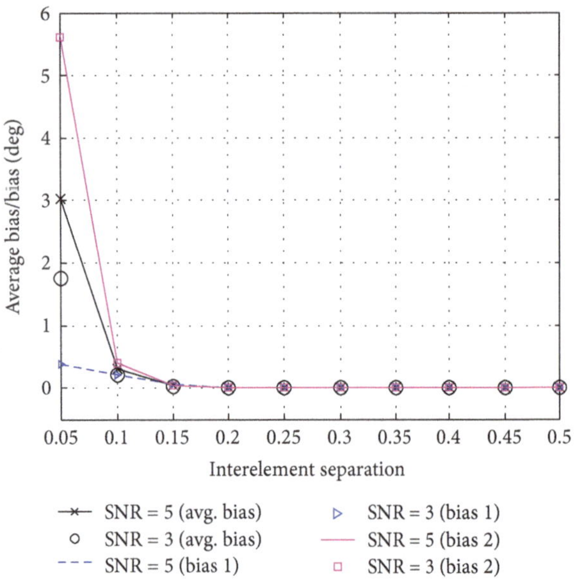

FIGURE 1: Summary of DOA estimations using MPM for a four-element ULA with interelement separation varying from 0.05λ to 0.5λ. Two equal-power incoming signals are from the directions of $\phi_1 = 5^\circ$ and $\phi_2 = 6^\circ$ with an SNR of 5 dB and 3 dB, respectively. The graph shows (i) average bias of the estimated directions and (ii) bias of the estimation of signals 1 and 2.

fails to resolve the two incoming signals as the difference between the two signals is only 1°. To probe further, the biases of the estimated directions of the two signals are also included. It is interesting to see that for the case of $\Delta = 0.05\lambda$, the bias of ϕ_1 is about 0.4° (i.e., the estimated direction is about 5.4°), which is close to the mean of the two incoming signals (5° and 6°). The bias of ϕ_2 is around 6° which is far from the actual incoming direction, which further confirms that MPM fails to resolve the two incoming signals under such an array configuration. To investigate if the performance varies at different SNR level, the same DOA problem is repeated except that the SNR level is 3 dB. It is found that there is no significant difference as compared to the case of SNR = 5 dB.

Next, we study the effect on the performance of MPM when the incoming signals are not of equal power and equal phase. The entire procedure is repeated for the DOA estimation of $\phi_1 = 5^\circ$ and $\phi_2 = 6^\circ$ except that the amplitudes of the two signals are different and they are 90° out of phase. For $\phi_2 = 6^\circ$, four different amplitudes of $A_2 = 2A_1, 3A_1, 5A_1$, and $10A_1$ are considered, where A_1 and A_2 are the amplitudes of the incoming signals at $\phi_1 = 5^\circ$ and $\phi_2 = 6^\circ$, respectively. The amplitude of the signal from $\phi_1 = 5^\circ$ remains to be the same for all cases. The average biases for the four cases are shown in Figure 2(a). The average biases in the range of 1.5° to 2° are found when $\Delta = 0.05\lambda$, which is considered as a fail estimate as the difference between the two incoming signals is only 1°. As Δ is larger than 0.2λ, the average biases decrease to the level of $10^{-3\circ}$ which agrees with the results for the equal-power and equal-phase case shown in Figure 1. The biases of the two signals are also included in Figure 2(b). At $\Delta = 0.05\lambda$, the biases of $\phi_1 = 5^\circ$ and $\phi_2 = 6^\circ$ are at the levels of 0.5° and 3°, respectively, which is similar to the results for the two signals with equal power in Figure 1. Such biases indicate that the two signals cannot be accurately resolved with this array aperture.

To investigate the performance of MPM for DOA estimations when the incoming signal direction changes, DOA estimation problems with the two signals coming from the directions of $\phi_1 = 10^\circ$ to 90° and $\phi_2 = \phi_1 + 1^\circ$ are considered. Here, only the angular range of 0° to 90° is considered as the angular range of 90° to 180° is simply the mirror image of that of 0° to 90°. The same DOA estimation calculation procedure is repeated for all cases independently. The step size of $\Delta\phi_1 = 5^\circ$ is considered, and the results are shown in Figure 3. Similarly, the average biases decrease as Δ increases for the same incoming signals. The results show that the average bias is below $10^{-3\circ}$ when the array element separation is greater than or equal to 0.2λ. Furthermore, we also note from the figure that as the signals are coming from a direction closer to the broadside direction of 90°, the accuracy of estimation increases quite significantly even when the element separation remains the same. For instance, when the element separation is 0.05λ with incoming signals of $\phi_1 = 10^\circ$ and $\phi_2 = 11^\circ$, an average bias of $\sim 0.7^\circ$ is resulted. When the incoming signals become $\phi_1 = 55^\circ$ and $\phi_2 = 56^\circ$, the average bias now becomes $< 10^{-4\circ}$ which is considered as accurate (with less than 0.01% error).

Case 2 (Fixed Aperture ULAs with Different Numbers of Elements (2 Incoming Signals)). In the previous case, when the two closely spaced emitters approach to the endfire directions, it was found that the array aperture of the four-element ULA has to be $\geq 0.60\lambda$ ($\Delta \geq 0.2\lambda$) for an accurate DOA estimation. Figure 4 shows the results of DOA estimations with different fixed array apertures that ranged from 0.25λ to 1λ, but with a different number of array elements varying from 4 to 20. Note that under a fixed array aperture, the interelement separation has to be changed with the change in the number of elements in the array. For example, for the four-element array with an aperture of 0.5λ, the interelement separation is $0.1667\lambda(= 0.5\lambda/(4-1))$.

The array apertures of 0.25λ and 0.5λ are first considered, and the results are shown in Figures 4(a) and 4(b). When the incoming signals are close to the endfire directions (between 5° and 15°), the average biases decrease when the number of elements increases from 4 to 20. As the incoming signals are moving towards the broadside direction (between 20° and 90°), the average biases decrease rapidly with an average bias less than $10^{-4\circ}$ regardless of the number of elements.

Now, the same estimation procedure is repeated with an array aperture of 0.75λ. When the aperture is increased from 0.5λ to 0.75λ, as shown in Figure 4(c), the average biases have been significantly reduced for all cases (below $2.2 \times 10^{-3\circ}$ for all cases). The maximum average bias is reduced from $1.56 \times 10^{-2\circ}$ (0.5λ aperture) to $2.2 \times 10^{-3\circ}$ (0.75λ aperture), which is almost a factor of 7. This indicates a significant improvement of the DOA estimations compared with the two cases of aperture size of 0.25λ and 0.5λ.

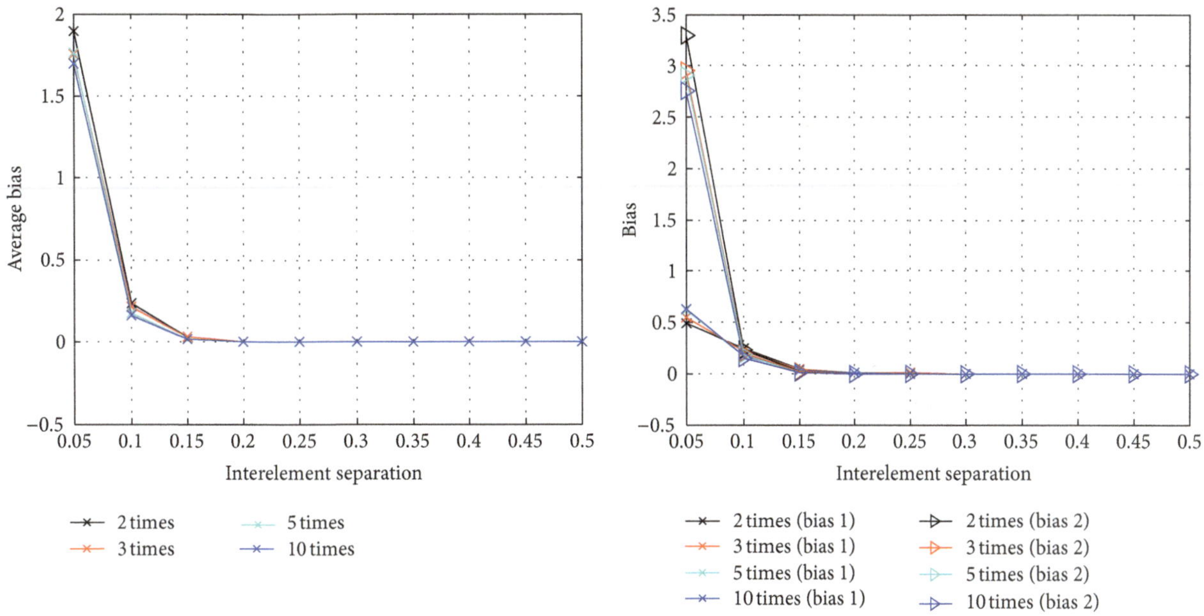

(a) Average Bias (b) Bias of Signal 1 and 2

FIGURE 2: Summary of DOA estimations using MPM for a four-element ULA with interelement separation varying from 0.05λ to 0.5λ. The two incoming signals are from the directions of $\phi_1 = 5^\circ (A_1 = 1, \gamma_1 = 90^\circ)$ and $\phi_2 = 6^\circ (A_2 = 2, 3, 5, \text{ and } 10, \gamma_2 = 180^\circ)$ with an SNR of 5 dB. (a) Average biases of the estimated directions; (b) biases of the estimation of signal 1 and signal 2.

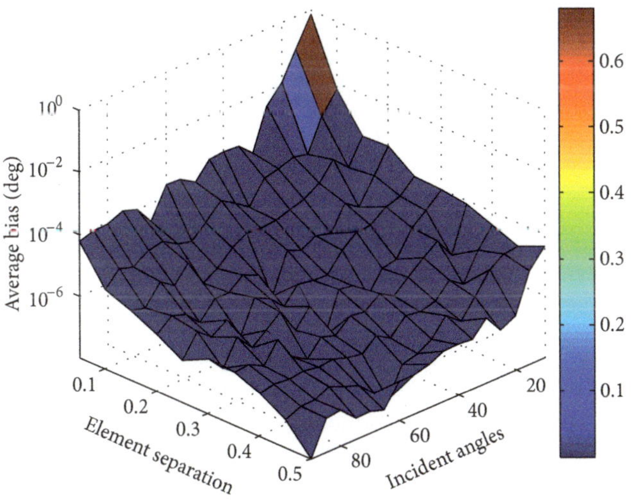

FIGURE 3: Summary of DOA estimations using MPM for a four-element ULA with interelement separation varying from 0.05λ to 0.5λ. Two equal-power incoming signals are from the directions of $\phi_1 = 10^\circ$ to 90° and $\phi_2 = \phi_1 + 1^\circ$ with an SNR of 5 dB. The step size of $\Delta\phi_1 = 5^\circ$ is considered. The figure shows the average biases of the estimated directions.

The previous results show that for a fixed small array aperture (especially for the cases of 0.25λ and 0.5λ), the accuracies of DOA estimation can be improved by increasing the number of elements when the incoming signals are close to the endfire directions. When the incoming signals are moving towards to the broadside direction, increasing the number of elements under a fixed aperture does not necessary improve the accuracies of the estimation. Comparing between signals coming from the broadside direction and the endfire directions, better accuracies are expected under the same array configuration.

When the array aperture is further increased to 1λ, as shown in Figure 4(d), it is not obvious if increasing the number of elements enhances the DOA performance under this aperture. On the other hand, it is worth noting that as the array aperture increases from 0.25λ to 1λ, the average biases for all cases (different number of elements and incoming signals) decrease. This indicates that the accuracy of DOA estimation depends more on the array aperture than on the number of elements when the aperture is sufficiently large. While not presented here, the results on the array apertures of 1.25λ and 1.5λ indicate that there is some improvement of DOA accuracy as the number of elements increases. These results show that further increase in the number of elements for large aperture arrays ($> 1\lambda$) only yields minor improvement to the accuracy of the estimated angles.

Case 3 (*Fixed Aperture ULAs with Different Number of Elements* (*3 Incoming Signals*)). In the last case, we consider DOA estimation problems of three closely spaced, equal-power, and equal-phase signals with an SNR of 5 dB come from the directions from $\phi_1 = 45^\circ$ to 90° with $\Delta\phi_1 = 5^\circ$, $\phi_2 = \phi_1 + 1^\circ$, and $\phi_3 = \phi_1 + 2^\circ$. First, the array aperture of 0.75λ is investigated. The number of array elements varies from 6 to 20 (the minimum number of sensors required for resolving 3 incoming signals using MPM is 6). The average biases of the DOA estimations are shown in Figure 5(a). It

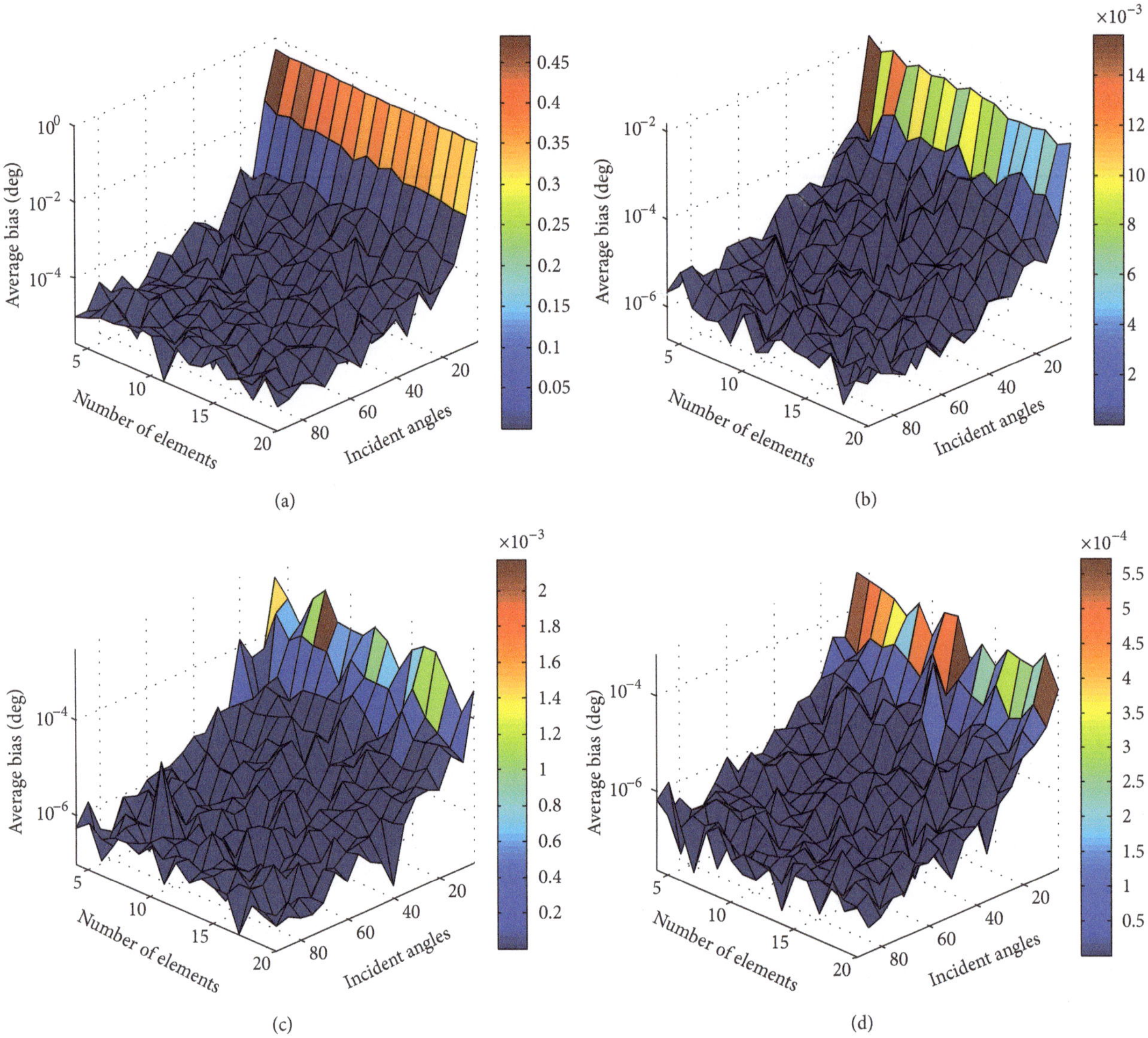

FIGURE 4: Summary of DOA estimations using MPM for a four-element ULA with the number of array elements varying from 4 to 20. Two equal-power incoming signals are from the directions of $\phi_1 = 10^\circ$ to 90° and $\phi_2 = \phi_1 + 1^\circ$ with an SNR of 5 dB. The step size of $\Delta\phi_1 = 5^\circ$ is considered. The figures show the average biases of the estimated direction. The apertures of the arrays are (a) 0.25λ, (b) 0.5λ, (c) 0.75λ, and (d) 1λ.

can be clearly seen that the average biases decrease with the number of the elements. Compared to the case with two signals of $\phi_1 = 45^\circ$, $\phi_2 = 46^\circ$ with the array aperture of 0.25λ (Figure 4(a)) and the case of $\phi_1 = 5^\circ$, $\phi_2 = 6^\circ$ with the array aperture of 0.75λ (Figure 4(c)) with 6 array elements, the average biases presented in Figure 5(a) (6°) are much higher than those in Cases 1 and 2 ($10^{-4\circ}$ and $10^{-3\circ}$, resp.). This is because resolving three incoming signals using MPM and resolving two incoming signals are two different mathematical problems with different CRLB, and they should be considered separately. As expected, when the incoming signals are moving towards the broadside directions, the average biases reduce to 2° for the case of 6 antenna elements. The average biases are also reduced as the number of elements increases. However, even with 20 elements, the average biases are between 0.5° and 3°. These bias values are considered as high as the actual separations between the three sources, only 1°, and thus it cannot be concluded that the sources are clearly resolved.

Lastly, the incoming signals of the same DOA problems and the array aperture are increased to 1λ and 1.25λ, and the average biases are shown in Figures 5(b) and 5(c), respectively. For the same DOA problem, the average biases decrease as the number of array element increases. At an array aperture of 1λ with 20 array elements, the average biases are at a level less than 0.1° for DOA when the incoming signals are more than 75°. For the DOA problem with incoming signals ($\phi_1 = 45^\circ$, $\phi_2 = 46^\circ$, and $\phi_3 = 47^\circ$) with 20 array elements, the average bias is 0.628°. However, when the array aperture is further increased to 1.25λ, the average bias reduces to smaller than 0.1° with the 20 array elements (as shown in Figure 5(c)). The average biases for the same DOA estimation problem

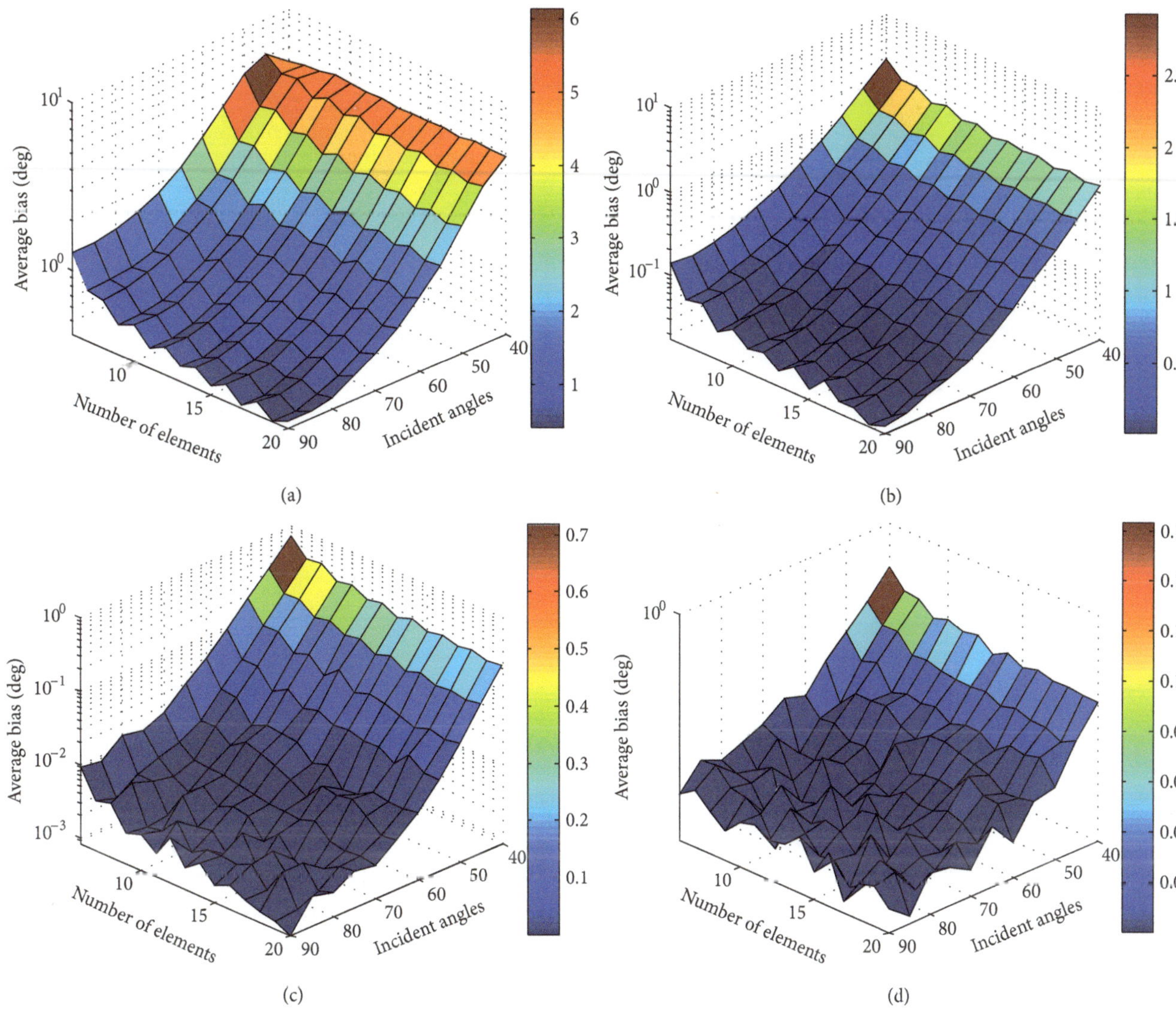

FIGURE 5: Summary of DOA estimations using MPM for a four-element ULA with the number of array elements varying from 6 to 20. Three equal-power incoming signals are from the directions of $\phi_1 = 40^\circ$ to 90°, $\phi_2 = \phi_1 + 1^\circ$, and $\phi_3 = \phi_1 + 2^\circ$ with an SNR of 5 dB. The step size of $\Delta\phi_1 = 5^\circ$ is considered. The figures show the average biases of the estimated directions with array aperture (a) 0.75λ, (b) 1.00λ, (c) 1.25λ, and (d) 1.50λ.

but with an array aperture of 1.5λ are shown in Figure 5(d). As can be seen, when the signal sources are closer to the broadside direction, the average biases are smaller. However, an increasing number of array elements do not necessarily increase the accuracy of DOA estimation in this aperture. On the other hand, when the signal sources are closer to the endfire direction, the accuracy of DOA estimation does increase substantially when the number of array elements is increased (see Figure 5(d)).

4. Discussions and Conclusions

In this paper, we have investigated the performance of MPM in DOA estimations using ULAs with different array apertures and different numbers of array elements through Monte Carlo simulations. The results show that minimum linear array apertures of 0.6λ and 1λ are required to resolve two and three incoming signals, respectively (i.e., element separation of $\leq 0.2\lambda$). The performance could be improved by increasing the number of elements. In practice, the emitters are less closely spaced which would be easier to resolve, and potentially, the size of the array and/or the number of elements can be further reduced. The finding provides us with the fundamental limit of DOA estimation in compact array. Arguably, the conclusions from this study are not limited to DOA estimations using compact antenna arrays (with interelement separation of 0.25λ [35] down to 0.125λ [37]) with MPM but are also applicable to acoustic or sonar arrays, as well as to other subspace methods.

Acknowledgments

This work was supported in part by the National University of Singapore (NUS) under the Grant no. R-263-000-469-112 and in part by the US ONR research fund under the Project no. 09PR03332-01.

References

[1] J. C. Liberti and T. S. Rappaport, *Smart Antennas for Wireless Communications, IS-95 and Third Generation CDMA Applications*, Prentice Hall, Upper Saddle River, NJ, USA, 1999.

[2] H. Krim and M. Viberg, "Two decades of array signal processing research: the parametric approach," *IEEE Signal Processing Magazine*, vol. 13, no. 4, pp. 67–94, 1996.

[3] L. C. Godara, "Application of antenna arrays to mobile communications, part II: beam-forming and direction-of-arrival considerations," *Proceedings of the IEEE*, vol. 85, no. 8, pp. 1195–1245, 1997.

[4] A. Paulraj, B. Ottersten, R. Roy, A. Swindlehrst, G. Xu, and K. Kailath, "Subspace methods for directions-of-arrival esitmation," in *Handbook of Statistics*, vol. 10, chapter 16, pp. 693–739, Elsevier Science, Amsterdam, The Netherlands, 1993.

[5] R. O. Schmidt, "Multiple emitter location and signal parameter estimation," *IEEE Transactions on Antennas and Propagation*, vol. 34, no. 3, pp. 276–280, 1986.

[6] B. Friedlander and A. J. Weiss, "Direction finding in the presence of mutual coupling," *IEEE Transactions on Antennas and Propagation*, vol. 39, no. 3, pp. 273–284, 1991.

[7] A. L. Swindlehurst and T. Kailath, "A performance analysis of subspace-based methods in the presence of model errors—part I: the MUSIC algorithm," *IEEE Transactions on Signal Processing*, vol. 40, no. 7, pp. 1758–1774, 1992.

[8] D. Madurasinghe, "Super resolution technique for source power and bearing estimation," *Electronics Letters*, vol. 39, no. 17, pp. 1232–1233, 2003.

[9] A. Paulraj, R. Roy, and T. Kailath, "Subspace rotation approach to signal parameter estimation," *Proceedings of the IEEE*, vol. 74, no. 7, pp. 1044–1046, 1986.

[10] R. Roy and T. Kailath, "ESPRIT—estimation of signal parameters via rotational invariance techniques," *IEEE Transactions on Acoustics, Speech, and Signal Processing*, vol. 37, no. 7, pp. 984–995, 1989.

[11] T. K. Sarkar and O. Pereira, "Using the matrix pencil method to estimate the parameters of a sum of complex exponentials," *IEEE Antennas and Propagation Magazine*, vol. 37, no. 1, pp. 48–55, 1995.

[12] L. H. S. Antony and N. Shuley, "Consequence of incorrect sampling procedures in resonance-based radar target identification," *Electronics Letters*, vol. 40, no. 8, pp. 507–508, 2004.

[13] H. S. Lui and N. V. Z. Shuley, "Sampling procedures for resonance based radar target identification," *IEEE Transactions on Antennas and Propagation*, vol. 56, no. 5, pp. 1487–1491, 2008.

[14] H. Ouibrahim, D. D. Weiner, and T. K. Sarkar, "Generalized approach to direction finding," *IEEE Transactions on Acoustics, Speech, and Signal Processing*, vol. 36, no. 4, pp. 610–612, 1988.

[15] C. K. E. Lau, R. S. Adve, and T. K. Sarkar, "Minimum norm mutual coupling compensation with applications in direction of arrival estimation," *IEEE Transactions on Antennas and Propagation*, vol. 52, no. 8, pp. 2034–2041, 2004.

[16] T. K. Sarkar, M. C. Wicks, M. Salazar-Palma, and R. J. Bonneau, *Smart Antennas*, John Wiley & Sons, 2003.

[17] Y. Hua, "Estimating two-dimensional frequencies by matrix enhancement and matrix pencil," *IEEE Transactions on Signal Processing*, vol. 40, no. 9, pp. 2267–2280, 1992.

[18] T. K. Sarkar, M. Salazar-Palma, and E. L. Mokole, *Physics of Multiantenna Systems and Broadband Processing*, John Wiley, 2008.

[19] S. Hwang, S. Burintramart, T. K. Sarkar, and S. R. Best, "Direction of Arrival (DOA) estimation using electrically small tuned dipole antennas," *IEEE Transactions on Antennas and Propagation*, vol. 54, no. 11, pp. 3292–3301, 2006.

[20] N. Yilmazer, J. Koh, and T. K. Sarkar, "Utilization of a unitary transform for efficient computation in the matrix pencil method to find the direction of arrival," *IEEE Transactions on Antennas and Propagation*, vol. 54, no. 1, pp. 175–180, 2006.

[21] Y. Hua and T. K. Sarkar, "Matrix pencil method for estimating parameters of exponentially damped/undamped sinusoids in noise," *IEEE Transactions on Acoustics, Speech, and Signal Processing*, vol. 38, no. 5, pp. 814–824, 1990.

[22] Y. Hua and T. K. Sarkar, "Generalized pencil-of-function method for extracting poles of an EM system from its transient response," *IEEE Transactions on Antennas and Propagation*, vol. 37, no. 2, pp. 229–234, 1989.

[23] N. Yilmazer, S. Burintramart, and T. K. Sarkar, "Performance analysis of direct data domain approach and ESPRIT method for DOA estimation," in *Proceedings of the IEEE Antennas and Propagation Society International Symposium (AP-S '07)*, pp. 5295–5298, Honolulu, Hawaii, USA, June 2007.

[24] S. M. Kay, *Fundamentals of Statistical Signal Processing: Estimation Theory*, Prentice Hall, 1993.

[25] Y.-X. Yao and S. M. Pandit, "Cramer-Rao lower bounds for a damped sinusoidal process," *IEEE Transactions on Signal Processing*, vol. 43, no. 4, pp. 878–885, 1995.

[26] P. Stoica and A. Nehorai, "MUSIC, maximum likelihood, and Cramer-Rao bound," *IEEE Transactions on Acoustics, Speech, and Signal Processing*, vol. 37, no. 5, pp. 720–741, 1989.

[27] W. M. Steedly and R. L. Moses, "Cramer-Rao bound for pole and amplitude coefficient estimates of damped exponential signals in noise," *IEEE Transactions on Signal Processing*, vol. 41, no. 3, pp. 1305–1318, 1993.

[28] E. H. Djermoune and M. Tomczak, "Perturbation analysis of subspace-based methods in estimating a damped complex exponential," *IEEE Transactions on Signal Processing*, vol. 57, no. 11, pp. 4558–4563, 2009.

[29] K. R. Dandekar, H. Ling, and G. Xu, "Experimental study of mutual coupling compensation in smart antenna applications," *IEEE Transactions on Wireless Communications*, vol. 1, no. 3, pp. 480–487, 2002.

[30] H. T. Hui, "Improved compensation for the mutual coupling effect in a dipole array for direction finding," *IEEE Transactions on Antennas and Propagation*, vol. 51, no. 9, pp. 2498–2503, 2003.

[31] H.-S. Lui, H. T. Hui, and M. S. Leong, "A note on the mutual-coupling problems in transmitting and receiving antenna arrays," *IEEE Antennas and Propagation Magazine*, vol. 51, no. 5, pp. 171–176, 2009.

[32] H. S. Lui and H. T. Hui, "Improved mutual coupling compensation in compact antenna arrays," *IET Microwaves, Antennas and Propagation*, vol. 4, no. 10, pp. 1506–1516, 2010.

[33] H. S. Lui and H. T. Hui, "Mutual coupling compensation for direction-of-arrival estimations using the receiving-mutual-impedance method," *International Journal of Antennas and Propagation*, vol. 2010, Article ID 373061, 2010.

[34] Y. Yu, H. S. Lui, C. H. Niow, and H. T. Hui, "Improved DOA estimations using the receiving mutual impedances for mutual

coupling compensation: an experimental study," *IEEE Transactions on Wireless Communications*, vol. 10, no. 7, pp. 2228–2233, 2011.

[35] C. Volmer, J. Weber, R. Stephan, K. Blau, and M. A. Hein, "An eigen-analysis of compact antenna arrays and its application to port decoupling," *IEEE Transactions on Antennas and Propagation*, vol. 56, no. 2, pp. 360–370, 2008.

[36] B. K. Lau, J. B. Anderson, G. Kristensson, and A. F. Molisch, "Impact of matching network on bandwidth of compact antenna arrays," *IEEE Transactions on Antennas and Propagation*, vol. 54, no. 11, pp. 3225–3237, 2006.

[37] D. M. Browne, M. Manteghi, M. P. Fitz, and Y. R. Samii, "Experiments with compact antenna arrays for MIMO radio communications," *IEEE Transactions on Antennas and Propagation*, vol. 54, no. 11, pp. 3239–3250, 2006.

5

Electromagnetic Simulation of the Near-Field Distribution around a Wind Farm

Shang-Te Yang and Hao Ling

Department of Electrical and Computer Engineering, The University of Texas at Austin, Austin, TX 78712, USA

Correspondence should be addressed to Shang-Te Yang; shangte.yang@utexas.edu

Academic Editor: David de la Vega

An efficient approach to compute the near-field distribution around and within a wind farm under plane wave excitation is proposed. To make the problem computationally tractable, several simplifying assumptions are made based on the geometry problem. By comparing the approximations against full-wave simulations at 500 MHz, it is shown that the assumptions do not introduce significant errors into the resulting near-field distribution. The near fields around a 3 × 3 wind farm are computed using the developed methodology at 150 MHz, 500 MHz, and 3 GHz. Both the multipath interference patterns and the forward shadows are predicted by the proposed method.

1. Introduction

Due to the rapid development of wind energy around the globe, the adverse effect of large wind turbines on radar and communication systems is raising serious concerns [1–7]. Both theoretical and experimental studies have been conducted to assess the impact of wind turbines on various systems including weather radar [1], air traffic control radar [2], air defense radar [3, 4], and marine radar [5], as well as radio communication and navigation systems [6, 7]. Two types of electromagnetic phenomenology can give rise to interference effects : radar clutter and electromagnetic transmission interference. Wind turbine clutter arises from the backscattered signal from the turbine structures and is of concern mainly to radar systems. Due to the spinning turbine blades, the resulting Doppler clutter can interfere with the detection of moving targets such as weather and aircraft. This topic has been well studied to date. For example, the United States Air Force Research Laboratory collected in situ measurement data of several utility class wind turbines in Fenner, NY, USA and corroborated the measurements with high-frequency ray-tracing simulation [8, 9]. The detailed phenomenology of the backscattering from wind turbines is fairly well understood.

The issue of electromagnetic transmission interference, on the other hand, is more subtle and therefore more difficult to assess. There exists a shadow region directly behind a turbine structure where the scattered electromagnetic field cancels destructively with the incident electromagnetic field. Outside the shadow region, the scattered field and incident field interfere due to their phase difference, forming a multipath region characterized by a rapid spatial oscillation pattern [10]. If a receiver in a communication link is positioned sufficiently close to a wind turbine (or a cluster of turbines), the receiver could experience signal fades. Similarly for a radar system, if a target of interest comes sufficiently close to a wind farm, the received radar signal from the target will experience a two-way signal transmission loss through the wind farm. The aforementioned effects are strongly dependent on the locations of the transmitter and receiver with respect to the wind turbine structure. Therefore, this issue also deserves careful consideration.

Both theoretical and experimental works have been conducted on the characterization of the transmission interference problem. A simple, approximate geometrical blockage estimate can be derived based on the Fresnel zone argument [11]. This is the standard methodology used to estimate the shadowing effect due to wind turbine structures by the United States Federal Aviation Administration obstruction evaluation process [12, 13]. The radar cross section (RCS) concept was applied to address the multipath interference problem in [14] and to estimate the loss of radar detection range in [15]. Full-wave analysis was used to analyze the shadow due to a turbine tower under a pencil beam excitation in [16]. Measurements of the shadow behind a wind turbine were carried out in [17, 18]. The effect of an offshore wind farm on radio communication and navigation systems was measured in-situ in [6]. The wind farm transmission interference problem has received less attention thus far in comparison to the radar clutter issue. This may be due to several reasons. From the measurement perspective, the transmission problem involves a bistatic transmitter/receiver configuration, which is logistically more difficult to carry out. From the computational perspective, near-field simulation around an entire wind farm poses a large computational burden. As a result, the interference effect of an entire wind farm on signal transmission is not yet well characterized.

In this paper, we set out to develop an efficient but approximate electromagnetic approach to compute the received field strength within and around a wind farm. Our approach is based on several simplifying assumptions. First, the turbine scattering is assumed to be dominated by the tower structure of the turbine, while additional scattering from the blades and nacelle structures are assumed to be of secondary importance. Second, due to the large length-to-cross section ratio of typical tower structures, the scattering process is assumed to be predominately two-dimensional (2D) in nature for an observer located close to the turbine. Third, we assume that the scattered fields can be predicted by using the far-field complex echo width (EW) of the tower. Lastly, the individual turbines in a wind farm are assumed to be excited under the incident excitation while neglecting the interaction terms. This is also known as the Born approximation. Using these assumptions, we formulate an approximate prediction methodology that leads to a significant reduction in the computational complexity of the problem and allows us to predict the shadow and multipath interference within and around a large wind farm.

This paper is organized as follows. In Section 2, we establish the validity of the three assumptions that form the basis of our prediction methodology. Full-wave simulation based on the multilevel fast multipole method (MLFMM) in the commercial electromagnetic package FEKO [19] is used to check the validity of the assumptions. In Section 3, we apply our prediction approach to two or more turbines by ignoring the interturbine interactions. It is again shown by comparison with full-wave simulation that such approach leads to a reasonable approximation barring any occlusion effects among the turbines. Results for a three-by-three farm are then shown, and the behavior of the fields as a function of frequency is investigated. The savings in computation time are also summarized and discussed in Section 3. Discussions and conclusions are given in Section 4.

FIGURE 1: Illustration of the problem statement.

2. EM Simulation Methodology: Single-Turbine Case

The problem of interest is illustrated in Figure 1. The transmitter is assumed to be located far away, so that the incident field is a plane wave. The field at an arbitrary observation point OB in close proximity of the wind farm is desired. Of particular interest to us is the case where both the transmitter and the receiver are located along the same horizontal plane, as our research is motivated by the impact on marine radar and communication systems from offshore wind farms. The more general configuration of arbitrary transmitter and receiver elevation angles, although interesting, is outside the scope of the present work. Clearly, a full-wave solution to this problem at frequencies above 1 GHz is computationally too exhaustive to carry out even using MLFMM, which already requires much less memory and time than the traditional method of moments (MoM). Here, we describe several simplifying approximations to arrive at the desired field distribution for a single turbine. In the next section, the Born approximation will be further applied to deal with multiple turbines that comprise a farm.

2.1. Geometry Simplification. While a turbine consists of multiple components including the tower, the blades, and the nacelle, it is believed that the tower gives rise to the strongest scattering and shadowing effects. Angulo et al. simulated the RCS of individual components of a turbine using physical optics (PO) and showed that the tower gives the strongest contribution [20]. To test the validity of this assumption under a near-field observation, we first simulate a single turbine at 500 MHz and compute the electric field distribution around the structure. The turbine model consists of a long cone-shaped tower, a rectangular-shaped nacelle, and turbine blades which are modeled by thin plates. The tower is 64.5 m tall with a 3.8 m bottom diameter and 2.8 m top diameter. The nacelle is 15 m long and has a 3 m × 3 m cross section. The three turbine blades are modeled as 34.3 m long and 3 m wide plates with a 15° pitch angle with respect to the rotation plane. All three components are assumed to be perfect conductors for simplicity, although different materials could be modeled without too much additional effort by using an impedance boundary condition approximation.

The turbine is excited with a plane wave at zero elevation and 30° with respect to hub-on incidence in azimuth. The incident wave is vertically polarized with an electric field strength of 1 V/m. The full-wave simulation is carried out using the MLFMM solver in FEKO. The resulting total electric field in the vertical polarization is plotted at a height of $z = 34.25$ m with a sampling interval of 5 m in the xy plane in Figure 2(a). The turbine is located at $x = y = 0$. The field strength is plotted in dB with respect to 1 V/m, and the dynamic range is 10 dB. A shadow region behind the turbine is clearly observed. The shadow is the darkest immediately behind the turbine but gradually disappears as the distance increases. Outside the shadow region, rapidly oscillating interference patterns are observed. It should be noted that the sampling interval of 5 m does not fully capture the fine oscillation of the field pattern at 500 MHz.

To test the modeling fidelity required of the turbine structure, we simulate only the cone-shaped tower using MLFMM and plot the near-field distribution in Figure 2(b). Comparing Figure 2(b) to Figure 2(a), it is noted that the main difference is in the region at the bottom of the figures. In addition, the shadow region in Figure 2(a) has more fluctuation as compared to that in Figure 2(b), which may be due to the contribution from the nacelle on the top of the tower. To provide a more quantitative measure, the error between the two plots is computed. We define the percentage RMS error as

$$\%\text{ of RMS error} = \sqrt{\frac{\sum_{\text{pixels}} \left|I_1 - I_0\right|^2}{\sum_{\text{pixels}} \left|I_0\right|^2}} \times 100\%, \quad (1)$$

where I_0 and I_1 are, respectively, the field magnitude of each pixel in the two figures on a linear scale. The RMS error between Figures 2(a) and 2(b) is 3.34%. This result confirms that most of the near-field oscillations and the forward shadow region are contributed by the tower structure.

Next, we simulate the total field around a 64.5 m tall perfect conducting cylinder with the mean diameter of the cone-shaped tower, 3.3 m, at 500 MHz using FEKO. The field in the same observation plane is computed and plotted in Figure 3(a). Comparing Figure 3(a) to 2(b), it is observed that the two figures bear the same interference patterns across most of the regions of the plot. The main difference is again in the backscattering region. The RMS error between Figure 2(b) and Figure 3(a) is 3.42%. This result indicates that the cone-shaped tower can be modeled by a finite cylinder with an acceptable error. Lastly, the RMS error between Figures 2(a) and 3(a) is 4.35%. Therefore, the features of the turbine scattering can be mostly captured by that from a simplified tower alone.

2.2. 2D Modeling. Since a simple cylinder can model the tower structure with acceptable error and since the cylinder is very long compared to its cross section, we next investigate the use of a 2D simulation to model the same problem. This amounts to using an infinitely long cylinder to model the finite one, neglecting the effect of the end truncation, and in the process turning the problem into a 2D one. To carry out the 2D simulation in FEKO, we use a short (a half wavelength in height) cylinder with a 3.3 m diameter bounded by periodic boundary conditions (PBCs) on the top and the bottom. The full-wave simulation is done using the MoM solver. This setup is equivalent to an infinitely long cylinder excited by a plane wave. The observation plane is at the center of the short cylinder, and the result is plotted in Figure 3(b). The interference pattern and strength are almost identical to those of the 3D finite cylinder. The only noticeable difference is that the shadow gradually fills out without the fluctuation observed in Figure 3(a), which is due to the two ends of the finite cylinder. The RMS error between the 3D full model MLFMM simulation and the 2D simulation (i.e., Figures 2(a) and 3(b)) is 4.38%. The RMS error between the 3D finite cylinder simulation and the 2D simulation (Figures 3(a) and 3(b)) is only 0.09%. The 2D approach dramatically reduces the number of unknowns and accelerates the simulation time. Detailed simulation time savings will be reported later in Section 3.

The previously mentioned approximation is valid because the finite cylinder is very long compared to its cross section, and we only observe the field at a very close distance. To see the maximum range for the validity of the 2D approximation, the following investigation is carried out. We study the radiation from a uniform, finite line current of length 64.5 m. The result from numerical integration is shown as the solid blue curve in Figure 4. The figure is plotted on a log-log scale, and the curve shows a very clear break in slope at about 7000 m. The break in slope suggests that the field changes from a 2D ($1/\sqrt{r}$) decay to a 3D ($1/r$) decay. Also plotted in the dotted red line is the field due to an infinitely long line current [21]:

$$E_z = -I\frac{\beta^2}{4\omega\varepsilon}H_0^{(2)}(\beta r) \overset{\beta r \gg 0.25}{\approx} -\eta I\sqrt{\frac{2j}{\pi\beta r}}e^{-j\beta r}, \quad (2)$$

where I is the current strength, β is the free space wave number, ω is the angular frequency, ε is the permittivity of free space, $H_0^{(2)}$ is the Hankel function of the second kind with order zero, r is the observation distance, and η is the wave impedance of free space. It is seen that the field from the finite line current tracks this line quite well until the break in slope occurs. This implies that the 2D scattering model is a valid approximation for the field of a finite line current within a certain observation distance. Plotted in the dashed black line is the far field due to an infinitesimal-length point current given by

$$E_\theta^{FF} = -\frac{j\beta\eta}{4\pi r}Ile^{-j\beta r}\sin\theta, \quad (3)$$

where I is the current strength, l is the length of the current, and θ is the angle between the position vector of the observer and the direction of the current. In Figure 4, θ is set to 90°. It is seen that the field of a finite-length current tracks this line after the break in slope. The distance at which the break in slope occurs can therefore be determined by equating (2) and (3) and solving for r. The result is l^2/λ (or 6,939 m for the case under consideration). Note that this expression is different

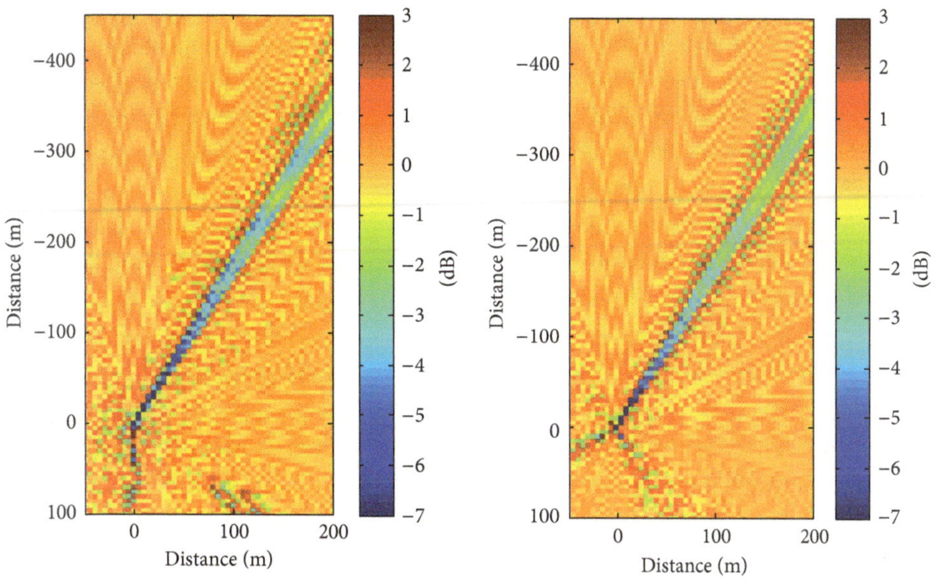

(a) 3D turbine model

(b) 3D finite tower

FIGURE 2: Results of the single-turbine simulation at 500 MHz. (a) Near-field distribution around a 3D turbine model including the tower, nacelle, and blades. (b) Near-field distribution around the cone-shaped tower.

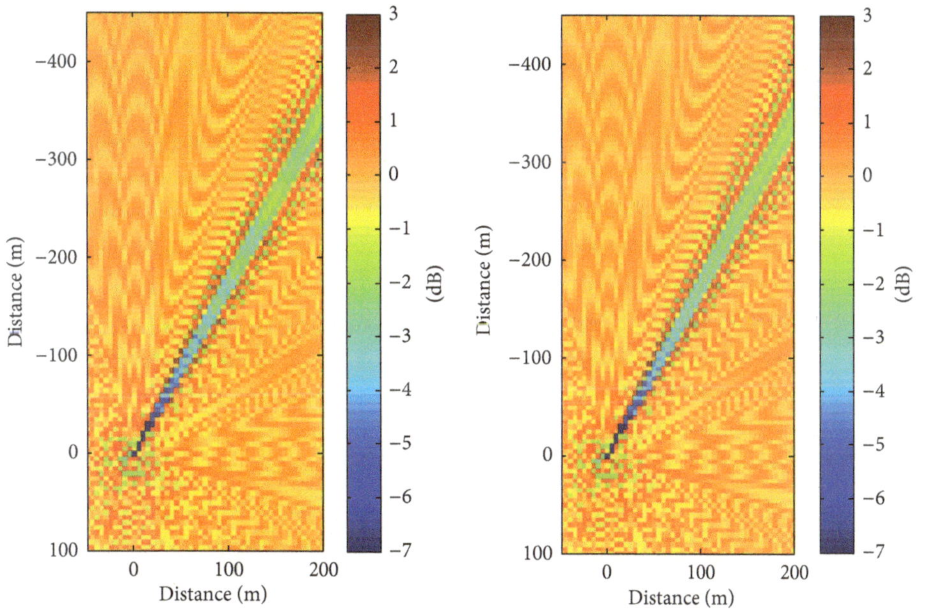

(a) 3D finite cylinder

(b) infinite 2D cylinder

FIGURE 3: Results of the single-turbine simulation at 500 MHz. (a) Near-field distribution around a 3D finite cylinder. (b) Near-field distribution around an infinite 2D cylinder.

by a factor of two from the standard far-field criterion, $2l^2/\lambda$, which is derived based on a phase consideration. As long as the region of interest is within l^2/λ of the turbine, the 2D scattering model should be fairly accurate. Another observation we can make from Figure 4 is that the field predicted by the 3D far-field approximation is higher than the field computed by the exact numerical integral at close range. This implies that the scattered field estimated using the RCS, which is defined based on the far field, will be too high when the observation position is close to the structure. Van Lil et al. observed this effect and proposed the use of an equivalent height instead of the physical height of the tower when computing the scattered field in the near zone [22, 23]. Since their proposed equivalent size is distance dependent, it

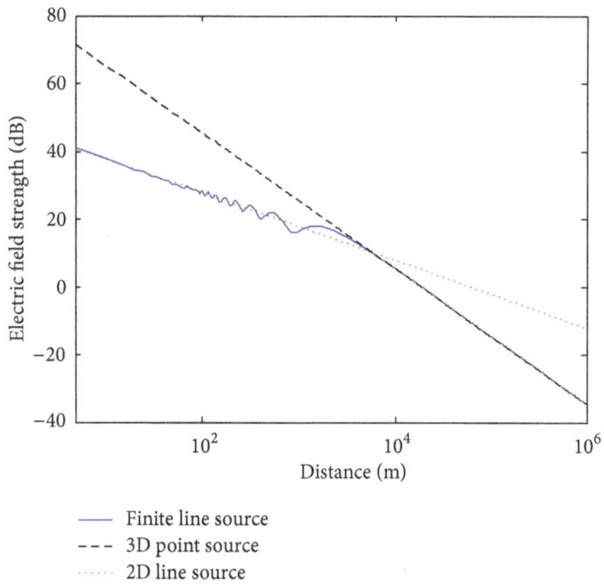

FIGURE 4: Electric field strength versus distance from a finite 64.5 m line current at 500 MHz. The field strengths of an infinite line current and an infinitesimal point current are overlaid, respectively, in red dotted and black dashed lines for comparison.

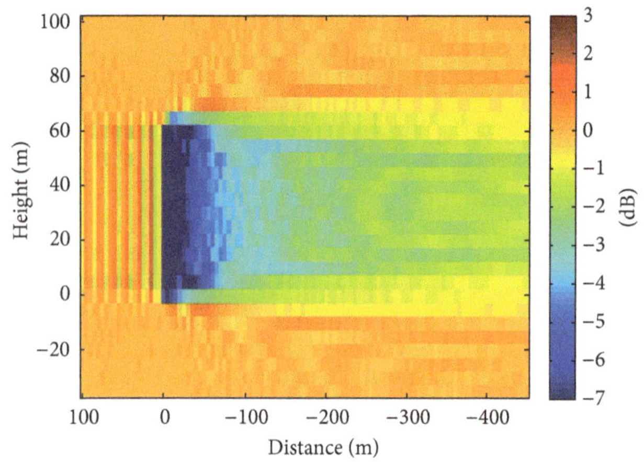

FIGURE 5: Near-field distribution around a 3D finite cylinder in the vertical cut plane along the incident field direction computed using MLFMM. The field inside the shadow region is relatively uniform for different observation heights.

in effect converts the decay of the scattered field from $(1/r)$ to $(1/\sqrt{r})$.

So far, the observation height has been restricted at the middle of the tower. Next, we simulate the same 3D finite cylinder using MLFMM and observe the near-field distribution in the vertical cut plane along the incident field direction. The result is shown in Figure 5. Strong shadow boundaries are observed at the bottom and the top edges of the cylinder. This agrees with the optical shadow intuition. More importantly, the field strength is relatively uniform at different observation heights. For example, the RMS difference between the horizontal cut at a height of 33.3 m (Figure 3(a)) and 48.4 m (not shown) is only 1.39%. This result indicates that the 2D scattering model can be used to approximate the horizontal cut at various observation heights as long as the observation height stays sufficiently away from the top and the bottom ends of the cylinder.

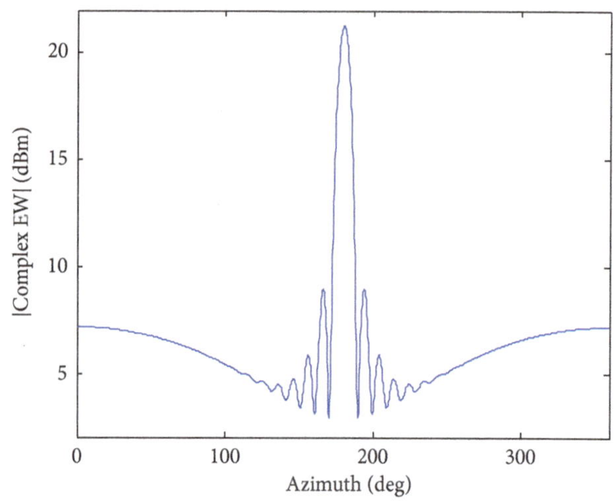

FIGURE 6: Magnitude plot of the complex echo width (EW) of an infinite 2D cylinder.

2.3. Complex Echo Width Approximation. As discussed previously, the near field of a turbine can be readily modeled by a 2D cylinder. However, computing the near field at many sampling points at 1 GHz and above is still computationally demanded since the computation time scales as the product of the number of observation positions and the number of current basis functions. To further reduce the computation time for the near field, we use the 2D far-field approximation to compute the 2D scattered field. Shown in Figure 6 is the magnitude of the 2D bistatic EW of a 2D circular cylinder with a 3.3 m diameter simulated via FEKO. The incident azimuth angle is 0°. Although only the magnitude is plotted in Figure 6, the phase front information on the scattered far field is also stored as the phase of the echo width. We then compute the total field by adding the incident plane wave to the scattered field, which is constructed by the complex EW with the proper space decay and phasing as follows:

$$E^s = E^i \frac{e^{-jkr}}{\sqrt{2\pi r}} \sqrt{\mathrm{EW}}\, e^{j[\mathrm{phase(EW)}]}. \qquad (4)$$

This approach is similar to that used earlier in [24] for simulating human radar returns. However, a well-defined phase center is assumed for each target part in [24], while no such assumption is needed in the approach presented here. Since the phase of the EW can be regarded as the phase compensation relative to a geometrical reference point, this approach works for all observation directions, including the forward direction. The resulting near-field distribution computed using this approach is shown in Figure 7(b). The same region in Figure 3(b) is copied to Figure 7(a) for

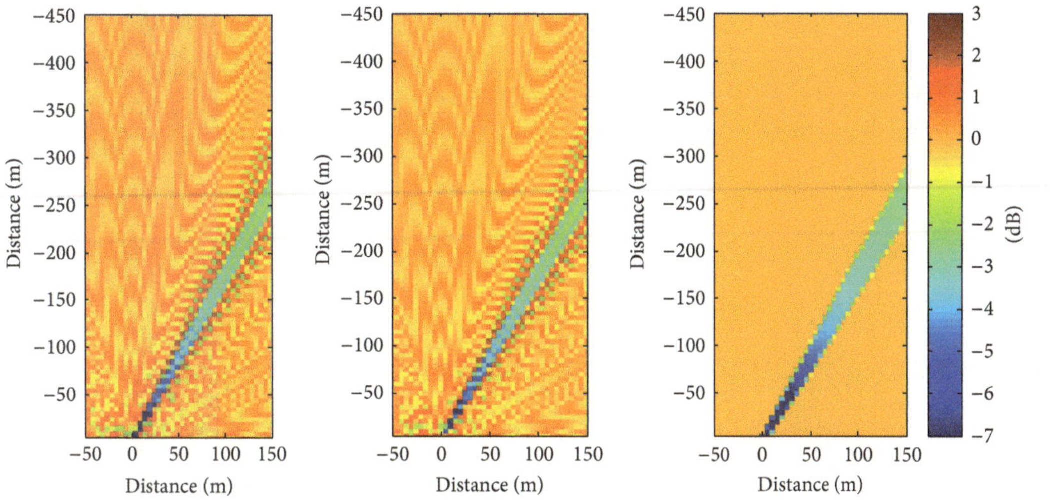

(a) Infinite 2D cylinder (b) Complex EW approx (c) Fresnel approx

FIGURE 7: Computed near-field distribution around an infinite 2D cylinder at 500 MHz. (a) Exact numerical integration. (b) Complex echo width (EW) approximation. (c) Fresnel approximation.

comparison. The two results agree well, and the RMS error between the brute-force near-field calculation in Figure 7(a) and the complex EW approximation in Figure 7(b) is 1.36%. Noted that the complex EW approximation requires the distance to be farther than $2D^2/\lambda$, where D is the cylinder diameter. In this case, that distance is 36.5 m, which is quite close to the structure.

For comparison, we also compute the forward shadow using the Fresnel zone blockage formula described in [11–13]. The attenuation of the incident field at a position directly behind the turbine is estimated by $1D/F_1$ in linear scale, where D is the diameter of the cylinder and F_1 is the first Fresnel zone radius, which is given by

$$F_1 = \sqrt{\frac{\lambda d_1 d_2}{d_1 + d_2}}, \tag{5}$$

where λ is the operating wavelength, d_1 is the distance from the transmitter to the cylinder, and d_2 is the distance from the cylinder to the observation position. To extend the previously mentioned formula for an arbitrary position, we solve the geometry problem to find the width of the turbine protruding into the first Fresnel zone, and the protrusion width is used as the diameter D in the blockage formula. It should be noted that the previously mentioned blockage formula will give erratic results when the observation distance is very close since the Fresnel radius can become smaller than the diameter. We use a zero total field when such situation happens. The Fresnel shadow is presented in Figure 7(c). It is observed that the shadow width is properly predicted, but the shadow is too dark. Also, this approach cannot be extended to predict the interference pattern outside the shadow region. The RMS error between Figures 7(a) and 7(c) is 8.24%.

3. Multiturbine Results under the Born Approximation and Computation Time

In the previous section, we established an approximate approach to efficiently compute the field distribution near a single-turbine structure. To extend the approach to a wind farm consisting of tens or hundreds of turbines, we apply the Born approximation and assume each turbine is fully illuminated by the incident field. This approximation is expected to be the least accurate when the turbines are lined up, so that one turbine casts a shadow over subsequent turbines. However, this scenario exists only at very few incident angles, considering the slenderness of the turbine structure and the large spacing between turbines found in offshore wind farms (600 m–1000 m).

To test whether the Born approximation is reasonable, two 64.5 m long, finite cylinders, each with a 3.3 m diameter, and spaced 600 m apart, are simulated rigorously using FEKO's MLFMM solver at 500 MHz. The near-field result is plotted in Figure 8(a). The Born approximation in conjunction with the 2D modeling and the complex echo width approximation discussed in the last section are applied to generate the result plotted in Figure 8(b). It is seen that both the shadow region and the interference pattern agree quite well with those in Figure 8(a). The RMS error between Figures 8(a) and 8(b) is 3.89%.

Based on the proposed methodology, the near field around a 3 × 3 wind farm with a 600 m spacing between turbines is computed at 150 MHz, 500 MHz, and 3 GHz. We use a 2D cylinder to model each turbine, the complex echo width to compute its near field, and the Born approximation to account for multiple turbines. The results are shown in Figures 9, 10, and 11, respectively. At 150 MHz, it can be observed that dark shadow occurs only when the observation position is within 100 m of the back of the turbine with

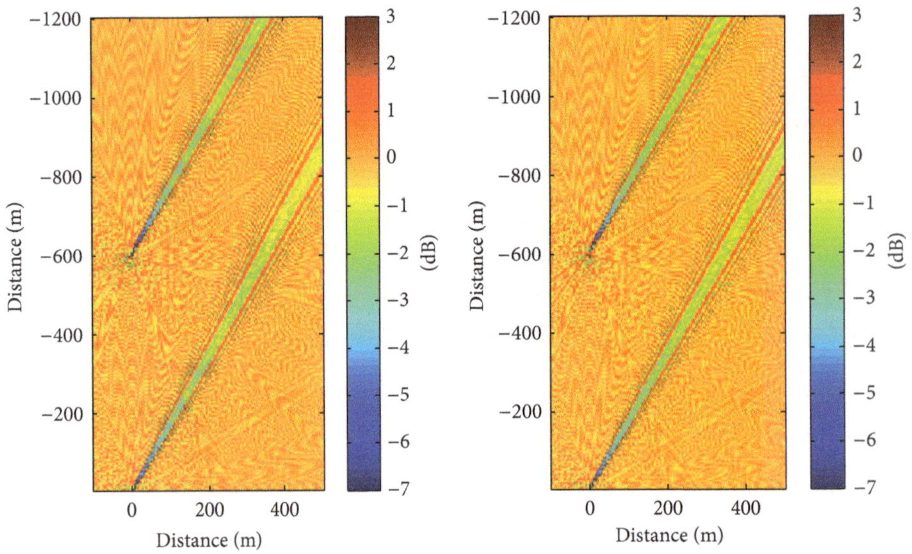

(a) Two finite cylinders

(b) Born approx. + complex EW approx

FIGURE 8: (a) Near-field distribution around two finite 3D cylinders at 500 MHz computed using the full-wave solver. (b) Near-field distribution around two infinite cylinders generated using the complex EW and the Born approximations.

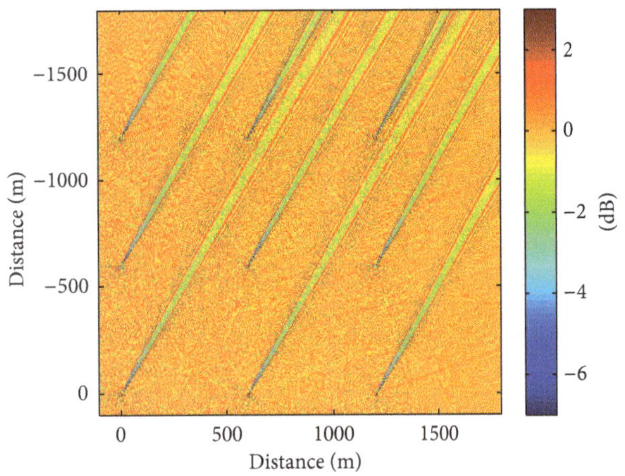

FIGURE 9: Near-field distribution around a 3 × 3 wind farm at 150 MHz.

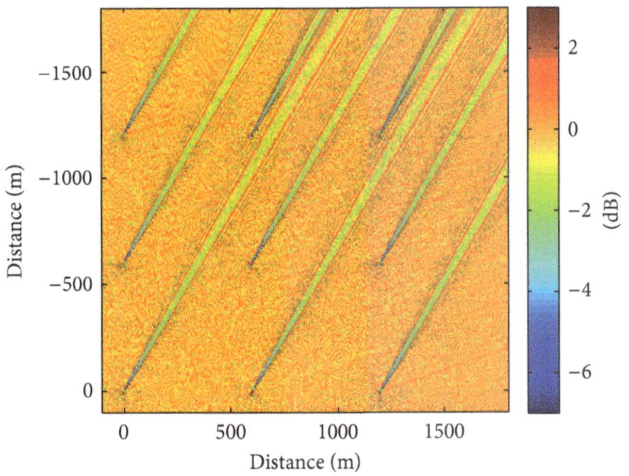

FIGURE 10: Near-field distribution around a 3 × 3 wind farm at 500 MHz.

respect to the incident direction. The field strength in the shadow region as we move farther away from the turbine becomes almost the same as the field strength outside the shadow region. As the frequency is increased to 500 MHz, the shadow becomes darker and extends farther in range. The interference pattern also changes more rapidly as a function of position due to the decrease in wavelength. Figure 11 shows the near-field distribution at 3 GHz. It shows a dark and narrow shadow region extending beyond 2 km behind each turbine. This follows the expected trend since the shadow should eventually approach the geometry optics limit in the very high-frequency limit. The multipath interference pattern also becomes even more rapidly changing as a function of position.

Finally, we discuss the computation time savings from the proposed methodology. The computation time to generate Figures 3(a), 3(b), and 7(b) is summarized in Table 1. All simulations are done on a desktop computer with an Intel Core i7-2600 CPU and 32 GB of memory. It is noted that the MLFMM utilizes an iterative solver; thus the number of iterations is also included for reference. Comparing the statistics for Figures 3(a) and 3(b), the 2D model reduces the solver time for the induced current by a factor of 779 and reduces the time for the near-field computation by a factor of 10. The overall saving is about a factor of 12. Comparing the statistics for Figures 3(b) and 7(b), the complex EW approach further reduces the time for the near-field computation to a negligible amount. Table 2 summarizes the computation

TABLE 1: Computation time summary for Figures 3(a), 3(b), and 7(b).

	3D finite cylinder – MLFMM	2D cylinder – MoM + PBC	2D cylinder + complex EW
Matrix elements	276 sec	81 sec	81 sec
Precondition	171 sec	0.063 sec	0.063 sec
Solving	779 sec (79 iterations)	1 sec	1 sec
Near field	692 sec	67 sec	0.027 sec
Total time	31 min	2.5 min	1.4 min

TABLE 2: Computation time summary for Figures 8(a) and 8(b).

	Two 3D finite cylinders – MLFMM	Two cylinders – complex EW + Born
Matrix elements	1588 sec	81 sec
Precondition	185 sec	0.063 sec
Solving	5905 sec (100 iterations)	1 sec
Near field	17466 sec	0.132 sec
Total time	7 hrs	1.4 min

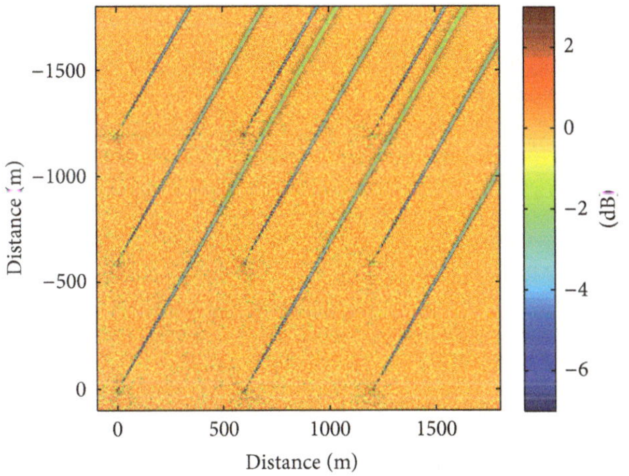

FIGURE 11: Near-field distribution around a 3 × 3 wind farm at 3 GHz.

time for the two-turbine case shown in Figures 8(a) and 8(b). Comparing the computation time of Figure 8(a) to 3(a), which are both computed using MLFMM, it is noted that the solver time and the near-field computation time are both increased significantly. Lastly, as seen by the time for Figure 8(b), the complex echo width approach reduces the near-field computation time in the two-turbine case from 17466 sec to a negligible amount. Overall, there is a time saving of a factor of 300 between the brute-force approach in Figure 8(a) and the proposed method in Figure 8(b). Such dramatic reduction in computation time means that this method could be quite useful to generate large amounts of data for radio frequency interference assessment.

4. Conclusions

In this paper, we have studied the near-field distribution around and within a wind farm under plane wave incidence. To make the problem computationally tractable, the following four assumptions were made. First, we assumed that the electromagnetic scattering is dominated by the tower structure and that the effect of the nacelle and blades is of secondary importance. Second, we assumed that the scattering from the tower can be regarded as a 2D problem when the observation distance is sufficiently close. Third, the 2D echo width concept was further applied to simplify the near-field computation. Lastly, we assumed that the interactions between turbines can be neglected and that the Born approximation can be applied when considering a whole wind farm. It was shown through a series of numerical tests that these assumptions can be made without introducing significant errors into the resulting near-field distribution. The near-field distributions of a 3 × 3 wind farm were then generated using the proposed methodology at different frequencies. The results showed the spatial extent and depth of the shadows behind the turbine towers as well as the multipath interference between the scattered and incident fields.

It should be noted that the plane wave incidence considered in this paper does not include any phase variations across the scatterers or any realistic decay as a function of distance from the transmitter. Thus, the distance from the transmitter to the wind farm must be sufficiently large to ensure the validity of this assumption. An appropriate modeling methodology for the close-in transmitter case is still needed. It is also important to mention that, while we are initially motivated by offshore wind farms and their impacts on marine radar/communication systems, where both the transmitter and receiver are located along the same plane, changing the elevation angle of the incident field will also be a very interesting study. However, additional formulation will be needed to account for such scenarios. Lastly, when the turbines on a farm are lined up with respect to the incident direction, the Born approximation must be modified to account for the mutual occlusion among the turbines. These will be topics of further study.

Acknowledgments

This work is supported by the Department of Energy under Grant DE-EE0005380, by the National Science Foundation under Grant ECCS-1232152, and in part by the Texas Norman

Hackerman Advanced Research Program under Grant no. 003658-0065-2009.

References

[1] R. J. Vogt, T. D. Crum, W. Greenwood, E. J. Ciardi, and R. G. Guenther, *New Criteria for Evaluating Wind Turbine Impacts on NEXRAD Weather Radars*, WINDPOWER 2011, Anaheim, Calif, USA, 2011.

[2] Air Warfare Center, *The Effects of Wind Turbine Farms on ATC RADARS*, Air Warfare Center, Royal Air Force, London, UK, 2005.

[3] Air Warfare Center, *Further Evidence of the Effects of Wind Turbine Farms on Air Defense RADARS*, Air Warfare Center, Royal Air Force, London, UK, 2005.

[4] "The effect of Windmills on military readiness," United States Department of Defense Report To the Congressional Defense Committees, 2006.

[5] "*U.S. Coast Guard Assessment of Potential Impacts to Marine Radar as it Relates to Marine Navigation Safety from the Nantucket Sound Wind Facility as Proposed by Cape Wind*, LLC," 2009.

[6] QinetiQ, *Results of the Electromagnetic Investigations and Assessments of Marine Radar, Communications and Positioning Systems Undertaken at the North Hoyle Wind Farm*, QinetiQ and the Maritime and Coastguard Agency, Salisbury, UK, 2004.

[7] D. Vega, C. Frenandez, O. Grande I Angulo et al., "Software tool for the analysis of potential impact of wind farms on radiocommunication services," in *Proceedings of the IEEE International Symposium on Broadband Multimedia Systems and Broadcasting (BMSB '11)*, Erlangen, Germany, 2011.

[8] B. M. Kent, K. C. Hill, A. Buterbaugh et al., "Dynamic radar cross section and radar doppler measurements of commercial general electric windmill power turbines part 1: predicted and measured radar signatures," *IEEE Antennas and Propagation Magazine*, vol. 50, no. 2, pp. 211–219, 2008.

[9] A. B. Buterbaugh, B. M. Kent, K. C. Hill et al., "Dynamic radar cross section and radar Doppler measurements of commercial General Electric windmill power turbines Part 2-Predicted and Measured Doppler Signatures," in *Proceedings of the 2007 Symposium of the Antenna Measurement Technology Association*, St. Louis, Mo, USA, 2007.

[10] E. F. Knott, J. F. Shaeffer, and M. T. Tuley, *Radar Cross Section: Its Prediction, Measurement and Reduction*, Artech House Radar Library, Norwood, Mass, USA, 1985.

[11] European Organisation for the Safety of Air Navigation, "Assessment methodology to determine the impact of wind turbines on ATC surveillance systems," Eurocontrol Document 0.3, European Organisation for the Safety of Air Navigation, Brussels, Belgium, 2007.

[12] J. J. Lemmon, J. E. Carroll, F. H. Sanders, and D. Turner, "Assessment of the effects of wind turbines on air traffic control radars," National Telecommunications and Information Administration Technical Report 08-454, Department of Commerce, Boulder, Colo, USA, 2008.

[13] S. Hawk, "Impact study of 130 offshore wind turbines in Nantucket Sound," FAA Western Service Area Engineering Services AJW-W15B, 2009.

[14] Meridian Energy Corp, *Application Documents of Project Hurunui Wind-Appendix N: Compatibility with Radio Services*, Meridian Energy Corp, Wellington, New Zealand, 2011.

[15] A. Theil and L. J. Van Ewijk, "Radar performance degradation due to the presence of wind turbines," in *Proceedings of the IEEE Radar Conference*, pp. 75–80, April 2007.

[16] G. Greving and M. Malkomes, "Weather radar and wind turbines-an update of the theoretical and numerical analysis of effects," in *Proceedings of the 6th European Conference in Meteorology and Hydrology*, pp. 1–5, 2010.

[17] S. Jordan, "Test report for evaluation of wind turbines impact on long-range radar surveillance at King Mountain, Texas," Radar Evaluation Squadron Report 84th, Hill Air Force Base, North Davis, Utah, USA, 2008.

[18] A. Frye, "Theoretical and measurement results of the effects of wind turbines of military radar systems and technical assessment methods," in *Proceedings of the 60th IEA Topical Expert Meeting*, Senternovem, The Netherlands, November 2009.

[19] EM Software & Systems, *FEKO Version 6. 2*, EM Software & Systems Inc., Stellenbosch, South Africa, 2012.

[20] I. Angulo, D. De La Vega, O. Rodríguez, O. Grande, D. Guerra, and P. Angueira, "Analysis of the mast contribution to the scattering pattern of wind turbines in the UHF band," in *Proceedings of the 5th European Conference on Antennas and Propagation (EuCAP '11)*, pp. 707–711, Rome, Italy, April 2011.

[21] C. A. Balanis, *Advanced Engineering Electromagnetics, 2/e*, John Wiley & Sons, Hoboken, NJ, USA, 1989.

[22] E. Van Lil, D. Trappeniers, J. W. De Bleser, and A. Van De Capelle, "Computations of radar returns of wind turbines," in *Proceedings of the 3rd European Conference on Antennas and Propagation (EuCAP '09)*, pp. 3852–3856, Berlin, Germany, March 2009.

[23] E. Van Lil, J. W. De Bleser, and A. Van De Capelle, "Computations of the effects of wind turbines in the close near field of RF installations," in *Proceedings of the 5th European Conference on Antennas and Propagation (EuCAP '11)*, pp. 1362–1366, Rome, Italy, April 2011.

[24] S. S. Ram, C. Christianson, Y. Kim, and H. Ling, "Simulation and analysis of human micro-Dopplers in through-wall environments," *IEEE Transactions on Geoscience and Remote Sensing*, vol. 48, no. 4, pp. 2015–2023, 2010.

6

EBG Size Reduction for Low Permittivity Substrates

Gonzalo Expósito-Domínguez,[1] José Manuel Fernández-González,[1] Pablo Padilla,[2] and Manuel Sierra-Castañer[1]

[1] Radiation Group, Department of Signals, Systems and Radio Communications, University of Madrid, 28040 Madrid, Spain
[2] Department of Signal Theory, Telematics and Communications, University of Granda, Granda 18071, Spain

Correspondence should be addressed to Gonzalo Expósito-Domínguez, gexposito@gr.ssr.upm.es

Academic Editor: Eric Lheurette

Double layer and edge-location via techniques are combined for electromagnetic band gap (EBG) size reduction. The study of the required number of elements and their dimensions is carried out in order to suppress the surface wave propagation modes and consequently to reduce the mutual coupling between radiating elements in low-permittivity substrates. By applying these techniques, the size of the EBG mushroom is reduced by 30%; however, the bandwidth operation maintains its value, and these structures can be integrated between radiating elements in broad bandwidth antennas.

1. Introduction

Low-profile integrated antennas are demanded to be small [1]. In order to reduce the size of the antenna or to increase the working frequency band, high-permittivity substrates are used. However, with this type of substrates, surface wave propagation modes are enhanced, and there is strong mutual coupling [2]. In array antennas, the separation between elements should not be higher than $0.8\lambda_0$ in order to avoid grating lobes and to get high directivity. This last condition regarding the separation between radiating elements increases the mutual coupling because the elements are closer and they have stronger interaction.

The most common mutual coupling reduction techniques are cavity structures [3], nonuniform feeding distribution [4], unequally space distribution [5], and defected ground plane (DGP) [6], but lately, (EBG) structures are being used. The introduction of several rows of EBG structures between two printed antennas has been proved to increase the isolation [7]. However, when low-permittivity substrates are used in order to enhance the radiation efficiency, the size of the printed antenna elements increases, and the available space between radiating elements is not sufficient to introduce the necessary number of EBG rows.

Patch antennas are found to have very strong mutual coupling due to the severe surface waves on thick and high-permittivity substrates. In the literature, it can be found a variety of works which apply metamaterials for the reduction of this effect. In [7], four rows of EBG mushrooms are inserted between the patch antennas in $\epsilon_r = 10.2$ and thickness ($h = 2$ mm) substrate. With this configuration, 8 dB of mutual coupling reduction is obtained. In [8], different substrates are combined; radiating elements are suspended over a thick foam layer in order to increase the bandwidth, and meanwhile, EBG structures are printed in a thin high-permittivity substrate for size reduction and surface wave suppression. In [9], by using edge-located vias, the size of mushroom-type EBG is reduced by 20%. Among other strategies, in [10], a fork shape is used. The area occupied by the fork-like structure is less than 25% of the mushroom-like structure. Besides, microelectromechanical systems (MEMS) are used, and reconfigurable stop band is obtained. Another studied technique is the use of metal strips. Basically, the idea is to combine the EBG concept with soft surfaces. A comparison of mushroom-type EBG surfaces and corrugated and strip-type soft surfaces is shown in [11]. This stripe-type is used in [12, 13] to reduce mutual coupling. Finally, dual band planar soft surfaces are developed in [14]; two sizes of

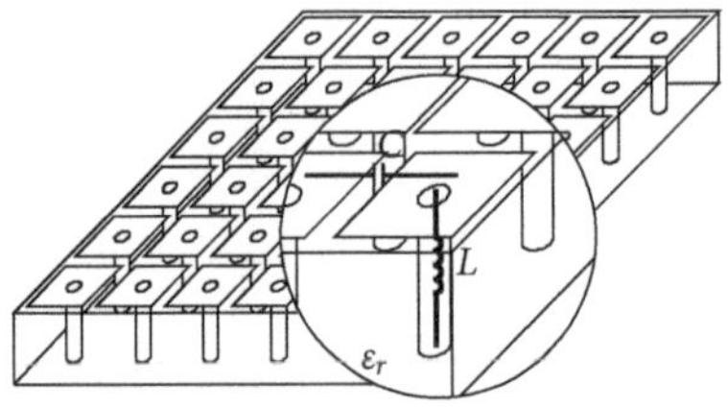

FIGURE 1: High impedance surface and its model with parallel resonant LC circuit. The substrate is transparent in order to get better visualization of metallic vias.

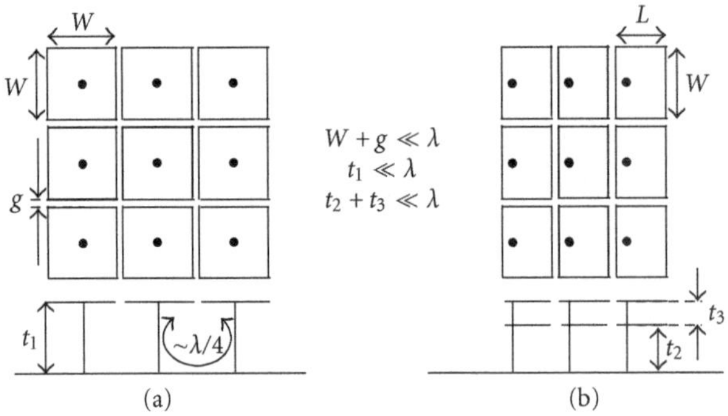

FIGURE 2: Layer and top views of traditional EBG structures (a) and multilayered F structure (b).

strips are mixed in order to get dual forbidden band. Thanks to its frequency selective feature, these EBG structures can be used as a filter. By applying three elements as a ground plane for a microstrip line, in [15], isolation levels higher than 55 dB in the first frequency band (2.1 GHz) and 40 dB in the second one (2.45 GHz) are achieved.

The main goal of this work is to combine multilayer technique with edge-location via [16] to find a new solution for EBG size reduction when using thick and low-permittivity substrates (in this case mushroom type [17]). By applying multilayered structure, different shapes, and edge-location via, the effective size is reduced by 30%, and EBG structures fit in between printed antennas. This paper is organized as follows. An overview of the EBG materials theory is provided in Section 2. Section 3 is devoted to study different solutions to obtain surface wave suppression characteristics. In Section 4, different EBG topologies, sizes, and number of periods are simulated and constructed in order to reduce the size of the mushroom to place it between radiating elements. Finally, in Section 5, the conclusions are drawn.

2. EBG Theory Fundamentals

EBG technique appears as an application of truncated frequency selective surfaces (FSSs) [17]. These structures consist of array patches printed in a substrate, which are short circuited to the ground plane with vias and can be visualized as mushrooms protruding from the surface, as shown in Figure 1.

When the period is small compared to the wavelength of interest, it is possible to analyze the material as an effective medium with a surface impedance (metamaterial with electrical properties [18]). These "mushrooms" present very high impedance values for vertical and horizontal propagation modes at certain frequencies. In Figure 2, on the left hand side, traditional mushrooms are shown, and meanwhile, on the right hand side, a multilayered mushroom structure is presented.

The behavior of this structure is similar to an LC circuit in (1). Below the resonance frequency, the surface is inductive, while above resonance frequency, the surface is capacitive. Consider

$$\omega_0 = \frac{1}{\sqrt{LC}}. \tag{1}$$

Nearby ω_0, the surface impedance (Z_s) is much higher than the impedance of free space, as (2) depicts. Therefore, no vertical or horizontal propagation modes are allowed. Consider

$$Z_s = \frac{j\omega L}{1 - \omega^2 LC}, \tag{2}$$

where the capacitance C is provided by the proximity of the metal plates, according to (3) [19] as follows:

$$C = \left[\frac{w\varepsilon_0(\varepsilon_{\text{eff}})}{\pi}\right]\cosh^{-1}\left(\frac{2w}{g_0}\right), \tag{3}$$

and the inductance L (4) is related to the thickness of the structure, because its value is due to the length of the via as follows:

$$L = \mu_0 t. \tag{4}$$

Analytically, it can be noticed that the band gap operation band is proportional to $\sqrt{L/C}$. Thus, for a fixed separation of

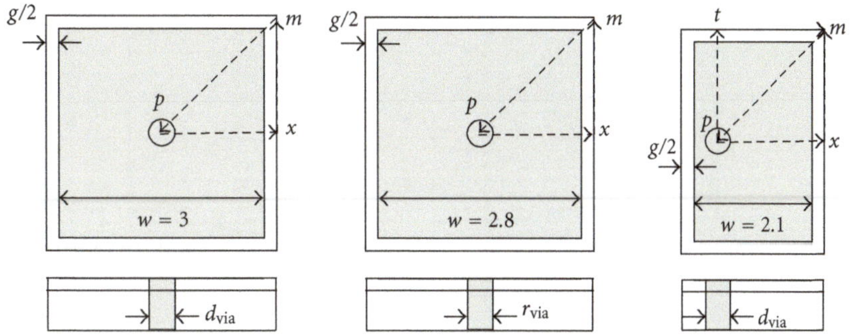

(a) Original mushroom (b) Double layer mushroom (c) Double layer and edge-location via mushroom (F shape)

FIGURE 3: Unit cell scheme for eigenmode solutions.

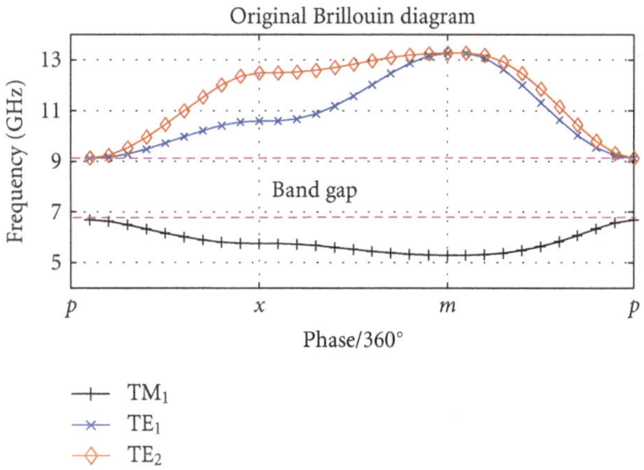

FIGURE 4: Brillouin diagram for original mushrooms.

the metal plates, the value of the capacitance is set, and the working bandwidth can be increased by increasing the length of the via and therefore the thickness of the substrate, which is adequate for wideband antennas.

Finally, EGB structures can be seen as bend corrugations, where the length of the corrugation is the sum of the via and the patch multiplied by two, as it is depicted in Figure 2. The band gap operation of the corrugations starts and finishes when the length is similar to $\lambda/4$ and $\lambda/2$, respectively, which yields a frequency octave [20].

3. Surface Wave Suppression

Thanks to the high surface impedance, horizontal or vertical modes are not allowed in the mushroom structures at certain frequencies. In order to find the allowed frequencies for each wave vector, a single unit cell with periodic conditions is simulated. The x-axis represents the dimensions inside the unit cell in which the boundary conditions are fulfilled at certain frequencies for the transmission modes of the structure. In Figures 3(a) and 3(b), the unit cell is symmetric with respect to the via; thus, only three traces have to be computed. However, for the F shape case in Figure 3(c), the EBG structure does not have the same dimensions in x and y. Therefore, five traces have to be analyzed to describe all the possible propagation modes for these boundary conditions.

All the studied structures are developed in a low-permittivity substrate $\varepsilon_r = 2.17$ and thickness of 1.143 mm. For the traditional mushroom, the size of the original patch in X band (7.25–8.4 GHz) is 3 mm × 3 mm, while on the other hand, for double layer mushroom 2.8 mm × 2.8 mm size is used. For this last configuration, two stacked substrates of 0.762 mm and 0.381 mm thickness are used.

The electric field is described in terms of an eigenvalue equation, which is solved numerically. In Figures 4, 5, and 6, the mode solutions that satisfy the boundary conditions are shown. The abscissa value represents the wave number which fulfils the requirements at a certain frequency. The lowest line is the TM mode, while the second and third lines are TE modes. A frequency band gap (dashed lines), in which the surface does not support surface wave propagation of either polarization, horizontal nor vertical, extends from the top of the TM band to the point where TE band crosses the light line. The comparison between original shape and double layer mushrooms yields the result of 2 GHz bandwidth for the original shape and 1.2 GHz for the multilayer solution.

In this work, the combination of multilayer structure [17] and edge-location via [16] for mushroom size reduction

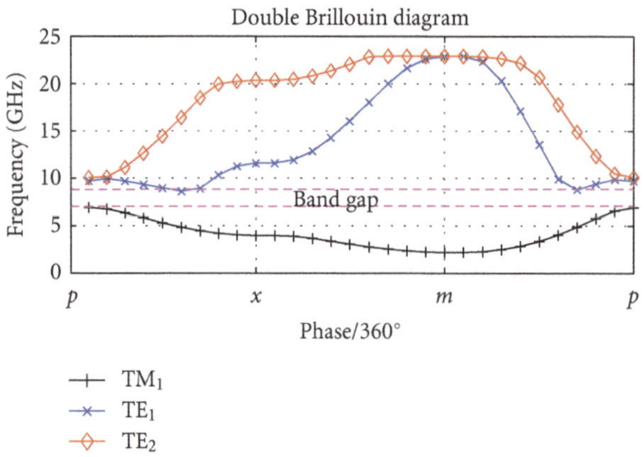

Figure 5: Brillouin diagram for double layer mushrooms.

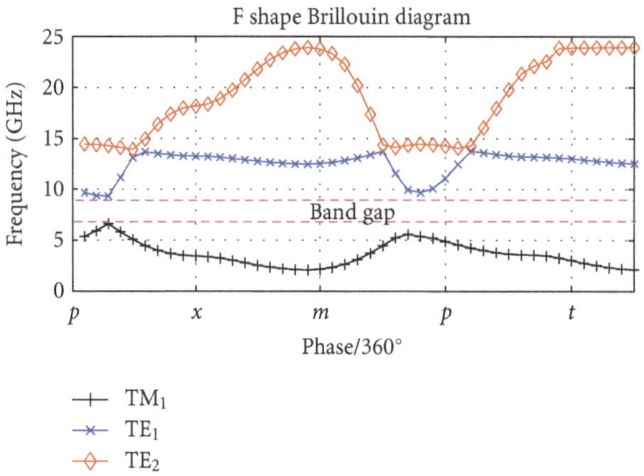

Figure 6: Brillouin diagram for F shape mushrooms.

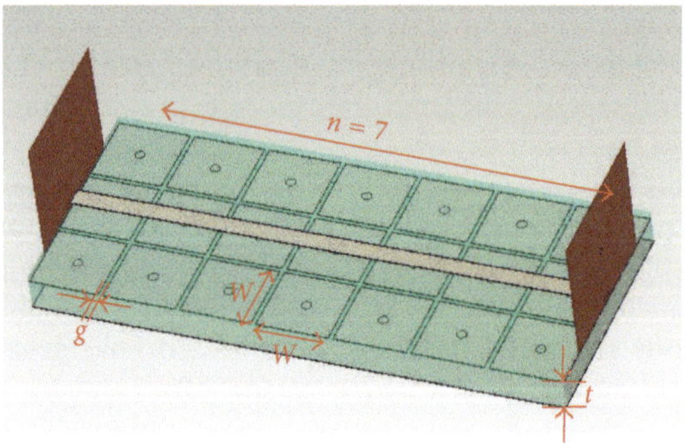

Figure 7: Simulation scheme for S_{21} analysis.

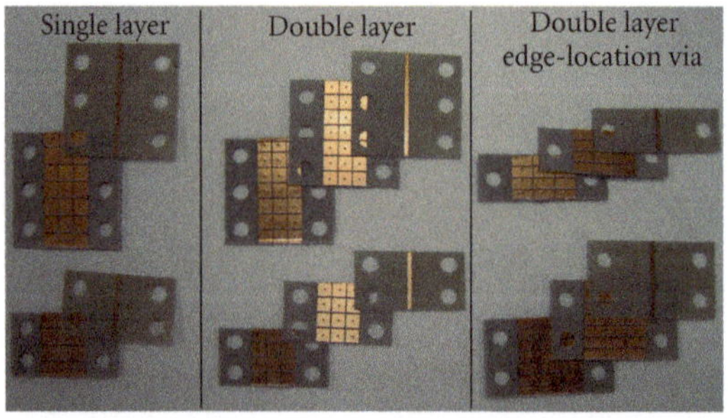

Figure 8: Samples of single and multilayered EBG mushrooms with different shapes and number of elements.

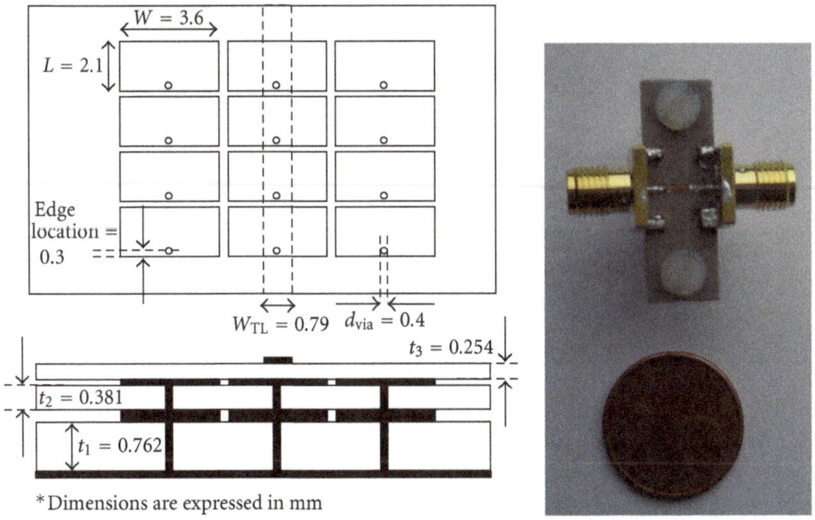

(a) Schematic (b) Prototype

FIGURE 9: Multilayered mushroom with rectangular shape, 4 elements and edge-located via (F-shape).

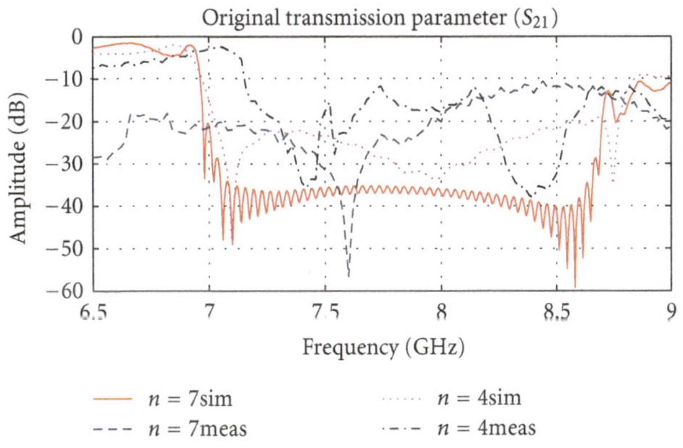

FIGURE 10: Transmission parameters (S_{21}) for original mushrooms.

is discussed. In order to keep the frequency working band (15%) and radiation efficiency, the same substrate ($\varepsilon_r = 2.17$ and thickness of 1.143 mm) is used. As long as the inductance depends on the substrate thickness, which is fixed already, the only available parameter is the capacitance C. In order to increase this parameter, a multilayered structure with edge-location via is presented. The dimensions of the patches are 2.1 mm × 3.6 mm (this value means 30% size reduction in the critical direction). The Brillouin diagram for this configuration is presented in Figure 6. In this case, the bandwidth keeps its value, but significant size reduction is noticed.

4. Mutual Coupling Reduction

The size of the mushroom patches and the necessary number of periods for mutual coupling reduction are higher than the available space between radiating elements for low-permittivity substrates. For example, for steering arrays antennas, separation must not be higher than 0.6 λ_0 in order to avoid grating lobes. To study this effect, a simulation scheme for S_{21} analysis is proposed in Figure 7. In this simulation, different topologies (original, double, or F shape), size of the mushroom w, gap size g, and number of elements n are tested in order to obtain the surface wave suppression behavior.

In order to validate the whole process, prototypes of four and seven rows are built. Six prototype transmission lines (TLs) with EBG ground plane are mounted in Figure 8. On the left hand side, the two circuits are single layered, the two circuits in the middle are double layered, and the last two circuits on the right hand side combine double layer with edge-location via (F shape).

All the substrates used have permittivity of $\varepsilon_r = 2.17$. However, four different thickness values are used. The TLs impedance is 50 Ω, and those TLs are printed in a 0.254 mm thick substrate. Mushroom in single layer case are printed in a 1.143 mm thick substrate, while in double layer case is printed in 0.762 mm (bottom layer) and 0.381 mm thick (upper layer) substrates. Therefore, total thickness value maintains its value as it can be seen in Figures 9(a) and 9(b).

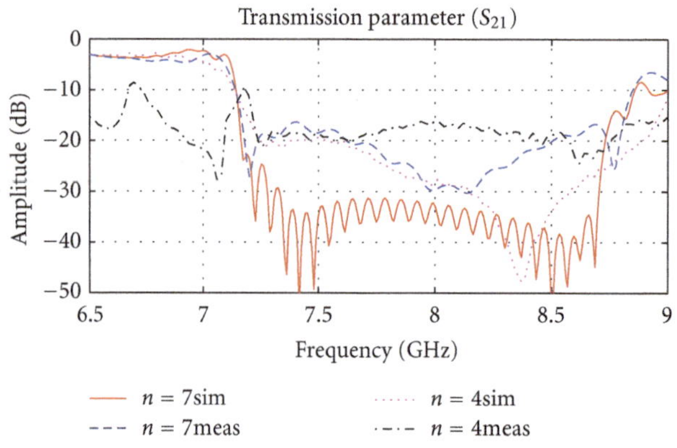

FIGURE 11: Transmission parameters (S_{21}) for double layer mushrooms.

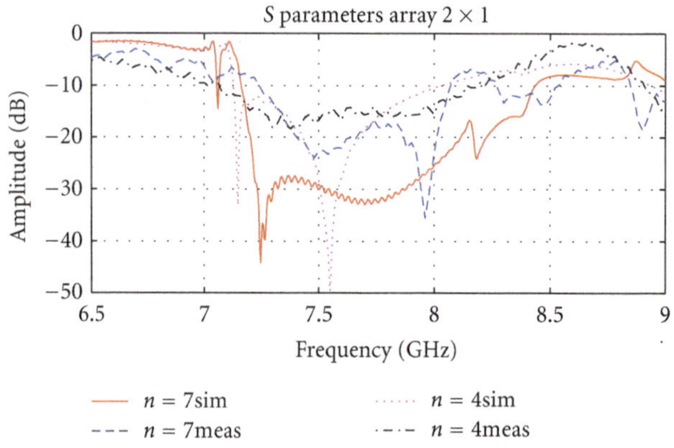

FIGURE 12: Transmission parameters (S_{21}) for F shape mushrooms.

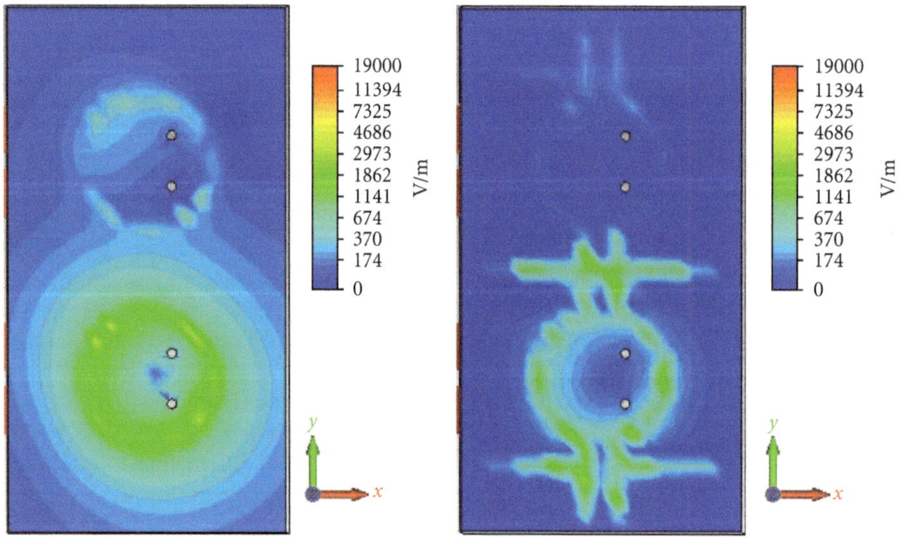

(a) Without EBG structures

(b) With EBG structures

FIGURE 13: $|E|$ field simulation of two round patches with dual circular polarization.

In Figures 10, 11, and 12, comparisons between the simulated structures and the measured prototypes are shown. Difficulties due to the small size of the circuits (size circuit approximately 1 cm) add some differences between simulations and measurements. However, the overall behavior of an LC filter in a certain frequency band is observed. It can be noticed, comparing the figures, that traditional mushrooms have larger frequency operation band than F shape mushrooms. However, F shape ones keep fulfilling the bandwidth and isolation requirements.

Finally, the tradeoff solution between isolation and available space is carried out, being the necessary number of periods for surface wave suppression: $n = 4$. In Figures 9(a) and 9(b), the chosen topology is shown. With final dimensions of 2.1×3.6 mm, four elements, double-layered structure and edge-location via, fulfil the requirements of available space (10.2 mm) between 2 printed antennas separated $0.6\,\lambda_0$ in $\varepsilon_r = 2.17$ and 1.143 mm thick substrate. A bandwidth operation in X band, from 7.1 GHz to 8.2 GHz (15%) and 10 dB of isolation, is obtained.

In order to prove this solutions four rows of double-layered edge-located via, EBG mushrooms are introduced between two round patches with double circular polarization. The radiating elements are integrated in the same substrate, and they are circular polarized. The elements are fed by a 90°/3 dB branch line coupler in order to get the double circular polarization.

In Figure 13(a), $|E|$ field simulation for the two patches is shown for left-handed circular polarization (LHCP). In Figure 13(b), it can be seen graphically how $|E|$ field value decays quicker when using double layer edge-location via EBG structures.

5. Conclusion

This paper presents and proposes a combination of double layer and edge-location via techniques for EBG size reduction. With these techniques, a 30% of size reduction in comparison to the original mushroom shapes is achieved. This F shape mushrooms maintain the broad bandwidth operation, and these are used in order to suppress the surface wave modes and consequently to reduce the mutual coupling between radiating elements in low-permittivity substrates.

Acknowledgments

The simulations contained in this work have been carried out using CST Microwave Studio Suite 2011 under a cooperation agreement between Computer Simulation Technology (CST) and Universidad Politécnica de Madrid. The authors kindly thank the company NELTEC S.A. for giving samples of the substrates, in which the prototypes were built. This work has been supported by a UPM Grant no. CH/003/2011 and the sicomoro project with Reference no. TEC2011-28789-C02-01.

References

[1] M. A. Salas-Natera, A. García-Aguilar, J. Mora-Cuevas et al., *Satellite Communications*, InTech, 2011.

[2] J. R. James and P. S. Hall, *Handbook of Microstrip Antennas*, IEEE Waves Series, 1989.

[3] R. J. Mailloux, "On the use of metallized cavities in printed slot arrays with dielectric substrates," *IEEE Transactions on Antennas and Propagation*, vol. AP-35, no. 5, pp. 477–487, 1987.

[4] C. A. Balanis, *Antenna Theory, Analysis and Design*, Wiley, 3rd edition, 1997.

[5] B. Preetham Kumar, "Design of unequally spaced arrays for performance improvement," *IEEE Transactions on Antennas and Propagation*, vol. 47, no. 3, pp. 511–523, 1999.

[6] M. Salehi, A. Motevasselian, A. Tavakoli, and T. Heidari, "Mutual coupling reduction of microstrip antennas using defected ground structure," in *Proceedings of the 10th IEEE Singapore International Conference on Communications Systems (ICCS '06)*, Singapore, November 2006.

[7] F. Yang and Y. Rahmat-Samii, "Microstrip antennas integrated with Electromagnetic Band-Gap (EBG) structures: a low mutual coupling design for array applications," *IEEE Transactions on Antennas and Propagation*, vol. 51, no. 10, pp. 2936–2946, 2003.

[8] E. Rajo-Iglesias, O. Quevedo-Teruel, and L. Inclán-Sánchez, "Mutual coupling reduction in patch antenna arrays by using a planar EBG structure and a multilayer dielectric substrate," *IEEE Transactions on Antennas and Propagation*, vol. 56, no. 6, pp. 1648–1655, 2008.

[9] E. Rajo-Iglesias, L. Inclan-Sanchez, J. L. Vazquez-Roy, and E. Garcia- Muñoz, "Size reduction of mushroom-type EBG surfaces by using edge-located vias," *Microwave and Wireless Components Letters, IEEE*, vol. 17, no. 9, pp. 670–672, 2007.

[10] L. Yang, M. Fan, F. Chen, J. She, and Z. Feng, "A novel compact electromagnetic-bandgap (EBG) structure and its applications for microwave circuits," *IEEE Transactions on Microwave Theory and Techniques*, vol. 53, no. 1, pp. 183–189, 2005.

[11] E. Rajo-Iglesias, M. Caiazzo, L. Inclan-Sanchez, and P.-S. Kildal, "Comparison of bandgaps of mushroom-type EBG surface and corrugated and strip-type soft surfaces," *Microwaves, Antennas Propagation, IET*, vol. 1, no. 1, pp. 184–189, 2007.

[12] E. Rajo-Iglesias, O. Quevedo-Teruel, and L. Inclán-Sánchez, "Planar soft surfaces and their application to mutual coupling reduction," *IEEE Transactions on Antennas and Propagation*, vol. 57, no. 12, pp. 3852–3859, 2009.

[13] S. Quevedo-Teruel, L. Inclan-Sanchez, and E. Rajo-Iglesias, "Soft surfaces for reducing mutual coupling between loaded PIFA antennas," *IEEE Antennas and Wireless Propagation Letters*, vol. 9, pp. 91–94, 2010.

[14] E. Rajo-Iglesias, J. L. Vázquez-Roy, O. Quevedo-Teruel, and L. Inclán-Sánchez, "Dual band planar soft surfaces," *IET Microwaves, Antennas and Propagation*, vol. 3, no. 5, pp. 742–748, 2009.

[15] L. Inclan-Sanchez, J. L. Vaquez-Roy, and E. Rajo-Iglesias, "High isolation proximity coupled multilayer patch antenna for dual-frequency operation," *IEEE Transactions on Antennas and Propagation*, vol. 56, no. 4, pp. 1180–1183, 2008.

[16] F. Yang and Y. Rahmat-Samii, "Polarization-dependent electromagnetic band gap (PDEBG) structures: designs and applications," *Microwave and Optical Technology Letters*, vol. 41, no. 6, pp. 439–444, 2004.

[17] D. Sievenpiper, L. Zhang, R. F. Jimenez Broas, N. G. Alexöpolous, and E. Yablonovitch, "High-impedance electromagnetic surfaces with a forbidden frequency band," *IEEE Transactions on Microwave Theory and Techniques*, vol. 47, no. 11, pp. 2059–2074, 1999.

[18] C. Caloz and T. Itoh, *Electromagnetic Metamaterials: Transmission Line Theory and Microwave Applications: The Engineering Approach*, John Wiley and Sons, 2006.

[19] F. Yang and Y. Rahmat-Samii, *Electromagnetic Band Gap Structures in Antenna Engineering*, The Cambridge RF and Microwave Engineering Series, 2008.

[20] P.-S. Kildal, "Definition of artificially soft and hard surfaces for electromagnetic waves," *Electronics Letters*, vol. 24, no. 3, pp. 168–170, 1988.

On the Second-Order Statistics of Correlated Cascaded Rayleigh Fading Channels

Yawgeng A. Chau and Karl Y.-T. Huang

Department of Communications Engineering, Yuan Ze University, 135 Yuan-Tung Road, Chung-Li 32003, Taiwan

Correspondence should be addressed to Yawgeng A. Chau, eeyaw@saturn.yzu.edu.tw

Academic Editor: Matteo Pastorino

The second-order statistics of two correlated cascaded (double) Rayleigh fading channels are analyzed, where different relevant second-order cross-correlation functions of in-phase and quadrature components of the cascaded Rayleigh channels are derived. The level crossing rate (LCR) and average fade duration (AFD) of the cascaded channels are evaluated, and a single-integral form of the LCR is derived. Numerical results of the LCR and AFD are presented, and the effect of the correlation is illustrated.

1. Introduction

Recently, the cascaded fading channels have attracted a lot of research interests [1–12]. The cascaded fading channels can be used to model different wireless communication scenarios, such as the keyhole fading channel [4], the mobile-to-mobile transmission channel [5], dual-hop fading channels [8], and radio frequency identification (RFID) pinhole channels [9–12].

In the published works of [1–9], the cascaded processes are assumed to be independent. For cascaded keyhole or pinhole fading channels, since the faded signals terminate and originate at the same keyhole or pinhole location, the correlation between cascaded fading processes may have some impacts on the performance of wireless transmissions. In [10–12], the bit error rate (BER) of RFID with multiple tags has been examined for correlated cascaded (forward-backscatter) channels. In fact, the completely dependent or correlated cascaded Rayleigh fading channels have been studied in earlier works [13], where the corresponding BERs of different modulations have been evaluated. The scenarios of keyhole multiple-input multiple-output (MIMO) channels and RFID pinhole channels with a correlation coefficient are illustrated in Figure 1.

In this paper, in contrast to the evaluation of first-order performance for correlated cascaded Rayleigh channels, such as the bit error rate given in [10–12], second-order statistics of correlated cascaded Rayleigh processes are analyzed and evaluated. The second-order statistics such as the LCR and AFD are useful for the design of practical wireless communication systems [14–19]. The LCR of a faded signal is defined as the average rate at which the signal envelope crosses a given level, and the AFD is the average duration that the signal envelope becomes lower than the given threshold [14, 15]. As addressed in [20], the second-order statistics, LCR and AFD, are also important for the design of real RFID systems. In [8], the LCR and AFD were analyzed for independent cascaded Rayleigh fading channels. In [16], the LCR and AFD were applied to the burst-error analysis for independent cascaded Rayleigh fading channels. In [17], the LCR and AFD of cooperative selection diversity were studied, where independent cascaded Rayleigh fading channels were assumed. In [18, 19], the LCR and AFD were evaluated for independent cascaded Nakagami-m fading channels.

In the context, two completely correlated Rayleigh processes [13] in cascaded are modelled, corresponding second-order correlation functions are derived, and the LCR and AFD of the correlated cascaded Rayleigh processes are evaluated. The result of the paper can be applied to the analysis of second-order statistics of correlated cascaded Rayleigh fading channels, such as a path of keyhole or pinhole fading channels shown in Figure 1.

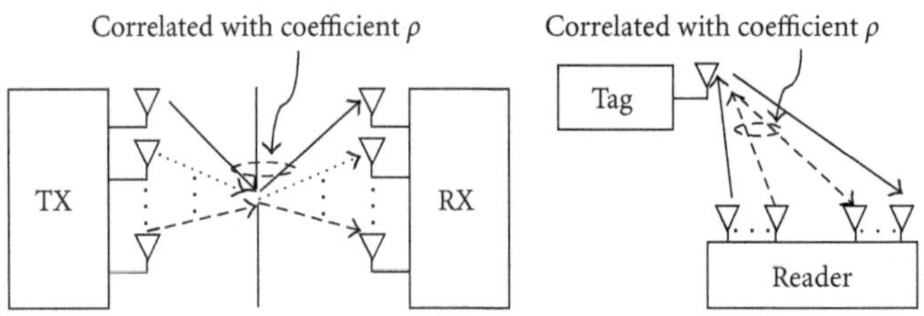

(a) Keyhole MIMO channels (b) RFID pinhole channels

FIGURE 1: The scenarios of keyhole MIMO channels and RFID pinhole channels.

In Section 2, the correlated cascaded Rayleigh random processes and their in-phase and quadrature processes are modelled, where various second-order correlation functions are derived. In Section 3, the LCR and AFD of the correlated cascaded processes are evaluated. In Section 4, numerical results of the LCR and AFD are given and the effect of the correlation is illustrated. Conclusions are drawn in Section 5.

2. Second-Order Correlation Functions of Correlated Cascaded Rayleigh Processes

2.1. Modeling. Let $\alpha_i(t)$ $(i = 1, 2)$ be a Rayleigh process with the probability density function (pdf) given by the following:

$$f_i(\alpha_i) = \frac{\alpha_i}{\sigma_i^2} e^{-\alpha_i^2/(2\sigma_i^2)}, \quad \alpha_i \geq 0, \tag{1}$$

where $2\sigma_i^2 = E[\alpha_i^2]$. Then, two correlated Rayleigh processes in cascade can be modelled by the following:

$$R(t, \tau) = \alpha_1(t)\alpha_2(t + \tau), \quad t, \tau \geq 0, \tag{2}$$

where τ is the time or phase offset between α_1 and α_2. The joint pdf of α_1 and α_2 is [21]

$$f_{12}(\alpha_1, \alpha_2) = \frac{\alpha_1\alpha_2}{\sigma_1^2\sigma_2^2(1-\rho^2)} e^{-(\sigma_2^2\alpha_1^2 + \sigma_1^2\alpha_2^2)/(2\sigma_1^2\sigma_2^2(1-\rho^2))} \times I_0\left(\frac{\alpha_1\alpha_2|\rho|}{\sigma_1\sigma_2(1-\rho^2)}\right), \quad \alpha_1, \alpha_2 \geq 0, \tag{3}$$

where $I_0(\cdot)$ is the zeroth order modified Bessel function of the first kind, and the correlation coefficient is in the range $-1 < \rho < 1$. Notice that when $\rho = 1$, α_1 and α_2 have a linear relation, and R reduces to the well-known chi-square distribution. With $f_{12}(\alpha_1, \alpha_2)$ given by (3), it is straightforward to show that the pdf of R is

$$f_R(r) = \int_0^\infty \frac{1}{\alpha_2} f_{12}\left(\frac{r}{\alpha_2}, \alpha_2\right) d\alpha_2 = \frac{r}{\sigma_1^2\sigma_2^2(1-\rho^2)} I_0\left(\frac{r|\rho|}{\sigma_1\sigma_2(1-\rho^2)}\right) K_0\left(\frac{r}{\sigma_1\sigma_2(1-\rho^2)}\right), \tag{4}$$

where $K_0(\cdot)$ is the zeroth order modified Bessel function of the second kind, and [22, (3.478.4)] is used to evaluate the integral. In Figure 2, the pdf with different correlation coefficients is shown. From Figure 2, when ρ increases, the cascaded Rayleigh processes has a larger probability of yielding smaller values.

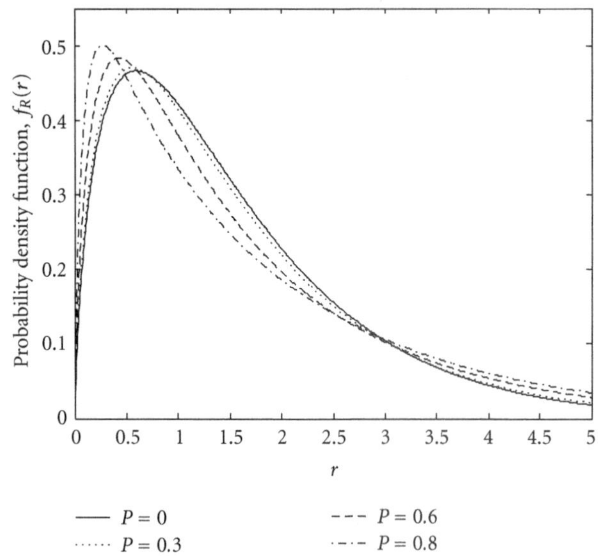

FIGURE 2: Probability density function of correlated cascade Rayleigh processes.

2.2. Second-Order Correlation Functions of In-Phase and Quadrature Processes. For $i = 1, 2$, let X_i and Y_i be the underlying in-phase and quadrature components of $\alpha_i(t)$, and

$$\alpha_i^2(t) = X_i^2(t) + Y_i^2(t). \tag{5}$$

The time derivatives of X_i, Y_i, and $\alpha_i(t)$ give the second-order information of cascaded Rayleigh processes.

Le f_m be the maximal Doppler frequency of the cascaded fading channels and θ_k be the uniformly distributed phase on $[0, 2\pi]$. Based on the modeling used in [13, 14], for $i = 1, 2$, we may write (X_i, Y_i) in the following forms [23]

$$X_i(t) = \lim_{n\to\infty} \sqrt{2\sigma_i^2} \sum_{k=1}^{n} a_k \cos(\omega_k t - \omega_{c,i} T_k),$$
$$Y_i(t) = \lim_{n\to\infty} \sqrt{2\sigma_i^2} \sum_{k=1}^{n} a_k \sin(\omega_k t - \omega_{c,i} T_k), \tag{6}$$

where $\omega_k = 2\pi f_m \cos\theta_k$, a_k is the corresponding fractional power for the kth component, $\omega_{c,i}$ is the frequency, and T_k is the time-delay [15]. In practice, T_k can be modelled by an exponential distribution with mean $\overline{T}$ [15]. The time derivatives of (X_i, Y_i) are

$$\dot{X}_i(t) = -\lim_{n\to\infty}\sqrt{2\sigma_i^2}\sum_{k=1}^{n} a_k\omega_k\cos\theta_k\sin(\omega_k t - \omega_{c,i}T_k),$$
$$\dot{Y}_i(t) = \lim_{n\to\infty}\sqrt{2\sigma_i^2}\sum_{i=1}^{n} a_k\omega_k\cos\theta_k\cos(\omega_k t - \omega_{c,i}T_k). \quad (7)$$

Let $\langle\cdot\rangle$ denote the average operator. With (7), we can directly obtain that the second-order autocorrelation functions are

$$\langle X_i(t)\dot{X}_i(t+\tau)\rangle = \langle Y_i(t)\dot{Y}_i(t+\tau)\rangle = 0, \quad i = 1,2. \quad (8)$$

Similarly, the second-order intercross-correlation functions of the in-phase and quadrature processes can be obtained as

$$\begin{aligned}\langle X_i(t)\dot{Y}_i(t+\tau)\rangle &= \langle \dot{X}_i(t)Y_i(t+\tau)\rangle \\ &= \langle \dot{X}_i(t)\dot{Y}_i(t+\tau)\rangle = 0, \quad i = 1,2.\end{aligned} \quad (9)$$

Let $\Delta\omega = \omega_{c,2} - \omega_{c,1}$ and $\omega_m = 2\pi f_m$. By using (7) and some manipulations, the second-order cross-correlation function $\langle \dot{X}_1(t)\dot{X}_2(t+\tau)\rangle$ and $\langle \dot{Y}_1(t)\dot{Y}_2(t+\tau)\rangle$ can be derived as

$$\begin{aligned}\langle \dot{X}_1(t)\dot{X}_2(t+\tau)\rangle &= \langle \dot{Y}_1(t)\dot{Y}_2(t+\tau)\rangle \\ &= \sigma_1\sigma_2\int_0^{2\pi}\int_0^{\infty}\frac{e^{-T_k/\overline{T}}}{2\pi\overline{T}} \\ &\quad\times\cos(\omega_m\tau\cos\theta_k - \Delta\omega T_k)d\theta_k dT_k \\ &\quad+\sigma_1\sigma_2\omega_m^2\int_0^{2\pi}\int_0^{\infty}\frac{e^{-T_k/\overline{T}}}{4\pi\overline{T}} \\ &\quad\times\cos 2\theta_k\cos(\Delta\omega T_k) \\ &\quad\times\cos(\omega_m\tau\cos\theta_k)d\theta_k dT_k \\ &\quad+\sigma_1\sigma_2\omega_m^2\int_0^{2\pi}\int_0^{\infty}\frac{e^{-T_k/\overline{T}}}{4\pi\overline{T}} \\ &\quad\times\cos 2\theta_k\sin(\Delta\omega T_k) \\ &\quad\times\sin(\omega_m\tau\cos\theta_k)d\theta_k dT_k \\ &= \frac{\sigma_1\sigma_2\omega_m^2}{2\left[1+\left(\Delta\omega\overline{T}\right)^2\right]} \\ &\quad\times[J_0(\omega_m\tau) - J_2(\omega_m\tau)],\end{aligned} \quad (10)$$

where $J_i(\cdot)$ $(i = 0,1,2,\ldots)$ denotes the ith-order Bessel function of the first kind, and [22, (3.715), (8.339), (3.895), and (8.473.1)] are applied to evaluate the integral.

When the two cascaded channels are using the same carrier frequency, we have $\Delta\omega = 0$. In Figure 3, this second-order cross-correlation of in-phase and quadrature processes with $\Delta\omega = 0$ is plotted for $\sigma_1^2 = \sigma_2^2 = 1$. From Figure 2, the vibration range of the cross-correlation may increase when f_m becomes higher.

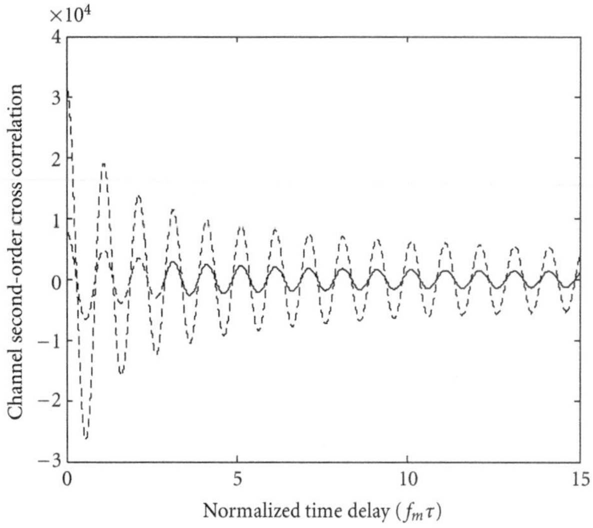

Figure 3: Cross-correlation function of in-phase and quadrature processes of the cascaded fading channels.

Other non-zero second-order correlation functions can be derived in a similar way. Following manipulations similar to those used for deriving (10), we can obtain the related second-order correlation functions as follows:

$$\begin{aligned}\langle \dot{X}_1(t)\dot{Y}_2(t+\tau)\rangle &= -\langle \dot{Y}_1(t)\dot{X}_2(t+\tau)\rangle \\ &= -\frac{\sigma_1\sigma_2\omega_m^2\Delta\omega\overline{T}}{2\left[1+\left(\Delta\omega\overline{T}\right)^2\right]} \\ &\quad\times[J_0(\omega_m\tau) - J_2(\omega_m\tau)], \\ \langle \dot{X}_1(t)X_2(t+\tau)\rangle &= -\langle X_1(t)\dot{X}_2(t+\tau)\rangle \\ &= -\langle \dot{Y}_1(t)Y_2(t+\tau)\rangle \\ &= -\langle Y_1(t)\dot{Y}_2(t+\tau)\rangle \\ &= -\frac{\sigma_1\sigma_2\omega_m J_1(\omega_m\tau)}{1+\left(\Delta\omega\overline{T}\right)^2}, \\ \langle X_1(t)\dot{Y}_2(t+\tau)\rangle &= \langle \dot{Y}_1(t)X_2(t+\tau)\rangle \\ &= -\langle \dot{X}_1(t)Y_2(t+\tau)\rangle \\ &= -\langle Y_1(t)\dot{X}_2(t+\tau)\rangle \\ &= \frac{\sigma_1\sigma_2\omega_m\Delta\omega\overline{T}J_1(\omega_m\tau)}{1+\left(\Delta\omega\overline{T}\right)^2}.\end{aligned} \quad (11)$$

The above second-order correlation functions can be applied to the derivation of the cross-correlation between $\dot{\alpha}_1(t)$ and $\dot{\alpha}_2(t+\tau)$ that will be useful for the LCR and AFD derivation in Section 3.

2.3. Cross-Correlation Function of $\dot{\alpha}_1(t)$ and $\dot{\alpha}_2(t)$. For $i = 1, 2$ the time derivative of $\alpha_i(t)$ has the form

$$\dot{\alpha}_i(t) = \frac{X_i(t)\dot{X}_i(t) + Y_i(t)\dot{Y}_i(t)}{\sqrt{X_i^2(t) + Y_i^2(t)}} = \frac{X_i(t)\dot{X}_i(t) + Y_i(t)\dot{Y}_i(t)}{\alpha_i(t)}. \quad (12)$$

The correlation function between $\dot{\alpha}_1(t)$ and $\dot{\alpha}_2(t+\tau)$ under fixed $\alpha_1(t)$ and $\alpha_2(t+\tau)$ can be written as follows:

$$\langle \dot{\alpha}_1(t)\dot{\alpha}_2(t+\tau)\rangle = \frac{\langle X_1\dot{X}_1X_2\dot{X}_2\rangle + \langle Y_1\dot{Y}_1Y_2\dot{Y}_2\rangle + \langle X_1\dot{X}_1Y_2\dot{Y}_2\rangle + \langle X_2\dot{X}_2Y_1\dot{Y}_1\rangle}{\alpha_1(t)\alpha_2(t+\tau)}, \quad (13)$$

where the variables t in (X_1, Y_1) and $(t+\tau)$ in (X_2, Y_2) are omitted for simplification. To derive $\langle \dot{\alpha}_1(t)\dot{\alpha}_2(t+\tau)\rangle$, we need to obtain the four expectations in the numerator of (13). Based on the model of central limit theorem considered in [14, 15], [24, (8.111)], all $X_i, \dot{X}_i, Y_i$, and $\dot{Y}_i$ have a zero-mean Gaussian distribution. On the other hand, it is well known that if the random variables (Z_1, Z_2, Z_3, Z_4) have a jointly zero-mean Gaussian distribution, $\langle Z_1Z_2Z_3Z_4\rangle$ can be evaluated by [19, (8.61)] the following:

$$\langle Z_1Z_2Z_3Z_4\rangle = \langle Z_1Z_2\rangle\langle Z_3Z_4\rangle + \langle Z_1Z_3\rangle\langle Z_2Z_4\rangle + \langle Z_1Z_4\rangle\langle Z_2Z_3\rangle. \quad (14)$$

Thus, we can use the property given by (14) to analyze the numerator of (13). The four expectations in the numerator of (13) can be derived by using (14) and the expectations of the forms $\langle Z_iZ_j\rangle$ and $\langle Z_i\dot{Z}_j\rangle$ $(i, j = 1, 2)$. For $Z = X$ or Y in (14), we have

$$\langle X_i(t)Y_i(t+\tau)\rangle = 0,$$

$$\langle X_1(t)X_2(t+\tau)\rangle = \langle Y_1(t)Y_2(t+\tau)\rangle = \frac{\sigma_1\sigma_2J_0(\omega_m\tau)}{1+\left(\Delta\omega\overline{T}\right)^2},$$

$$\langle X_1(t)Y_2(t+\tau)\rangle = -\langle X_2(t)Y_1(t+\tau)\rangle = \frac{-\Delta\omega\overline{T}\sigma_1\sigma_2J_0(\omega_m\tau)}{1+\left(\Delta\omega\overline{T}\right)^2}. \quad (15)$$

Using the result obtained in Section 2.2, and (13)–(15), we can simplify (13) into the form

$$\langle \dot{\alpha}_1(t)\dot{\alpha}_2(t+\tau)\rangle = \frac{\sigma_1^2\sigma_2^2\omega_m^2}{\alpha_1(t)\alpha_2(t+\tau)\left[1+\left(\Delta\omega\overline{T}\right)^2\right]} \times \left[J_0^2(\omega_m\tau) - 2J_1^2(\omega_m\tau) - J_0(\omega_m\tau)J_2(\omega_m\tau)\right]. \quad (16)$$

To evaluate (16), we employ the following relation derived in [15, (1.5.20)] and [23, (4.31b)] for ρ and $\Delta\omega\overline{T}$

$$\rho^2(\omega_m\tau) = \frac{J_0^2(\omega_m\tau)}{1+\left(\Delta\omega\overline{T}\right)^2}. \quad (17)$$

From (17), for $\Delta\omega = 0$, the correlation coefficient is a function of the maximum Doppler frequency and the phase offset between the cascaded processes. Substituting (17) with $\Delta\omega = 0$, we can simplify (16) into the form

$$\begin{aligned}\langle \dot{\alpha}_1(t)\dot{\alpha}_2(t+\tau)\rangle &= \frac{-\sigma_1^2\sigma_2^2}{\alpha_1(t)\alpha_2(t+\tau)}\frac{d^2}{d\tau^2}J_0^2(\omega_m\tau) \\ &= \frac{-\sigma_1^2\sigma_2^2\omega_m^2(\rho^2)''}{\alpha_1(t)\alpha_2(t+\tau)} \\ &= \frac{-\sigma_1^2\sigma_2^2\omega_m^2(\rho^2)''}{R(t,\tau)},\end{aligned} \quad (18)$$

where we use the recursive relations of Bessel functions given by [22, (8.473.1), (8.473.4)], and $(\rho^2)''$ is defined by

$$(\rho^2)'' = \left[\frac{d^2\rho^2(x)}{dx^2}\right]_{x=\omega_m\tau}. \quad (19)$$

For the derivation of the LCR and AFD of the correlated cascaded Rayleigh channels, in the next section, without loss of generality, $\Delta\omega = 0$ is considered.

3. Derivation of LCR and AFD

3.1. LCR. The average LCR of a random process R at $R = r$ can be evaluated by [14, 15]

$$N_R(r) = \int_0^\infty \dot{r}f_{R,\dot{R}}(r,\dot{r})d\dot{r} = f_R(r)\int_0^\infty \dot{r}f_{\dot{R}|R}(\dot{r} \mid r)d\dot{r}, \quad (20)$$

where $\dot{r}$ denotes the time derivative of r, and $f_{\dot{R}|R}(\dot{r} \mid r)$ is the pdf of the time derivative $\dot{R}$ conditioned on $R = r$.

To evaluate the LCR given by (20), we first characterize the probability distribution of $\dot{R}$ below, where

$$\dot{R}(t,\tau) = \alpha_2(t+\tau)\dot{\alpha}_1(t) + \alpha_1(t)\dot{\alpha}_2(t+\tau). \quad (21)$$

Based on [15], $\dot{\alpha}_i$ is Gaussian distributed with zero mean and variance $\sigma_{\dot{\alpha}_i}^2 = 2\pi^2f_m^2\sigma_i^2$. Consequently, conditioned on fixed (α_1, α_2), $\dot{R}$ is Gaussian distributed with zero mean and the variance

$$\dot{\sigma}_R^2 = \alpha_2^2\sigma_{\dot{\alpha}_1}^2 + \alpha_1^2\sigma_{\dot{\alpha}_2}^2 + 2\alpha_1\alpha_2\operatorname{cov}(\dot{\alpha}_1, \dot{\alpha}_2), \quad (22)$$

where $\operatorname{cov}(\dot{\alpha}_1, \dot{\alpha}_2)$ denotes the covariance of $\dot{\alpha}_1(t)$ and $\dot{\alpha}_2(t+\tau)$ under fixed (α_1, α_2), and is given by (18) for $\Delta\omega = 0$. Let $f_{\dot{R}|r,\alpha}(\dot{r} \mid \alpha_1, \alpha_2)$ denote the conditional pdf of $\dot{R}$ under fixed (α_1, α_2). Thus,

$$f_{\dot{R}|r,\alpha}(\dot{r} \mid \alpha_1, \alpha_2) \sim \text{Gaussian}(0, \dot{\sigma}_R^2). \quad (23)$$

Let $f_{R|\alpha}(r \mid \alpha_1, \alpha_2)$ denote the conditional pdf of R under fixed (α_1, α_2). Then, for the correlated cascaded processes, the LCR given by (20) can be rewritten in the form

$$\begin{aligned}N_R(r) &= \int\int\int_0^\infty \delta(r-\alpha_1\alpha_2)\dot{r}f_{\dot{R}|r,\alpha}(\dot{r} \mid r, \alpha_1, \alpha_2)d\dot{r} \\ &\quad \times f_{R|\alpha}(r \mid \alpha_1, \alpha_2)f_{12}(\alpha_1, \alpha_2)d\alpha_1d\alpha_2 \\ &= \int_0^\infty\left(\int_0^\infty \dot{r}f_{\dot{R}|\alpha}\left(\dot{r} \mid \frac{r}{\alpha_2}, \alpha_2\right)d\dot{r}\right)\frac{1}{\alpha_2}f_{12}\left(\frac{r}{\alpha_2}, \alpha_2\right)d\alpha_2,\end{aligned} \quad (24)$$

where $\delta(\cdot)$ is the Dirac delta function, $f_{12}(r/\alpha_2, \alpha_2)$ has the form given by (3) with α_1 replaced by r/α_2, and $f_{\dot{R}|\alpha}(\dot{r} \mid r/\alpha_2, \alpha_2)$ is given by (23) also with α_1 replaced by r/α_2. Substituting (18) into (22) and (3) into (24), we can express the normalized LCR in the single-integral form

$$\begin{aligned}\frac{N_R(r)}{f_m} &= \frac{\sqrt{\pi} r}{\sigma_1^2\sigma_2^2(1-\rho^2)} I_0\left(\frac{r|\rho|}{\sigma_1\sigma_2(1-\rho^2)}\right)\\ &\times \int_0^\infty \sqrt{\sigma_1^2 - \frac{4\sigma_1^2\sigma_2^2(\rho^2)''}{\alpha_2^2} + \frac{r^2\sigma_2^2}{\alpha_2^4}}\\ &\times e^{-((\sigma_1^2\alpha_2^2+\sigma_2^2 r^2/\alpha_2^2)/2\sigma_1^2\sigma_2^2(1-\rho^2))}\, d\alpha_2\end{aligned} \tag{25}$$

that can be easily evaluated with a computing software, such as MATLAB. If $\rho = 0$, it can be easily check that (25) will reduce to the form given in [8, 17, 18] for independent cascaded Rayleigh channels.

3.2. AFD. Let $P(R \le r)$ be the cumulative distribution function (cdf) of R. With the normalized LCR given by (25), the normalized AFD can be evaluated by [14]

$$T_R(r) = f_m \cdot \frac{P(R \le r)}{N_R(r)}. \tag{26}$$

By using the pdf of R given by (4), the cdf in the above AFD can be evaluated as

$$\begin{aligned}P(R \le r) &= \int_0^r f_R(y)\,dy\\ &= \frac{1}{\sigma_1^2\sigma_2^2(1-\rho^2)} \int_0^r y I_0\left(\frac{y|\rho|}{\sigma_1\sigma_2(1-\rho^2)}\right)\\ &\quad\times K_0\left(\frac{y}{\sigma_1\sigma_2(1-\rho^2)}\right) dy\\ &= 1 - \frac{r|\rho|}{\sigma_1\sigma_2(1-\rho^2)} I_1\left(\frac{r|\rho|}{\sigma_1\sigma_2(1-\rho^2)}\right)\\ &\quad\times K_0\left(\frac{r}{\sigma_1\sigma_2(1-\rho^2)}\right)\\ &\quad- \frac{r}{\sigma_1\sigma_2(1-\rho^2)} I_0\left(\frac{r|\rho|}{\sigma_1\sigma_2(1-\rho^2)}\right)\\ &\quad\times K_1\left(\frac{r}{\sigma_1\sigma_2(1-\rho^2)}\right),\end{aligned} \tag{27}$$

where $I_1(\cdot)$ is the first-order modified Bessel function of the first kind, and a formula modified from [22, (6.521.4)] with $I_i(z) = j^{-i}J_i(jz)$ $(j = \sqrt{-1})$ is used to write the integral into a closed-form.

4. Numerical Results of LCR and AFD

From (17), due to the property of Bessel functions, ρ^2 is not a monotone function of τ. However, as τ increases, the vibration range of $J_0^2(\omega_m\tau)$ tends to be smaller, and ρ^2 becomes smaller.

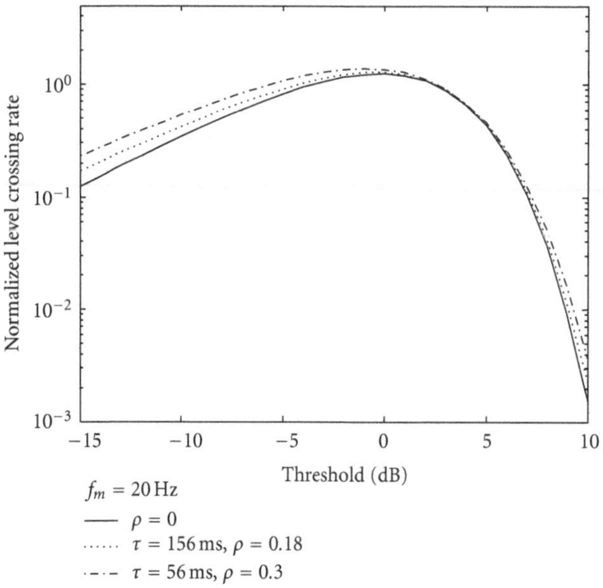

FIGURE 4: Normalized LCR for $\sigma_1^2 = \sigma_2^2 = 1$.

To demonstrate the impact of the correlation on the LCR and AFD, the normalized LCR of the correlated cascaded Rayleigh channels for different values of τ is plotted in Figure 4. The result in Figure 4 implies that when correlation increases, the cascaded envelope R has more chances to enter a deep fading area. For example, as shown in Figure 4, the correlated cascaded channels may have a 5% ~ 10% more crossing rate than independent cascaded channels to go into deep fade below −10 dB.

In Figure 5, the numerical AFD is plotted. According to Figure 5, as the correlation increases, the average duration when R temporarily stays in the deep fading area can be shorter although R enters the area more frequently.

Thus, the correlation between cascaded channels not only affects the BER as shown in [10–13] but also changes the second-order statistics which are more serious in a deep fading area.

5. Conclusions

Different second-order correlation functions of completely correlated cascaded fading channels have been derived and the corresponding LCR and AFD are analyzed, which can facilitate the design of wireless communication or RFID systems on keyhole or pinhole channels. Under the consideration of the channel correlation, the LCR and AFD will depart from those obtained for independent cascaded channels. Numerical results show that the correlation has a higher impact on the second-order statistics in the deep fading region. The evaluation of LCR and AFD will also be helpful for the design of interleaving transmission and encoding schemes over correlated cascaded fading channels.

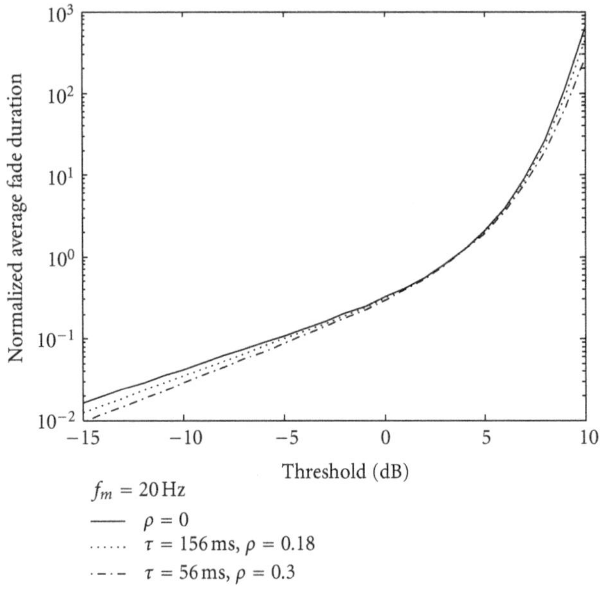

FIGURE 5: Normalized AFD for $\sigma_1^2 = \sigma_2^2 = 1$.

Further research on related topics may consider the second-order statistics of diversity schemes over correlated cascaded channels.

Acknowledgment

This paper was supported by the National Science Council, Taiwan, under Grant nos. NSC100-2220-E-155-001 and NSC100-2220-E-155-005.

References

[1] D. Gesbert, H. Bölcskei, D. A. Gore, and A. J. Paulraj, "Outdoor MIMO wireless channels: models and performance prediction," *IEEE Transactions on Communications*, vol. 50, no. 12, pp. 1926–1934, 2002.

[2] M. Uysal, "Diversity analysis of space-time coding in cascaded Rayleigh fading channels," *IEEE Communications Letters*, vol. 10, no. 3, pp. 165–167, 2006.

[3] G. K. Karagiannidis, N. C. Sagias, and P. T. Mathiopoulos, "N∗Nakagami: a novel stochastic model for cascaded fading channels," *IEEE Transactions on Communications*, vol. 55, no. 8, pp. 1453–1458, 2007.

[4] D. Chizhik, G. J. Foschini, and R. A. Valenzuela, "Capacities of multi-element transmit and receive antennas: correlations and keyholes," *Electronics Letters*, vol. 36, no. 13, pp. 1099–1100, 2000.

[5] V. Erceg, S. J. Fortune, J. Ling, A. J. Rustako, and R. A. Valenzuela, "Comparisons of a computer-based propagation prediction tool with experimental data collected in urban microcellular environments," *IEEE Journal on Selected Areas in Communications*, vol. 15, no. 4, pp. 677–684, 1997.

[6] J. Salo, H. M. El-Sallabi, and P. Vainikainen, "The distribution of the product of independent Rayleigh random variables," *IEEE Transactions on Antennas and Propagation*, vol. 54, pp. 639–643, 2006.

[7] C. Zhu, J. Mietzner, and R. Schober, "On the performance of non-coherent transmission schemes with equal-gain combining in generalized K-fading," *IEEE Transaction on Wireless Communications*, vol. 9, no. 4, pp. 1337–1349, 2010.

[8] C. S. Patel, G. L. Stüber, and T. G. Pratt, "Statistical properties of amplify and forward relay fading channels," *IEEE Transactions on Vehicular Technology*, vol. 55, no. 1, pp. 1–9, 2006.

[9] C. He and Z. J. Wang, "Closed-form BER analysis of non-coherent FSK in MISO double Rayleigh fading/RFID channel," *IEEE Communications Letters*, vol. 15, no. 8, pp. 848–850, 2011.

[10] J. Griffin and G. Durgin, "Link envelope correlation in the backscatter channel," *IEEE Communications Letters*, vol. 11, no. 9, pp. 735–737, 2007.

[11] J. D. Griffin and G. D. Durgin, "Gains for RF tags using multiple antennas," *IEEE Transactions on Antennas and Propagation*, vol. 56, no. 2, pp. 563–570, 2008.

[12] C. He and Z. J. Wang, "Impact of the correlation between forward and backscatter channels on RFID system performance," in *Proceedings of the IEEE International Conference on Acoustics, Speech and Signal Processing (ICASSP '11)*, pp. 3540–3543, May 2011.

[13] H. D. Goldman and R. C. Sommer, "An analysis of cascaded binary communication links," *IRE Transactions on Communications Systems*, vol. 10, no. 3, pp. 291–299, 1962.

[14] S. O. Rice, "Mathematical analysis of random noise," *Bell System Technical Journal*, vol. 23, pp. 282–332, 1944, pp. 46–156, 1945.

[15] W. C. Jakes, *Microwave Mobile Communications*, IEEE Press, Piscataway, NJ, USA, 1994.

[16] Y. A. Chau and K. Y. Huang, "Burst-error analysis of dual-hop fading channels based on the second-order channel statistics," *IEEE Transactions on Vehicular Technology*, vol. 59, no. 6, pp. 3108–3115, 2010.

[17] Y. A. Chau and K. Y. Huang, "Channel statistics and performance of cooperative selection diversity with dual-hop amplify-and-forward relay over Rayleigh fading channels," *IEEE Transactions on Wireless Communications*, vol. 7, no. 5, pp. 1779–1785, 2008.

[18] N. Slatanov, Z. H. Velkov, and G. K. Karagiannidis, "Level crossing rate and average fade duration of the double Nakagami-*m* random process and application in MIMO keyhole fading channels," *IEEE Communications Letters*, vol. 12, no. 11, pp. 822–824, 2008.

[19] Y. A. Chau and K. Y.-T. Huang, "Evaluation of level crossing rates and fade durations for multihop Nakagami fading channels with an amplify-and-forward relay," in *Proceedings of the International Symposium on Intelligent Signal Processing and Communications Systems (ISPACS '07)*, pp. 732–735, Xiamen, China, November 2007.

[20] C. Floerkemeier and S. Sarma, "RFIDSim—a physical and logical layer simulation engine for passive RFID," *IEEE Transactions on Automation Science and Engineering*, vol. 6, no. 1, pp. 33–43, 2009.

[21] M. K. Simon, *Probability Distributions Involving Gaussian Random Variables: A Handbook For Engineers and Scientists*, Kluwer Academic, Norwell, Mass, USA, 2002.

[22] I. S. Gradshteyn and I. M. Ryzhik, *Table of Integrals, Series, and Products*, Academic Press, New York, NY, USA, 6th edition, 2000.

[23] M. Yacoub, *Foundations of Mobile Radio Engineering*, CRC Press, Boca Raton, Fla, USA, 1993.

[24] A. Papoulis, *Probability, Random Variables, and Stochastic Processes*, McGraw-Hill, New York, NY, USA, 3rd edition, 1991.

8

Characterizing the Statistical Properties of SAR Clutter by Using an Empirical Distribution

Gui Gao, Gongtao Shi, Huanxin Zou, and Shilin Zhou

School of Electronics Science and Engineering, National University of Defense Technology, Changsha, Hunan 410073, China

Correspondence should be addressed to Gui Gao; dellar@126.com

Academic Editor: Deren Li

The performances on the applications of synthetic aperture radar (SAR) data strongly depend on the statistical characteristics of the pixel amplitudes or intensities. In this paper, a new empirical model, called simply $\mathscr{H}\mathscr{G}^o$, has been proposed to characterize the statistical properties of SAR clutter data over the wide range of homogeneous, heterogeneous, and extremely heterogeneous returns of terrain classes. A particular case of the $\mathscr{H}\mathscr{G}^o$ distribution is the well-known $\mathscr{G}^o$ distributions. We also derived analytically the estimators of the presented $\mathscr{H}\mathscr{G}^o$ model by applying the "method of log cumulants" (MoLCs). The performance of the proposed model is verified by using some measured SAR images.

1. Introduction

Nowadays, synthetic aperture radar (SAR) has become an advanced tool as compared to optical sensors for monitoring land or sea surfaces because of its advantages regardless of weather conditions [1, 2]. The interpretation of statistical properties of SAR clutter with various terrain classes is a crucial task for designing filtering [3], detection [4], segmentation [5], and classification [6–8] of algorithms to optimally exploit the contents of the processed SAR images. These analyses result in a growing interest in developing precise models for the statistics of the pixel amplitudes or intensities.

It is known that SAR images are strongly affected by speckle noise due to its coherent imaging mechanism [1, 2, 9, 10]. This noise effect degrades the information of SAR images and limits the surveillance performances. To overcome this drawback, much work has been done to match amplitude or intensity statistics [3, 5, 6, 8, 11]. Among the existing statistical models, the parametric distributions have been intensively investigated because of their high accuracy and flexibility, which are usually obtained by two approaches.

The first is to consider the physical mechanism of the backscattering from the land surfaces. The multiplicative model [1, 2] by combining the speckle noise and the terrain backscatter is commonly used in this class. Generally, multi-look SAR speckle noise intensity is assumed to obey a gamma distribution. Therefore, a central task for establishing this type of statistical models is to aim at the backscatter modeling of different terrain classes [1, 2, 11]. Two important distributions, consisting of $\mathscr{K}$ as well as $\mathscr{G}^0$, which, in turn, in relation to the Gamma and inverse Gamma distribution for the intensity backscatter, have received a great deal of attention [2, 5–7, 9, 11, 12]. As compared to $\mathscr{K}$, $\mathscr{G}^0$ agrees reasonably better with the heavy tail behavior coming from the extremely heterogeneous clutter like the cases of urban areas or other man-made structures [11].

The second is to conduct the distributions from a purely mathematics view irrespective with physical property of radar clutter backscatter. Some known examples are the lognormal, weibull, and more recently the Fisher [6] (completely identical to the $\mathscr{G}^0$) distributions. The lognormal and weibull fit sometimes well with the SAR histogram of some heterogeneous and ocean regions [9]. However, they tend to occur a large deviation estimating the histograms with the heavy tail behavior.

This paper is devoted to report an empirical model (denoted simply as $\mathscr{H}\mathscr{G}^0$) for characterizing the statistical properties of SAR clutter data to obtain the modeling ability of more heterogeneous clutter. The proposed model has the $\mathscr{G}^0$ distribution as a special case. Furthermore, using the second-kind statistics theory developed by Nicolas [13], which relies on the Mellin transform [14], that is, "method-of-log-cumulants" (MoLC), we derive the parameter estimators of the new distribution model.

In the rest of this paper, the proposed distribution is first given in Section 2. Section 3 derives the corresponding parameter estimators based on the MoLC. We provide the experimental results of the $\mathscr{HG}^0$ model using measured SAR data in Section 4, the comparisons with that $\mathscr{G}^0$ fits are also discussed in this section. The last section concludes this paper and give a perspective in the future work.

2. The Proposed Distribution

The proposed amplitude distribution is defined as

$$p_{Z_A}(z) = -\frac{2\sigma\lambda^{\upsilon\alpha}n^n\Gamma((\alpha-n)/\upsilon)}{\upsilon\Gamma(\upsilon\alpha)\Gamma(n)} \times z^{2n-1}\left(\lambda+nz^{-2\upsilon}\right)^{(n-\alpha)/\upsilon}, \quad -\alpha,-\upsilon,\lambda,\sigma,n,z>0, \tag{1}$$

where σ, α, and λ are the power, shape, scale parameters, respectively. υ indicates the stretching parameter. $\Gamma(\cdot)$ represents the Gamma function. n is the number of looks. The corresponding intensity expression of (1) is further given by

$$p_{Z_I}(z) = -\frac{\sigma\lambda^{\upsilon\alpha}n^n\Gamma((\alpha-n)/\upsilon)}{\upsilon\Gamma(\upsilon\alpha)\Gamma(n)} \times z^{n-1}\left(\lambda+nz^{-\upsilon}\right)^{(n-\alpha)/\upsilon}, \quad -\alpha,-\upsilon,\lambda,\sigma,n,z>0. \tag{2}$$

We refer to this distribution characterized by (1) or (2) as the $\mathscr{HG}^0$ distribution. Specifically, we call the $\mathscr{HG}_I^0$ distribution and the $\mathscr{HG}_A^0$ distribution, correspond to (2) and (1), respectively, to distinct the intensity statistic as well as the amplitude statistic. Figure 1 gives some examples of the $\mathscr{HG}_A^0$ distribution with respect to the various parameters. From this figure, it can be seen clearly that the parameter α reflects the degree of homogeneity for the tested returns, which implies that the smaller value of $|\alpha|$ obtains, the more in-homogeneous (i.e., larger tails) they are. Meanwhile, υ corresponds to a stretching of the SAR image amplitude or intensity, showing a strong effect for low values of the return. Moreover, as an independent parameter, σ indicates the whole fluctuation (contains magnifying or shrinking) of power of densities along the vertical axis. The parameter λ controls the peak value of the density.

Although the probability density functions (PDFs) characterized by (1) or (2) are shown as the empirical models, the $\mathscr{HG}^0$ model also can be derived within the structure of the multiplicative model [1, 2, 11]. Herein, we account for the intensity distribution as an example, the following theorem is established.

Theorem 1. *Letting X_I and Y_I indicates the backscattering RCS component and speckle noise one, respectively, Z_I denotes the observed intensity of SAR data. Therefore, the relationship of this three variables is expressed by the multiplicative model as*

$$Z_I = X_I \cdot Y_I. \tag{3}$$

If X_I obeys

$$p_{X_I}(x) = \frac{\sigma\lambda^{\upsilon\alpha}}{\Gamma(\upsilon\alpha)}x^{\alpha-1}\exp\left(-\lambda x^{\upsilon}\right), \quad -\alpha,-\upsilon,\lambda,\sigma,x>0 \tag{4}$$

and the PDF of Y_I is

$$p_{Y_I}(y) = \frac{n^n}{\Gamma(n)}y^{n-1}\exp\left(-ny^{-\upsilon}\right), \quad y,n,-\upsilon>0. \tag{5}$$

Then the distribution of the intensity return is characterized by the density shown in (2), that is, $Z_I \sim \mathscr{HG}_I^0(\alpha,\upsilon,\lambda,\sigma,n)$.

Proof. Combining (4) and (5) via (3), the PDF of Z_I can be easily derived as

$$\begin{aligned} p_{Z_I}(z) &= \int_0^{\infty}\frac{1}{x}\cdot p_{X_I}(x)\cdot p_{Y_I}\left(\frac{z}{x}\right)dx \\ &= \frac{\sigma\lambda^{\upsilon\alpha}n^n}{\Gamma(\upsilon\alpha)\Gamma(n)}z^{n-1}\int_0^{\infty}x^{\alpha-n-1}\exp\left\{\left(-\lambda-nz^{-\upsilon}\right)x^{\upsilon}\right\}dx. \end{aligned} \tag{6}$$

A variable change of $t = 1/x$ leads to $x = 1/t$ and $dx = -t^{-2}dt$, (6) turns out be

$$p_{Z_I}(z) = \frac{\sigma\lambda^{\upsilon\alpha}n^n}{\Gamma(\upsilon\alpha)\Gamma(n)}z^{n-1}\int_0^{\infty}t^{-\alpha+n-1}\exp\left\{\left(-\lambda-nz^{-\upsilon}\right)t^{-\upsilon}\right\}dt. \tag{7}$$

Applying the integral formula $\int_0^{\infty}x^{a-1}\exp(-\mu x^p)dx = (1/p)\mu^{-a/p}\Gamma(a/p)$, $\operatorname{Re}\mu>0$, $\operatorname{Re}a>0$, $p>0$ [15, 3.478, Equation 1], one can easily obtain the result that (7) is equal to (2).

Furthermore, $\mathscr{HG}^0$ has the charming property that the well-known $\mathscr{G}^0$, presented by Frery et al. [11] to model homogeneous, heterogeneous, and extremely heterogeneous terrains, is a special case of this proposed model when $\sigma = 1$ and $\upsilon = -1$. Thus, the proposed model exhibits higher fitting ability as compared to $\mathscr{G}^0$.

As derived in Appendix A, the mth order moments of the $\mathscr{HG}_I^0$ are given by

$$E\left(z_I^m\right) = \frac{\sigma\lambda^{\upsilon\alpha-(m+\alpha)/\upsilon}n^{n+(m+n)/\upsilon}\Gamma(-(m+n)/\upsilon)\Gamma((\alpha+m)/\upsilon)}{\upsilon^2\Gamma(\upsilon\alpha)\Gamma(n)}. \tag{8}$$

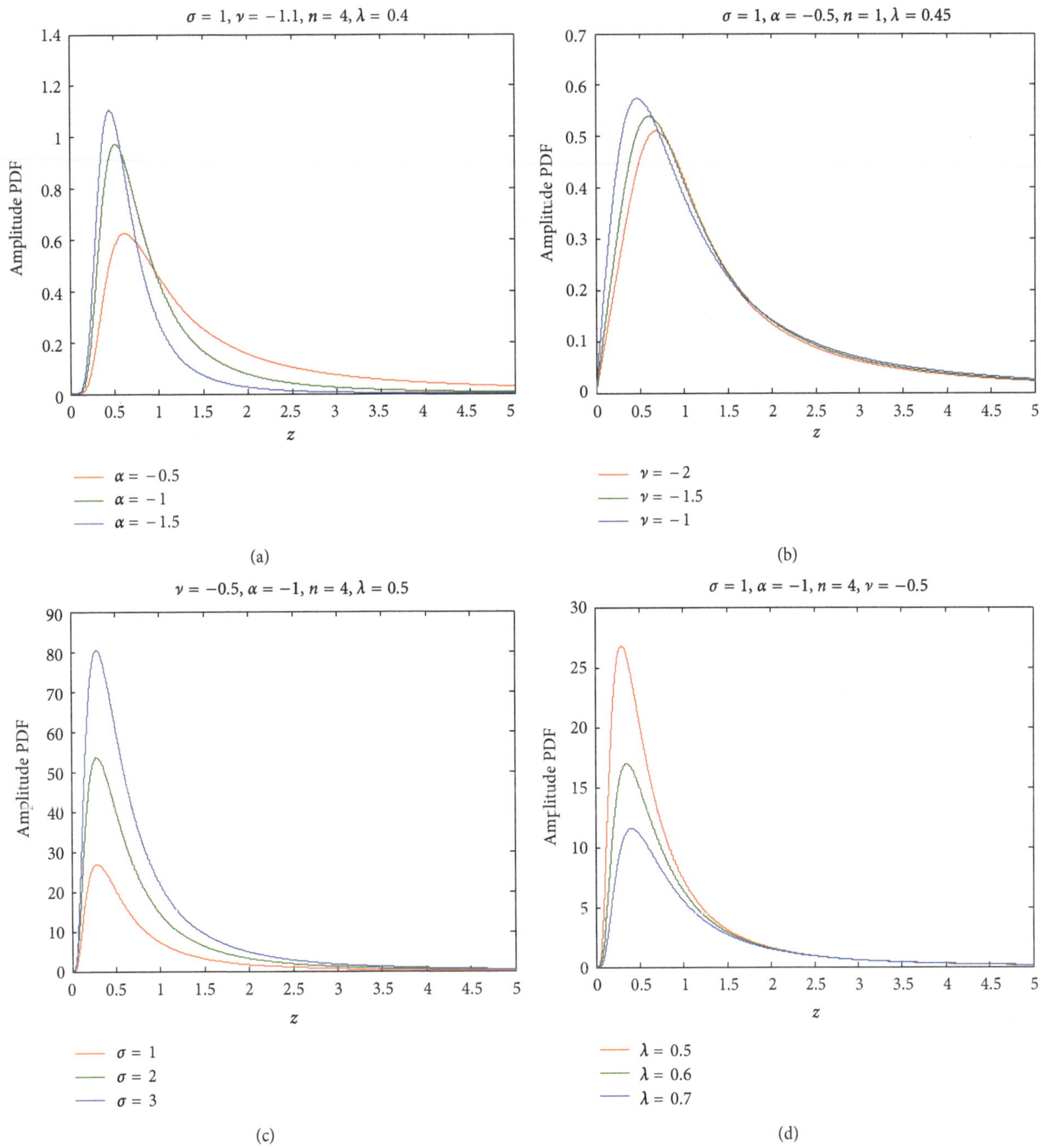

FIGURE 1: Plots of $\mathscr{HG}_A^0$ versus different parameters.

The mth order moments of the corresponding amplitude random variable can easily be given by $E(z_A^m) = E(z_I^{m/2})$. ⊠

3. The Parameter Estimators of the Proposed Distribution

3.1. The Log-Cumulants of $\mathscr{HG}^0$. Nicolas [13] has proposed a parametric PDF estimation technique based on the MoLC for a function defined over $\mathbb{R}^+$. Given a positive-valued random variable X with the PDF $p_X(x)$, the second-kind first and second characteristic functions are, respectively, defined as [13]

$$\phi_X(s) = \mathscr{M}[p_X(x)](s) = \int_0^{\infty} x^{s-1} p_X(x)\,dx,$$
$$\varphi_X(s) = \log(\phi_X(s)), \tag{9}$$

where $\mathscr{M}$ is the Mellin transform operator. The rth order derivative of $\varphi_X(s)$ at $s = 1$ is the log-cumulants of order r, that is,

$$k_r = \left. \frac{d^r \varphi_X(s)}{ds^r} \right|_{s=1}. \quad (10)$$

Hereafter, we take the intensity distribution $\mathscr{H}\mathscr{G}_I^0$ as an example to estimate its parameters υ, α, and λ. The processing of the amplitude one is similar. Owing to $\phi_X(s) = E(X^{s-1})$, consequently, via (8) and (9), the second-kind second characteristic functions of the $\mathscr{H}\mathscr{G}_I^0$ distribution yields

$$\begin{aligned}\varphi_{\mathscr{H}\mathscr{G}_I^0}(s) &= \log(\sigma) + \log\left(\lambda^{\upsilon\alpha-(\alpha/\upsilon)}\right) \\ &+ \frac{1-s}{\upsilon}\log(\lambda) + \log\left(n^{n+(n/\upsilon)}\right) + \frac{s-1}{\upsilon}\log(n) \\ &+ \log\left(\Gamma\left(-\frac{s-1+n}{\upsilon}\right)\right) + \log\left(\Gamma\left(\frac{\alpha+s-1}{\upsilon}\right)\right) \\ &- \log\left(\upsilon^2\Gamma(\upsilon\alpha)\Gamma(n)\right)\end{aligned} \quad (11)$$

which leads to that the log-cumulants of the $\mathscr{H}\mathscr{G}_I^0$ are expressed as

$$\begin{aligned} k_1 &= \frac{\log(n/\lambda) - \Psi(-n/\upsilon) + \Psi(\alpha/\upsilon)}{\upsilon}, \\ k_r &= \frac{\Psi(r-1,-n/\upsilon)}{(-\upsilon)^r} + \frac{\Psi(r-1,\alpha/\upsilon)}{\upsilon^r}, \quad r \geq 2,\end{aligned} \quad (12)$$

where $\Psi(\cdot)$ represents the digamma function, and $\Psi(r,\cdot)$ is the rth order polygamma function.

Given a sample set $\{z_i\}$, $i \in [1,N]$, the log-cumulants can be estimated directly by

$$\begin{aligned}\widehat{k}_1 &= \frac{1}{N}\sum_{i=1}^{N}[\ln(z_i)], \\ \widehat{k}_1 &= \frac{1}{N}\sum_{i=1}^{N}\left[\left(\ln(z_i) - \widehat{k}_1\right)^r\right], \quad r \geq 2.\end{aligned} \quad (13)$$

3.2. The Parameter Estimators of $\mathscr{H}\mathscr{G}^0$. We notice that (12) is independent of the parameter σ and the rth log-cumulants shown in (12) are irrespective with the parameter λ on the condition that $r \geq 2$. In addition, we regard the parameter n as a known constant, which can be replaced by the equivalent number of looks (*ENLs*) [4, 11] or obtained from some prior knowledge about processed SAR images, hence, allowing us to divide the parameter estimates to three stages.

First, the estimates $\widehat{\upsilon}$ and $\widehat{\alpha}$ of the parameters υ as well as α are obtained by solving the following equations resorting to the numerical computation:

$$\begin{aligned}\widehat{k}_2 &= \frac{\Psi(1,-n/\widehat{\upsilon}) + \Psi(1,\widehat{\alpha}/\widehat{\upsilon})}{\widehat{\upsilon}^2}, \\ \widehat{k}_3 &= \frac{\Psi(2,\widehat{\alpha}/\widehat{\upsilon}) - \Psi(2,-n/\widehat{\upsilon})}{\widehat{\upsilon}^3}.\end{aligned} \quad (14)$$

Second, according to the first equation of (12), the estimate $\widehat{\lambda}$ of the parameter λ is

$$\widehat{\lambda} = \frac{n}{\exp\left(\widehat{\upsilon}\widehat{k}_1 + \Psi(-n/\widehat{\upsilon}) - \Psi(\widehat{\alpha}/\widehat{\upsilon})\right)}. \quad (15)$$

We assume the observed amplitude PDF is $\widehat{p}_{Z_A}(z)$ from the actual data, which corresponds to the histogram of tested data. The theoretical amplitude PDF denotes as $p_{Z_A}(z)$. As the previous analysis, since σ indicates the whole proportion fluctuation of $p_{Z_A}(z)$, $p_{Z_A}(z)/\sigma$ can be calculated by using $\widehat{\upsilon}$, $\widehat{\alpha}$, and $\widehat{\lambda}$. Given a sample amplitude set $\{z_i \mid i = 1\cdots N\}$, let the symbol $q_{Z_A}(z)$ be equal to $p_{Z_A}(z)/\sigma$, then the estimate $\widehat{\sigma}$ of the parameter σ is simply given by

$$\widehat{\sigma} = E\left(\frac{\widehat{p}_{Z_A}(z_i)}{q_{Z_A}(z_i)}\right), \quad z_i \in \mathbb{Z}, \quad (16)$$

where $\mathbb{Z}$ is a subset of $\{z_i \mid i = 1\cdots N\}$. Here, in order to facilitate the stable fitting, we choose empirically all points of histogram within 3 dB bandwidth to calculate σ, such that $\mathbb{Z} = \{z_i \mid \widehat{p}_{Z_A}(z_i) > \widehat{p}_{\max}/2,\ \widehat{p}_{\max} = \max(\widehat{p}_{Z_A}(z_i)),\ i \in [1,N]\}$.

4. Experimental Results

In this section, we aim at verifying the performance of the proposed $\mathscr{H}\mathscr{G}^0$. In order to assess how the $\mathscr{H}\mathscr{G}^0$ performs, several space-borne TerraSAR-X geocoded scenes with various land-over typologies as examples are reported. Figures 2(a)–2(d) show four selected typical patches from a large "Sanchagang" TerraSAR-X image with low resolution, which are the portions of water body, drying riverbed, mountain, and a town returns, respectively. The four types of scenes are related to the homogeneous and heterogeneous terrains. For simplicity, we denote them as "water-body", "drying riverbed", "mountain", and "town." Additionally, two urban areas shown in Figures 2(e)-2(f), extracted from a large "Beijing" TerraSAR-X scene with high resolution, are further carried out to demonstrate the effectiveness of the proposed model on the extremely heterogeneous terrains. Likewise, the two urban images are denoted by "urban1" and "urban2." The main parameters of TerraSAR-X systems for "Sanchagang" and "Beijing" data are listed in Table 1.

The estimated PDFs of the proposed $\mathscr{H}\mathscr{G}^0$ model for the histograms of six selected areas indicated in Figure 2 are shown in Figure 3. As the corresponding comparisons, the fitting results with the intensively used $\mathscr{G}^0$ distribution are also provided, where parameter estimates of this distribution are derived by Tison et al. [6] based on the MoLC. Herein, the estimate n in $\mathscr{H}\mathscr{G}^0$ is first replaced by the *ENL*, that is, [4]

$$\widehat{n} = ENL = n_a + n_r, \quad (17)$$

where n_a and n_r are the effective number of looks in the azimuth and range. For the previous "Sanchagang" and "Beijing" TerraSAR-X images used in this investigation, n_a and n_r are listed in the last two columns of Table 1 according

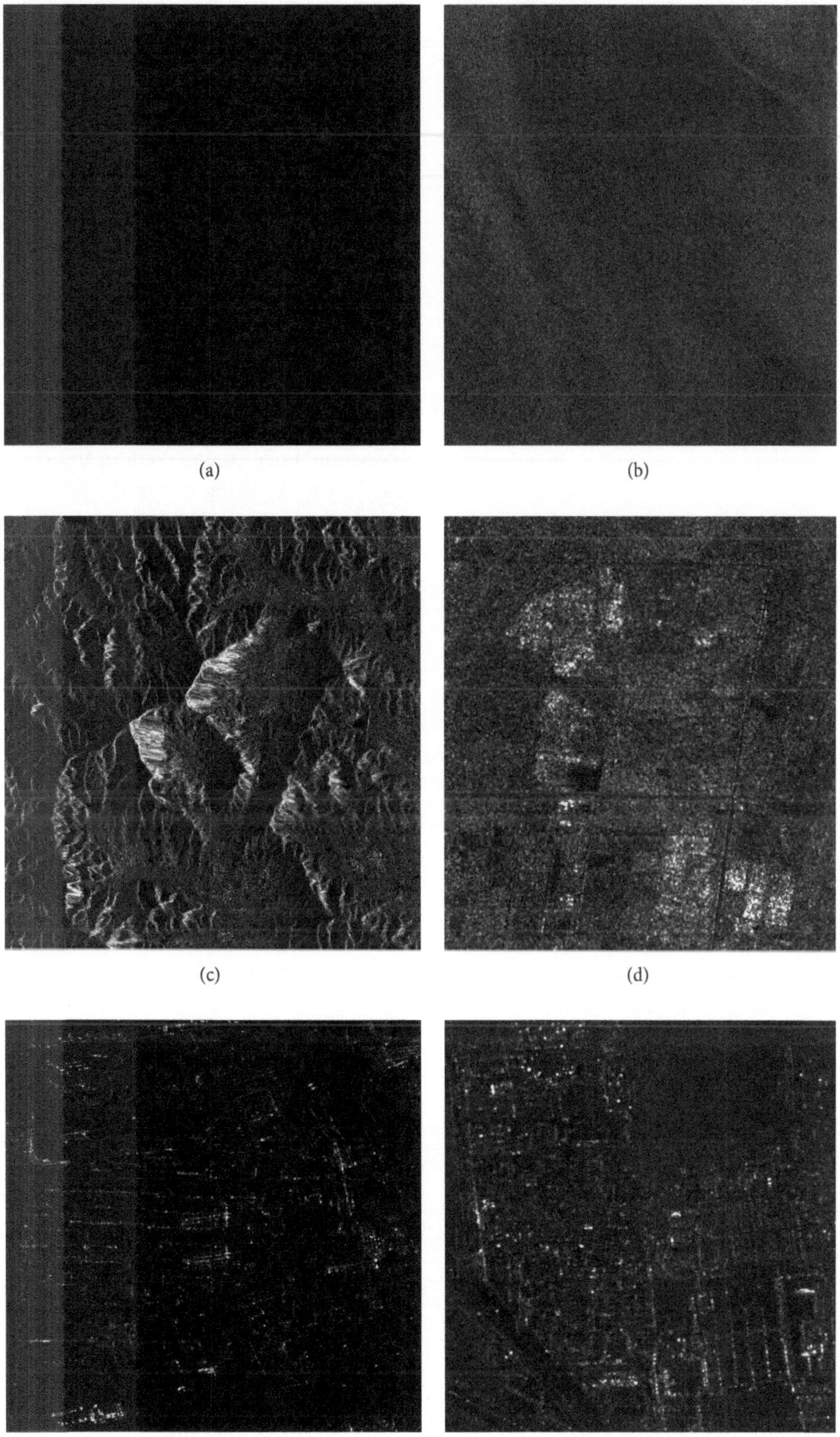

(a) (b) (c) (d) (e) (f)

Figure 2: The Tested TerraSAR-X SAR image. (a) "Water body" patch with a size of 332 px × 322 px, (b) "drying riverbed" image (size: 1597 × 1734 pixels), (c) "mountain" image (size: 1996 × 2034 pixels), (d) "town" image (size: 525 × 560 pixels), (e) "urban1" image (size: 1198 × 1342 pixels), and (e) "urban2" image (size: 705 × 728 pixels).

Table 1: The main parameters of TerraSAR-X space-borne systems for the data in this study.

Testing site	Acquisition mode	Polarization	Resolution (az × rg)	Date	Size	Azimuth looks	Range looks
Beijing, China	SpotLight	HH	1.92 m × 2.26 m	2008-01-31T10:03 UTC	21333 px × 20000 px	1.19	1.02
Sanchagang, China	StripMap	VV	6.11 m × 5.98 m	2008-01-01T22:20 UTC	11909 px × 22363 px	1.02	2.67

Table 2: Parameter estimations of noted areas in Figure 2.

Area	$\mathscr{HG}_A^0(\alpha, \upsilon, \lambda, \sigma, n)$
"Water body"	(−6.6029, −1.1001, 6.3048, 1.0709, 3.683)
"Drying riverbed"	(−6.0382, −1.0555, 5.6813, 1.1897, 3.683)
"Mountain"	(−2.9788, −0.5779, 2.4694, 0.1426, 3.683)
"Town"	(−5.062, −0.7271, 4.5872, 0.4861, 3.683)
"Urban1"	(−1.1774, −0.4895, 0.6905, 11.1945, 2.22)
"Urban2"	(−0.8641, −0.7065, 0.3227, 2.3517, 2.22)

to reading their metadata files. Next, the estimates of all other parameters in $\mathscr{HG}^0$ are accomplished by the estimators derived in Section 3, which are shown at Table 2.

Furthermore, in order to quantitatively assess the performances that different distributions fit, we define an error ratio factor (*ERF*) as

$$\delta = \frac{\sum_{i=1}^{N} \|p_c(z_i) - q(z_i)\|^2}{\sum_{i=1}^{N} \|p_b(z_i) - q(z_i)\|^2}, \quad (18)$$

where $\{z_i \mid i = 1 \cdots N\}$ is a sample set, $p_c(\cdot)$ represents the compared theoretical PDF, $p_b(\cdot)$ is the basic theoretical PDF, and $q(\cdot)$ indicates the actual PDF from the observed data. The symbol $\|\cdot\|$ denotes 2-norm. It is obvious that the numerator and denominator in (18) separately imply the total fitting errors with the p_c and p_b to approximate q. The better the performances of p_c related to p_b fit, the smaller δ is, and vice verse. Specifically, $\delta = 1$, if both p_c and p_b have the identical capability for fitting the measured data.

Let $\mathscr{HG}_A^0$ represent the basic amplitude PDF, and let $\mathscr{G}_A^0$ be the compared amplitude one. The δ values of the previous six areas in this study are given in Figure 3 (see the yellow textbox). It can be clearly seen that the proposed $\mathscr{HG}_A^0$ better agrees with the amplitude histograms of all six terrains than the $\mathscr{G}_A^0$, as expected, because the δ values are larger than 1 for all six areas. The same conclusion can be confirmed from a visual point of view and implies the higher fitting precision, using $\mathscr{HG}^0$ than using $\mathscr{G}^0$, over homogeneous, heterogeneous, and extremely heterogeneous regions.

5. Conclusion and Perspective

We have developed an empirical model, $\mathscr{HG}^0$, to exploit the knowledge of statistical characteristics of SAR amplitude or intensity images over the wide terrain classes with homogeneous, heterogeneous, and extremely heterogeneous backscattering properties. The parameter estimators of this model based on the MoLC are also provided. Consequently, we report the performances of different land-over typologies with $\mathscr{HG}^0$ distribution fits. The experimental results show that the $\mathscr{HG}^0$ distribution is a more advanced model compared with the known $\mathscr{G}^0$ distribution to characterize the multilook processed SAR data.

As we know, a preliminary statistical analysis of SAR clutter data is important for designing signal processing algorithms, such as speckle filtering, target detection, building extraction, image segmentation, and classification. In future, it is worth expecting to use the $\mathscr{HG}^0$ distribution in these fields. Herein, we firstly attempt to give an analytical derivation for constructing a constant false alarm rate (CFAR) detector to promote the upcoming studies of target detection in SAR images.

Given the propose amplitude model $\mathscr{HG}_A^0$ shown in (1), its cumulative distribution function (CDF) is written as (see Appendix B)

$$\begin{aligned} F_{\mathscr{HG}_A^0}(x) = &-\frac{\sigma\lambda^{\upsilon\alpha+(n-\alpha)/\upsilon} n^{n-1}\Gamma((\alpha-n)/\upsilon)}{\upsilon\Gamma(\upsilon\alpha)\Gamma(n)} x^{2n} \\ &\times {}_2F_1\left(\frac{\alpha-n}{\upsilon}, -\frac{n}{\upsilon}; 1-\frac{n}{\upsilon}; -\frac{n}{\lambda}x^{-2\upsilon}\right), \\ &-\alpha, -\upsilon, \lambda, \sigma, n, x > 0, \end{aligned} \quad (19)$$

where ${}_2F_1(\cdot,\cdot;\cdot;\cdot)$ is the Gauss hypergeometric function. For a given value of the false alarm probability, denoted by P_{fa}, the corresponding CFAR threshold T for the $\mathscr{HG}_A^0$ distribution can be obtained from

$$\begin{aligned} 1 - P_{fa} &= F_{\mathscr{HG}_A^0}(T) \\ &= -\frac{\sigma\lambda^{\upsilon\alpha+(n-\alpha)/\upsilon} n^{n-1}\Gamma((\alpha-n)/\upsilon)}{\upsilon\Gamma(\upsilon\alpha)\Gamma(n)} T^{2n} \\ &\quad \times {}_2F_1\left(\frac{\alpha-n}{\upsilon}, -\frac{n}{\upsilon}; 1-\frac{n}{\upsilon}; -\frac{n}{\lambda}T^{-2\upsilon}\right). \end{aligned} \quad (20)$$

Considering $F_{\mathscr{HG}_A^0}(T)$ is strictly monotonously increasing, the threshold T can be accurately calculated with the help of the numerical solution or a simple bisection method [16].

Our future work will focus on demonstrating the performances of the proposed CFAR detector using some measured SAR data and investigating how the $\mathscr{HG}^0$ distribution performs when extending it to other application fields.

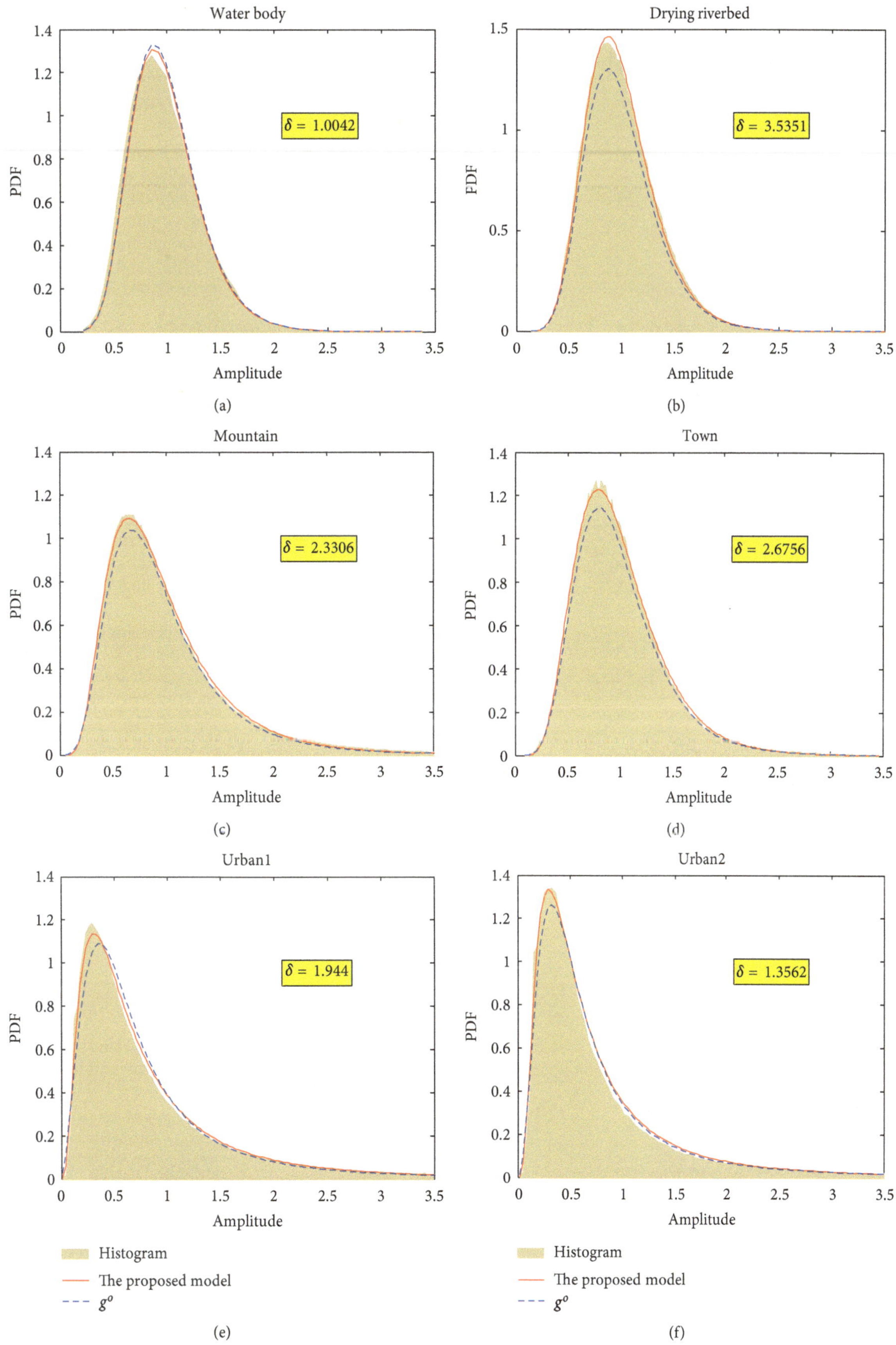

FIGURE 3: Plots of amplitude histograms and of the estimated PDFs: (a)–(f) are corresponded to the fitting results for the scenes shown in Figures 2(a)–2(f), respectively.

Appendices

A. The Derivation of mth Order Moments of the $\mathscr{H}\mathscr{G}_I^o$ Distribution

Via (2), the mth order moments of the $\mathscr{H}\mathscr{G}_I^o$ are expressed as

$$E\left(z_I^m\right) = \int_0^\infty z^m p_{Z_I}(z)\,dz = -\frac{\sigma\lambda^{\upsilon\alpha} n^n \Gamma\left((\alpha-n)/\upsilon\right)}{\upsilon\Gamma(\upsilon\alpha)\Gamma(n)} \times \lambda^{(n-\alpha)/\upsilon} \int_0^\infty z^{m+n-1}\left(1+\frac{n}{\lambda}z^{-\upsilon}\right)^{(n-\alpha)/\upsilon} dz. \tag{A.1}$$

A variable replacement $x = z^{-\upsilon}$ leads to $z = x^{-1/\upsilon}$ and $dz = -(1/\upsilon)x^{-(1/\upsilon)-1}dx$; thus (A.1) can be rewritten as

$$E\left(z_I^m\right) = -\frac{\sigma\lambda^{\upsilon\alpha} n^n \Gamma\left((\alpha-n)/\upsilon\right)}{\upsilon^2\Gamma(\upsilon\alpha)\Gamma(n)} \times \lambda^{(n-\alpha)/\upsilon} \int_0^\infty x^{((m+n)/\upsilon)-1}\left(1+\frac{n}{\lambda}x\right)^{(n-\alpha)/\upsilon} dx. \tag{A.2}$$

According to the integral formula $\int_0^\infty x^{\mu-1}dx/(1+\beta x)^\upsilon = \beta^{-\mu}\Gamma(\mu)\Gamma(\upsilon-\mu)/\Gamma(\upsilon)$, $|\arg\beta| < \pi$, $\operatorname{Re}\upsilon > \operatorname{Re}\mu > 0$ [15, 3.194, Equation 3], one can obtain (8).

B. The Derivation of the Cumulative Distribution Function of $\mathscr{H}\mathscr{G}_A^o$

Based on the definition of the CDF, the CDF of $\mathscr{H}\mathscr{G}_A^o$ is

$$F_{\mathscr{H}\mathscr{G}_A^o}(x) = \int_{-\infty}^x p_{Z_A}(z)\,dz = \int_0^x p_{Z_A}(z)\,dz = -\frac{2\sigma\lambda^{\upsilon\alpha} n^n \Gamma\left((\alpha-n)/\upsilon\right)}{\upsilon\Gamma(\upsilon\alpha)\Gamma(n)} \times \lambda^{(n-\alpha)/\upsilon} \int_0^x z^{2n-1}\left(1+\frac{n}{\lambda}z^{-2\upsilon}\right)^{(n-\alpha)/\upsilon} dz. \tag{B.1}$$

Similarly, utilizing a symbol change $t = z^{-2\upsilon}$, (B.1) turns out to be

$$F_{\mathscr{H}\mathscr{G}_A^o}(x) = \frac{\sigma\lambda^{\upsilon\alpha} n^n \Gamma\left((\alpha-n)/\upsilon\right)}{\upsilon^2\Gamma(\upsilon\alpha)\Gamma(n)} \times \lambda^{(n-\alpha)/\upsilon} \int_0^{x^{-2\upsilon}} t^{(-n/\upsilon)-1}\left(1+\frac{n}{\lambda}t\right)^{(n-\alpha)/\upsilon} dt. \tag{B.2}$$

Likewise, applying the integral formula $\int_0^u x^{\mu-1}dx/(1+\beta x)^\upsilon = (u^\mu/\mu)_2F_1(\upsilon,\mu;\ 1+\mu; -\beta u)$, $|\arg(1+\beta u)| < \pi$, $\operatorname{Re}\mu > 0$ [15, 3.194, Equation 1], we arrive at (19) by simplifying (B.2).

References

[1] C. J. Oliver and S. Quegan, *Understanding Synthetic Aperture Radar Images*, Artech House, Norwood, Mass, USA, 1998.

[2] G. Gao, "Statistical modeling of SAR images: a survey," *Sensors*, vol. 10, no. 1, pp. 775–795, 2010.

[3] A. Achim, E. E. Kuruo⊠lu, and J. Zerubia, "SAR image filtering based on the heavy-tailed rayleigh model," *IEEE Transactions on Image Processing*, vol. 15, no. 9, pp. 2686–2693, 2006.

[4] T. Esch, M. Thiel, A. Schenk, A. Roth, A. Müller, and S. Dech, "Delineation of Urban footprints from TerraSAR-X data by analyzing speckle characteristics and intensity information," *IEEE Transactions on Geoscience and Remote Sensing*, vol. 48, no. 2, pp. 905–916, 2010.

[5] A. C. Frery, J. Jacobo-Berlles, J. Gambini, and M. E. Mejail, "Polarimetric SAR image segmentation with B-splines and a new statistical model," *Multidimensional Systems and Signal Processing*, vol. 21, no. 4, pp. 319–342, 2010.

[6] C. Tison, J. M. Nicolas, F. Tupin, and H. Maître, "A new statistical model for Markovian classification of Urban areas in high-Resolution SAR images," *IEEE Transactions on Geoscience and Remote Sensing*, vol. 42, no. 10, pp. 2046–2057, 2004.

[7] A. C. Frery, A. H. Correia, and C. D. C. Freitas, "Classifying multifrequency fully polarimetric imagery with multiple sources of statistical evidence and contextual information," *IEEE Transactions on Geoscience and Remote Sensing*, vol. 45, no. 10, pp. 3098–3109, 2007.

[8] V. A. Krylov, G. Moser, S. B. Serpico, and J. Zerubia, "Supervised high-resolution dual-polarization SAR image classification by finite mixtures and copulas," *IEEE Journal on Selected Topics in Signal Processing*, vol. 5, no. 3, pp. 554–566, 2011.

[9] Y. Delignon, R. Garello, and A. Hillion, "Statistical modeling of ocean SAR image," *IEE Proceedings Radar, Sonar and Navigation*, vol. 144, no. 6, pp. 348–354, 1997.

[10] Y. Delignon and W. Pieczynski, "Modeling non-Rayleigh speckle distribution in SAR images," *IEEE Transactions on Geoscience and Remote Sensing*, vol. 40, no. 6, pp. 1430–1435, 2002.

[11] A. C. Frery, H.-J. Müller, C. D. C. F. Yanasse, and S. J. S. Sant'Anna, "A model for extremely heterogeneous clutter," *IEEE Transactions on Geoscience and Remote Sensing*, vol. 35, no. 3, pp. 648–659, 1997.

[12] K. L. P. Vasconcellos, A. C. Frery, and L. B. Silva, "Improving estimation in speckled imagery," *Computational Statistics*, vol. 20, no. 3, pp. 503–519, 2005.

[13] J. M. Nicolas, "Introduction to second kind statistic: application of log-moments and log-cumulants to SAR image law analysis," *Traitement du Signal*, vol. 19, no. 3, pp. 139–167, 2002.

[14] S. N. Anfinsen and T. Eltoft, "Application of the Matrix-variate mellin transform to analysis of polarimetric radar images," *IEEE Transactions on Geoscience and Remote Sensing*, vol. 49, no. 6, pp. 2281–2295, 2011.

[15] S. I. Gradshteyn and I. M. Ryzhik, *Table of Integrals, Series, and Products*, Academic Press, San Diego, Calif, USA, 7th edition, 2007.

[16] G. Moser, J. Zerubia, and S. B. Serpico, "SAR amplitude probability density function estimation based on a generalized Gaussian model," *IEEE Transactions on Image Processing*, vol. 15, no. 6, pp. 1429–1442, 2006.

A Rectangular Ring, Open-Ended Monopole Antenna with Two Symmetric Strips for WLAN and WiMAX Applications

Joong-Han Yoon,[1] Young-Chul Rhee,[2] and Woo-Su Kim[3]

[1] *Department of Electronic Engineering, Silla University, San 1-1 Gwaebeop-Dong, Sasang-Gu, Busan 617-736, Republic of Korea*
[2] *Department of Information and Communication Engineering, Kyungnam University, 449 Wolyong-Dong, Masan, Kyungnam 631-701, Republic of Korea*
[3] *School of Electrical and Computer Engineering, Georgia Institute of Technology, North Avenue, Atlanta, GA 30332, USA*

Correspondence should be addressed to Woo-Su Kim; wkim84@mail.gatech.edu

Academic Editor: Stefano Selleri

A triple-band rectangular ring, open-ended monopole antenna with symmetric L strips for wireless local area network (WLAN)/Worldwide Interoperability of Microwave Access (WiMAX) applications is proposed. The proposed antenna consists of two symmetric folded arms and L strips. Based on the concept, a prototype of the proposed triple antenna has been designed, fabricated, and tested. The numerical and experimental results demonstrated that the proposed antenna satisfied the −10 dB impedance bandwidth requirement while simultaneously covering the WLAN and WiMAX bands. Furthermore, this paper presented and discussed the 2D radiation patterns and 3D gains according to the results of the experiment. The proposed antenna's peak gain varied between 2.17 and 4.93 dBi, and its average gain varied between −2.97 and −0.53 dBi.

1. Introduction

The recent increasing demand for antennas with multiband operation capabilities in modern communication systems has attracted public attention. Likewise, the ability to integrate more than one communication standard into a single system has become a most sought-after feature of a modern wireless communication device. In the last few years, wireless local area networks (WLANs) with connections to portable terminals, such as personal digital assistants (PDAs), notebooks or laptop computers, smart phones, and other similar equipment, have been governed through the standardization of frequency bands that are defined by various standard protocols, namely, IEEE 802.11a (5.15–5.35 GHz and 5.725–5.85 GHz) and 802.11 a/g (2.4–2.484 GHz). Furthermore, Worldwide Interoperability of Microwave Access (WiMAX) has been deployed to provide mobile broadband Internet services with large coverage areas in many countries. Recently, WiMAX has primarily allocated three different frequency bands (WiMAX: 2.5–2.69, 3.3–3.7, and 5.25–5.85 GHz) [1, 2].

In the near future, these consumer electronic devices will be capable of integrating multiple frequency bands (WLAN, WiMAX, etc.) into one single system. Among the key components of these devices, the antenna plays an important role in determining the RF performance of the end product. The antenna used in these devices is required to operate at different frequency bands or broadband in WLAN and WiMAX systems. Furthermore, many monopole antennas with different structure have been reported such as the G-shaped monopole [3], the rectangular ring monopole [4], the C-shaped monopole [5], the W-shaped monopole [6], the Y-shaped monopole [7], F-shaped [8], double T-shaped [9], and double S-shaped [10]. However, these antennas can only cover one application. Thus, to make multiband and wideband WLAN/WiMAX antennas, studies on WLAN/WiMAX for triple-band applications have been made. In particular, the rectangular ring, monopole antenna for WLAN/WiMAX bands is proposed [3–7].

The dual rectangular ring, printed CPW-fed monopole antenna with inverted T-shaped strip [11], the rectangular

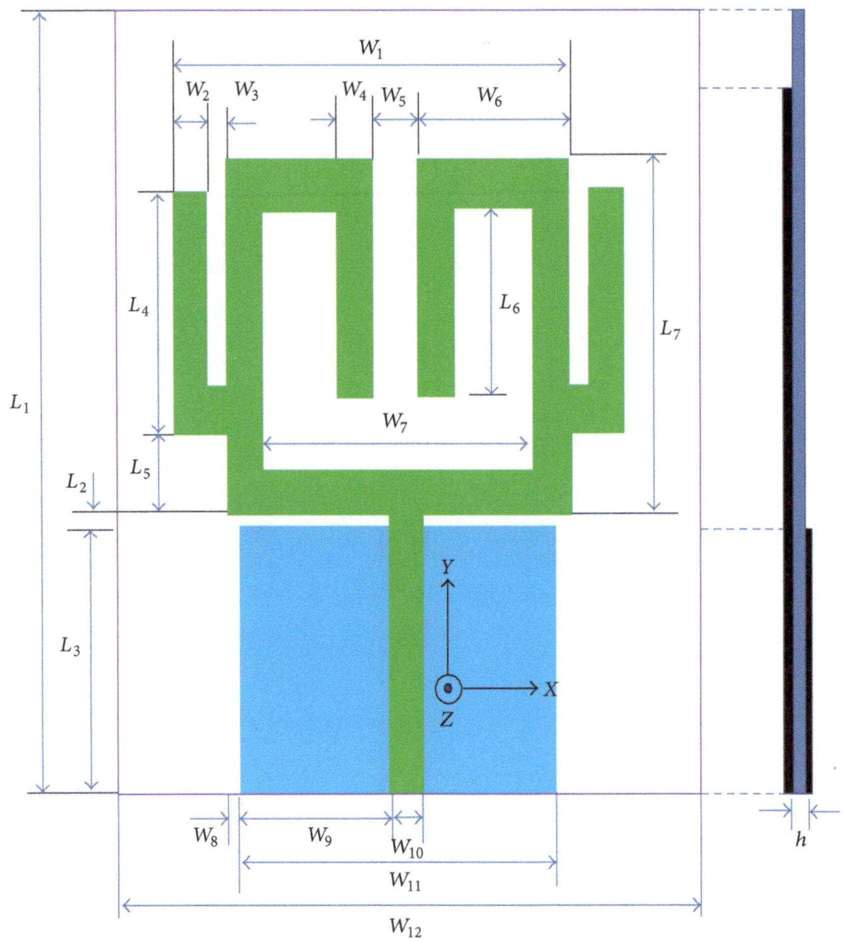

FIGURE 1: Configuration of the rectangular ring, open-ended monopole antenna with two symmetric folded arms and two symmetric L strips.

ring, monopole antenna with a straight strip and a meandered strip [12], and the rectangular ring, monopole antenna with symmetric L strips [13] have been suggested. These antennas are designed and fabricated at a substrate with relative permittivity of 3.5, 2.65, and 2.65, respectively. These proposed antennas are made of expensive materials, compared to conventional FR4 materials. Likewise, the dual rectangular ring with open-ended CPW-fed monopole antenna [14] is introduced. The total size of the proposed antenna is quite large. Next, the square ring, monopole antenna with a vertical and L-shaped strip [15] is proposed. The L strip is located on the bottom side and is connected to the square ring through a hole. This proposed antenna has a complicated structure because a hole has to be made.

In this paper, a rectangular ring, open-ended monopole antenna with two symmetric folded arms and L strips for WLAN and WiMAX band applications was designed. The proposed antenna was fabricated with the use of a conventional FR4 material that is often used to make printed circuit boards and is easy to manufacture. By properly selecting the dimensions of the proposed antenna, good triple broadband bandwidth capabilities, and suitable radiation characteristics for two multiband wireless communication systems, such as WLAN (2.4–2.483 GHz, 5.15–5.85 GHz) and WiMAX (2.5–2.69, 3.3–3.7, and 5.25–5.85 GHz) operations, this study's goals can be achieved. The antenna configuration and simulated data, as well as the constructed prototype and measured data, will be carefully examined and discussed in the succeeding sections.

2. Antenna Design

Figure 1 shows the specific configuration designed for the WLAN/WiMAX printed monopole antenna, which consists of two symmetric folded arms and L strips. The total sizes of the substrate and the ground plane of the proposed antenna that was used in this study were 30.0×40.0 mm ($W_{12}\times L_1$) and 19.0×17.2 mm ($W_{11}\times L_3$), respectively. The radiating element was placed on the same side as the feeding strip line, and the ground plane was placed on the other side of the substrate. The proposed antenna is fabricated on an inexpensive FR4 substrate with a relative permittivity of 4.4 and a thickness of 1.0 mm.

The proposed antenna had rectangular ring, open-ended strip lines that have a uniform width ($W_2 = W_4 = W_{10}$, 2 mm) for design convenience. In this design, the rectangular ring, open-ended antenna with symmetric, folded arms was resonant for 2.4 GHz/2.5 GHz WiMAX bands. Two symmetric,

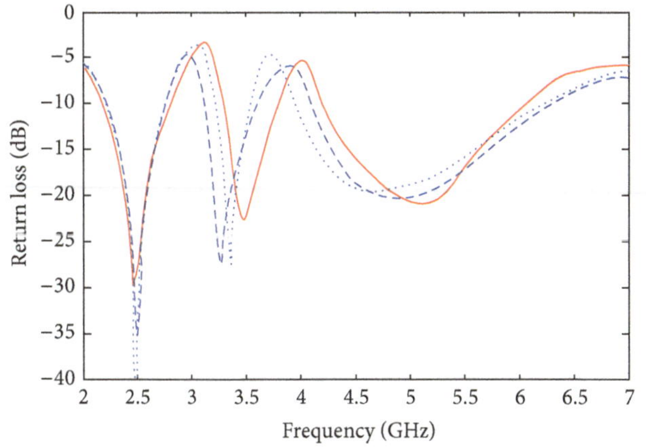

FIGURE 2: Simulated return loss of the proposed antenna with different values of gap width between the rectangular ring, open-ended antenna and the L strip (W_3). (Other parameters: L_1 = 40.0 mm; L_2 = 1.8 mm; L_3 = 17.2 mm; L_4 = 12.5 mm; L_5 = 3.0 mm; L_6 = 9.0 mm; L_7 = 16.0 mm; W_1 = 27.0 mm; W_2 = 2.0 mm; W_4 = 2.0 mm; W_5 = 3.0 mm; W_6 = 9.0 mm; W_7 = 17.0 mm; W_8 = 1.0 mm; W_9 = 8.5 mm; W_{10} = 2.0 mm; W_{11} = 19.0 mm; and W_{12} = 30.0 mm.)

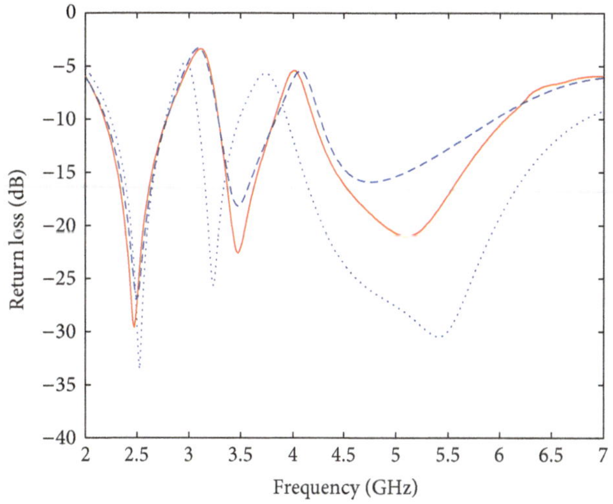

FIGURE 3: Simulated return loss of the proposed antenna with different values of the ground plane (W_9). (Other parameters: L_1 = 40.0 mm; L_2 = 1.8 mm; L_3 = 17.2 mm; L_4 = 12.5 mm; L_5 = 3.0 mm; L_6 = 9.0 mm; L_7 = 16.0 mm; W_1 = 27.0 mm; W_2 = 2.0 mm; W_3 = 1.0 mm; W_4 = 2.0 mm; W_5 = 3.0 mm; W_6 = 9.0 mm; W_7 = 17.0 mm; W_8 = 1.0 mm; W_{10} = 2.0 mm; W_{11} = 19.0 mm; and W_{12} = 30.0 mm.)

folded arms are attached to the end of the rectangular ring, open-ended antenna. To obtain good impedance matching and wide bandwidth for 3.5 and 5 GHz operating frequencies, we introduced the two symmetric L strips, 3.0 × 12.5 mm^2, (W_2 + W_3) × L_4. Two symmetric L strips are attached to the side of the rectangular ring, open-ended antenna. A 50 Ω microstrip feed line, W_{10} = 2 mm, is used for centrally feeding the antenna from the bottom edge of the symmetrical rectangular ring, open-ended antenna. To obtain the optimal parameters (lines and lengths) of the proposed antenna for WLAN and WiMAX applications, HFSS [16], a full-wave commercial EM software that is capable of simulating a finite substrate and a finite ground structure was used. The length and width of the two symmetric, folded arms and L strips control the input of the proposed antenna.

2.1. Effect of the Two Symmetric, Folded Arms (W_3). Figure 2 illustrates the return loss for the different values of gap width between the rectangular ring, open-ended antenna and the L strip (W_3). It can be seen in the figure that the impedance bandwidth and characteristics of the return loss slightly changed when W_3 changed from 0.5 to 1.5 mm in the 3.5 and 5 GHz bands. It can also be seen in the figure, however, that the impedance bandwidth and characteristics of the return loss did not change in the 2.4 GHz and 2.5 GHz bands. This implied that gap width is affected by the characteristics of the return loss in 3.5 GHz and 5 GHz bands. To design a good triple-band WLAN/WiMAX operation, W_3 was set at 1.0 mm.

2.2. Effect of the Ground Plane (W_9). Figure 3 illustrates the return loss for the different values of the ground plane (W_9). It can be seen in the figure that the impedance bandwidth and characteristics of the return loss did not change when W_9 changed from 7.5 to 9.5 mm in the 2.4 GHz WLAN and 2.5 GHz WiMAX bands. It can also be seen in the figure, however, that the impedance bandwidth and characteristics of the return loss greatly changed in the 3.5 GHz and 5 GHz bands. In particular, impedance bandwidth and characteristics of the return loss improved in the 5 GHz band, but required frequency bands deviated in the 3.5 GHz band when W_9 was 9.5 mm. To design a good triple-band WLAN/WiMAX operation, W_9 was set at 8.5 mm.

2.3. Effect of Gap Width between the Rectangular Ring, Open-Ended Antenna and the Ground Plane (L_2). Figure 4 illustrates the return loss for the different values of the gap width between the rectangular ring, open-ended antenna and the ground plane (L_2). It can be seen in the figure that the impedance bandwidth and characteristics of the return loss slightly changed when W_3 changed from 1.6 to 2.0 mm in the required frequency bands. This implied that the gap width is significantly affected by the characteristics of the return loss in the required frequency bands. To design a good triple-band WLAN/WiMAX operation, L_2 was set at 1.8 mm.

2.4. Effect of Two Symmetric L Strips (L_4). Figure 5 illustrates the return loss for the different values of L strips (L_4). It can be seen in the figure that the impedance bandwidth

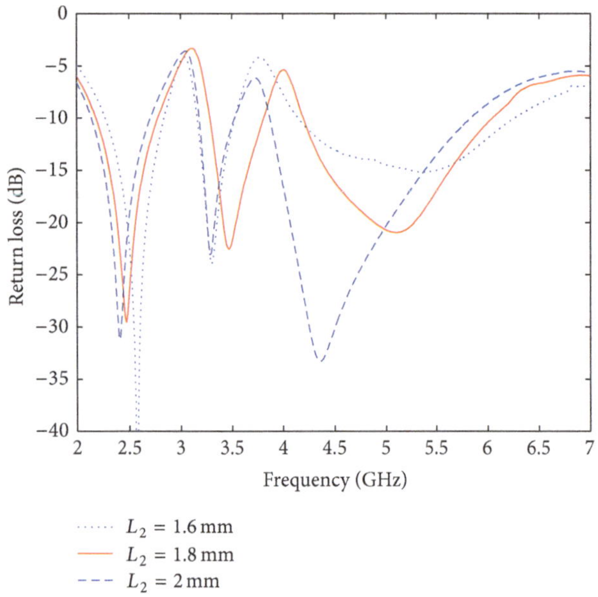

FIGURE 4: Simulated return loss of the proposed antenna with different values of gap length between the rectangular ring, open-ended antenna and the ground plane (L_2). (Other parameters: L_1 = 40.0 mm; L_3 = 17.2 mm; L_4 = 12.5 mm; L_5 = 3.0 mm; L_6 = 9.0 mm; L_7 = 16.0 mm; W_1 = 27.0 mm; W_2 = 2.0 mm; W_3 = 1.0 mm; W_4 = 2.0 mm; W_5 = 3.0 mm; W_6 = 9.0 mm; W_7 = 17.0 mm; W_8 = 1.0 mm; W_9 = 8.5 mm; W_{10} = 2.0 mm; W_{11} = 19.0 mm; and W_{12} = 30.0 mm.)

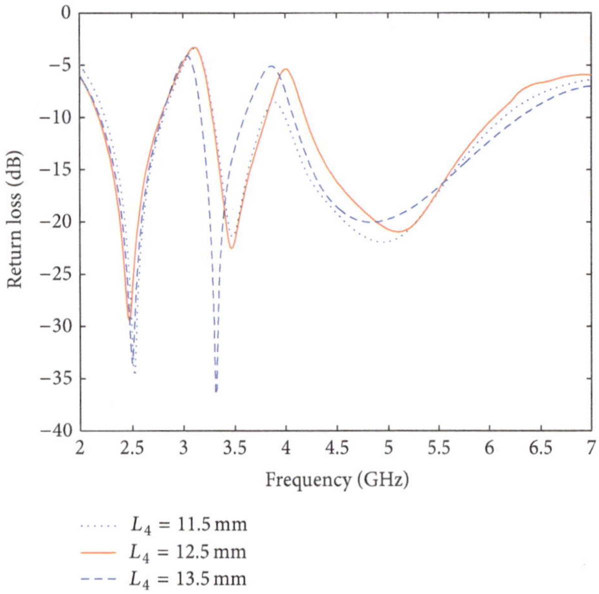

FIGURE 5: Simulated return loss of the proposed antenna with different values of two symmetric L strips (L_4). (Other parameters: L_1 = 40.0 mm; L_2 = 1.8 mm; L_3 = 17.2 mm; L_5 = 3.0 mm; L_6 = 9.0 mm; L_7 = 16.0 mm; W_1 = 27.0 mm; W_2 = 2.0 mm; W_3 = 1.0 mm; W_4 = 2.0 mm; W_5 = 3.0 mm; W_6 = 9.0 mm; W_7 = 17.0 mm; W_8 = 1.0 mm; W_9 = 8.5 mm; W_{10} = 2.0 mm; W_{11} = 19.0 mm; and W_{12} = 30.0 mm.)

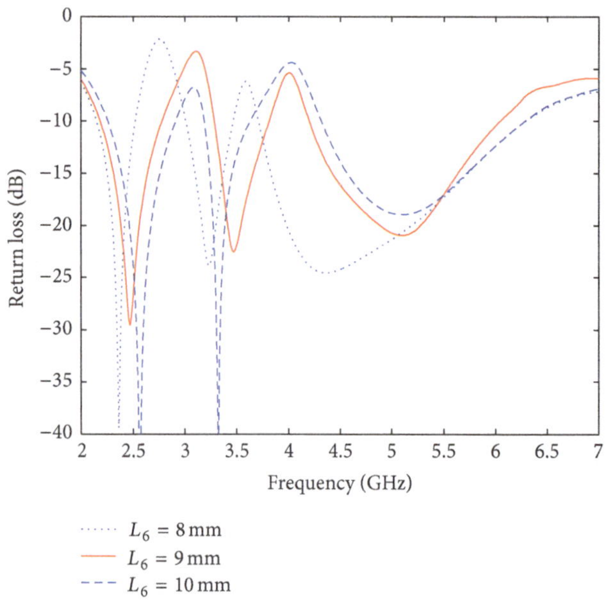

FIGURE 6: Simulated return loss of the proposed antenna with different values of two folded arms (L_6). (Other parameters: L_1 = 40.0 mm; L_2 = 1.8 mm; L_3 = 17.2 mm; L_4 = 12.5 mm; L_5 = 3.0 mm; L_7 = 16.0 mm; W_1 = 27.0 mm; W_2 = 2.0 mm; W_3 = 1.0 mm; W_4 = 2.0 mm; W_5 = 3.0 mm; W_6 = 9.0 mm; W_7 = 17.0 mm; W_8 = 1.0 mm; W_9 = 8.5 mm; W_{10} = 2.0 mm; W_{11} = 19.0 mm; and W_{12} = 30.0 mm.)

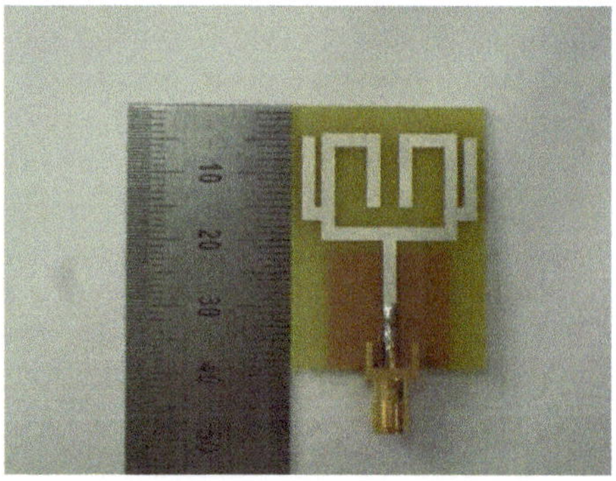

(a)

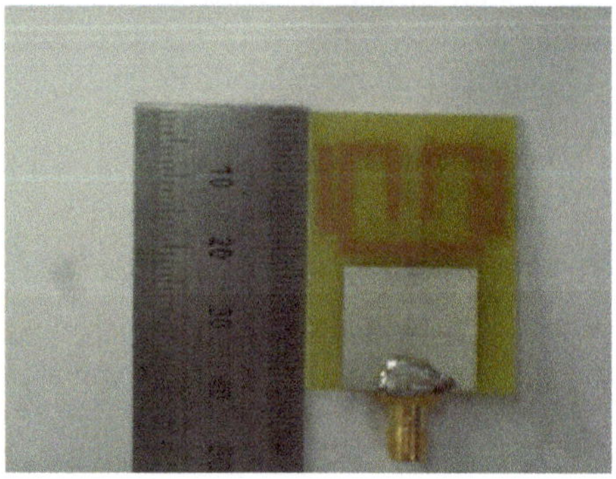

(b)

FIGURE 7: Prototype of the proposed dual-band antenna: (a) front view and (b) back view.

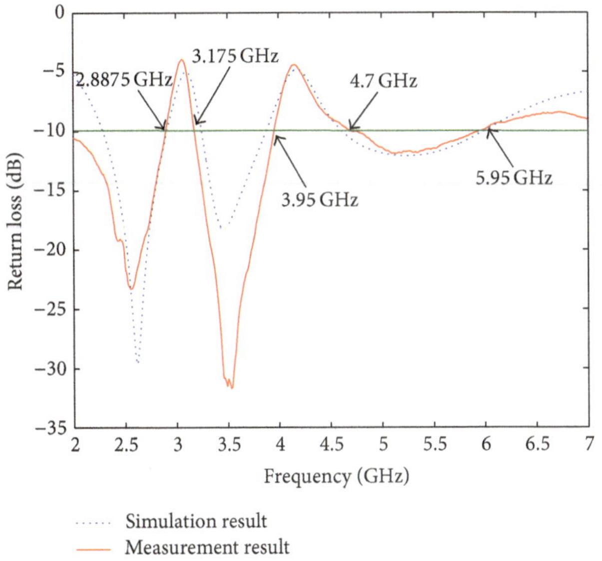

FIGURE 8: Simulated and measured return loss versus frequencies of the proposed antenna.

and characteristics of the return loss slightly changed when L_4 changed from 11.5 to 13.5 mm in the 3.5 and 5 GHz bands. It can also be seen in the figure, however, that the impedance bandwidth and characteristics of the return loss did not change in the 2 GHz bands. This implied that the two symmetric L strips are largely affected by the characteristics of the return loss in 3.5 and 5 GHz bands. To design a good triple-band WLAN/WiMAX operation, L_4 was set at 12.5 mm.

2.5. Effect of Two Symmetric Folded Arms (L_6). Figure 6 illustrates the return loss for the different values of the L strips (L_6). It can be seen in the figure that the impedance bandwidth and characteristics of the return loss slightly changed when L_6 changed from 8.0 to 10.0 mm in the required frequency bands. This implied that the two symmetric L strips are highly affected by the characteristics of the return loss in the required frequency bands. To design a good triple-band WLAN/WiMAX operation, L_6 was set at 9.0 mm.

The proposed antenna structure has several design parameters that can handle the resistance and reactance of the antenna input impedance. Therefore, the dimensions of the proposed antenna were set as follows: $L_1 = 40.0$ mm; $L_2 = 1.8$ mm; $L_3 = 17.2$ mm; $L_4 = 12.5$ mm; $L_5 = 3.0$ mm; $L_6 = 9.0$ mm; $L_7 = 16.0$ mm; $W_1 = 27.0$ mm; $W_2 = 2.0$ mm; $W_3 = 1.0$ mm; $W_4 = 2.0$ mm; $W_5 = 3.0$ mm; $W_6 = 9.0$ mm; $W_7 = 17.0$ mm; $W_8 = 1.0$ mm; $W_9 = 8.5$ mm; $W_{10} = 2.0$ mm; $W_{11} = 19.0$ mm; $W_{12} = 30.0$ mm; and $h = 1.0$ mm. Based on the design dimensions, the proposed compact antenna was constructed and studied. A prototype of the proposed antenna was fabricated with the aforementioned design parameters and is shown in Figures 7(a) and 7(b).

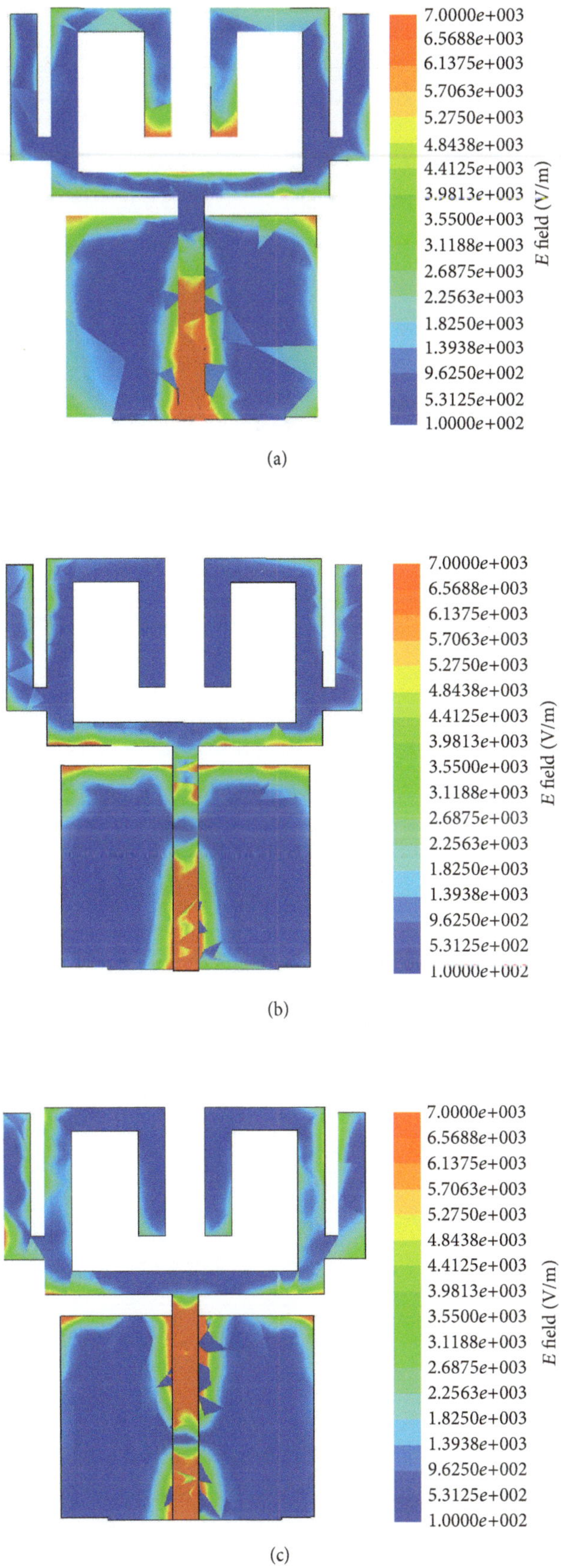

FIGURE 9: Surface current distributions at (a) 2.48 GHz, (b) 3.55 GHz, and (c) 5.06 GHz.

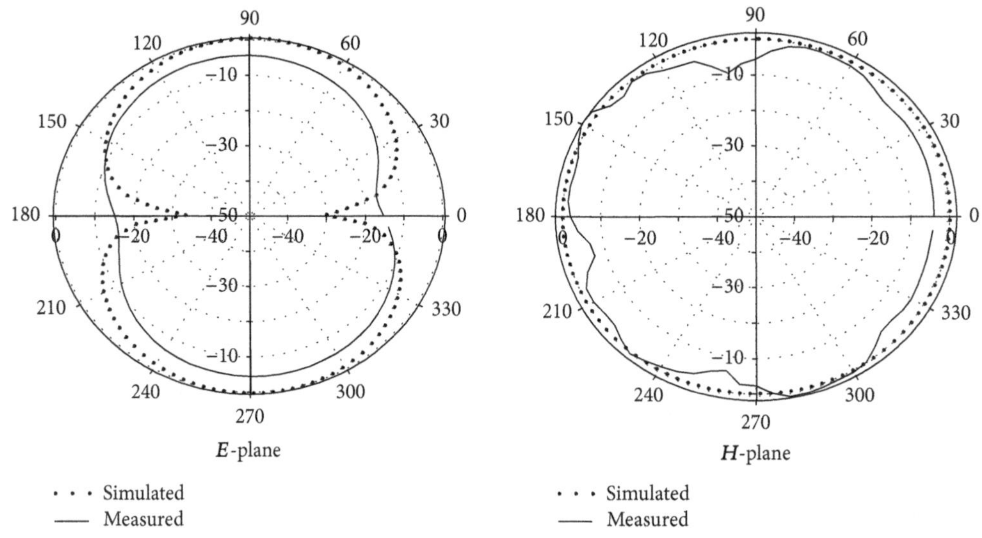

(a)

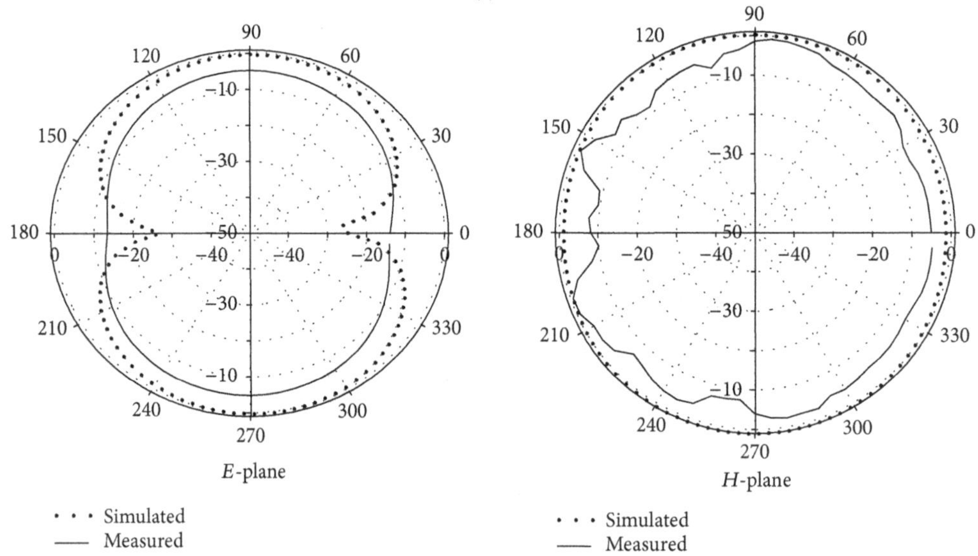

(b)

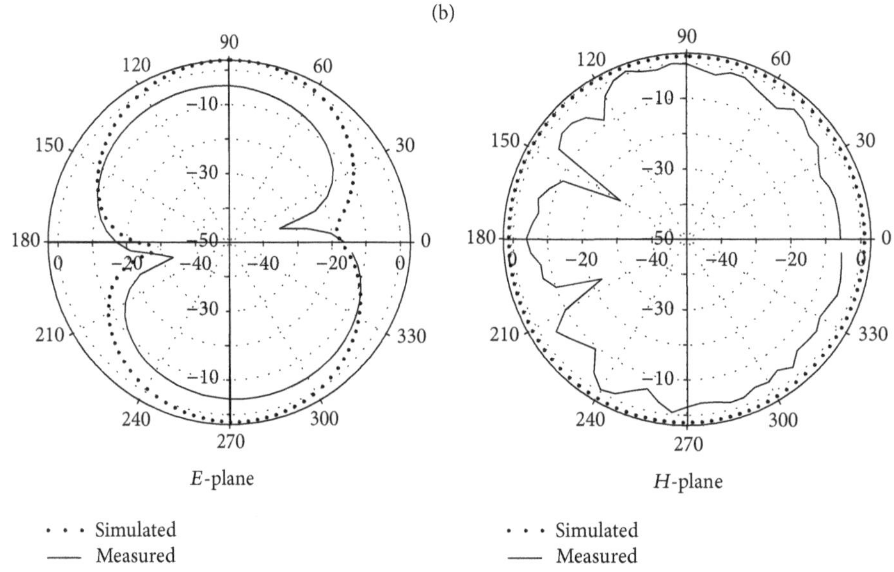

(c)

Figure 10: Continued.

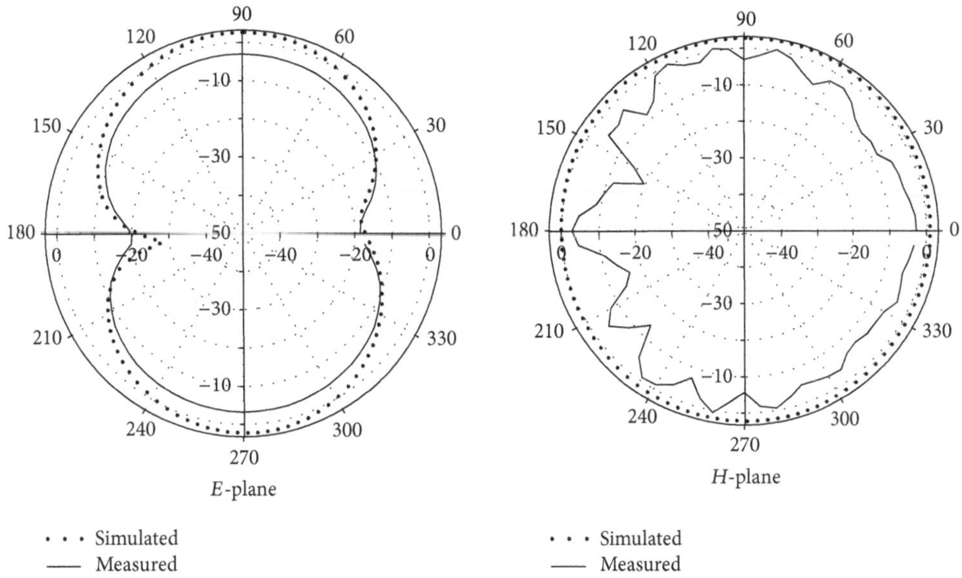

(d)

FIGURE 10: Radiation patterns of the proposed antenna for WLAN/WiMAX operation frequencies: in the (a) 2.5 GHz; (b) 3.5 GHz; (c) 5.3 GHz; and (d) 5.7 GHz bands.

3. Measurement

Based on the detailed dimensions given in the previous section, a prototype for the proposed triple-band antenna was fabricated and measured. The measured results were performed by using an Anritsu MS4644A vector network analyzer in Silla University. The far-field radiation patterns and gains were measured by using a far-field anechoic absorber obtained from Foxconn Corporation in Republic of Korea. Figure 8 describes the simulated and experimental return loss against the frequency for the proposed antenna, where generally acceptable agreements between them have been achieved. Because the simulation used a waveguide port and an SMA connector was used for measurement, there is a difference between the measured and the simulated results. To confirm the accurate return loss characteristics for the designed antenna, it is recommended that the manufacturing and the measurement process must be performed more carefully. Based on the measured results, three relative impedance bandwidths, with −10 dB return loss of about 887.5 MHz (36.32%, 2,000–2,887.5 MHz), 775 MHz (21.75%, 3,175–3,950 MHz), and 1,250 MHz (23.47%, 4,700–5,950 MHz), respectively, can satisfy the typical bandwidth requirement for the 2.4–2.484 and 5.15–5.825 GHz WLAN bands, as well as 2.5–2.69, 3.4–3.7, and 5.25–5.85 GHz WiMAX bands, respectively.

Theoretically, HFSS was used to evaluate and verify the three resonant frequencies 2.48, 3.55, and 5.06 GHz, which mainly depended on the lengths of the different current paths along the folded arms and L strips. As expected, it was obvious that the different surface currents were excited by the 2.48 and 3.55 GHz frequencies, respectively, through folded arms and L strips. Figures 9(a), 9(b), and 9(c) show the surface current density excitations along the folded arms and L strips in the cases of the three resonant frequencies, 2.48, 3.55, and 5.06 GHz, respectively. As shown in Figure 9(a), the lowest band surface current density excitations along the folded arms were observed when the resonant frequency was 2.48 GHz. As observed in the middle-band excitation in Figure 9(b), a larger surface current density flowing along the symmetric L strips was observed when the resonant frequency was 3.55 GHz. As shown in Figure 9(c), however, larger surface current densities flowing not only along the folded arms but also along the L strips were observed when the resonant frequency was 5.06 GHz. Thus, it is implied that the 5 GHz band excitation is contributed by the radiations of the folded arms and the L strips. Likewise, based on the current density of the proposed antenna, L_2, the gap length between the rectangular ring, open-ended antenna and the ground plane, and W_3, the gap width between the rectangular ring, open-ended antenna and the L strip, strongly affect antenna performance in all required frequency bands. The required frequency response characteristics can be obtained by optimizing the values of L_2 and W_3.

Figure 10 shows the measured 2D far-field radiation patterns in the E-plane (x-z plane) and H plane (y-z plane). Figures 10(a), 10(b), 10(c), and 10(d) show the 2D radiation patterns at 2.5, 3.5, 5.3, and 5.7 GHz, respectively. Good agreement is obtained between the simulated and measured results. Based on these radiation patterns, the proposed antenna displays nearly omnidirectional radiation characteristics in the H-plane and monopole-like radiation pattern characteristics in the E-plane at the considered frequencies.

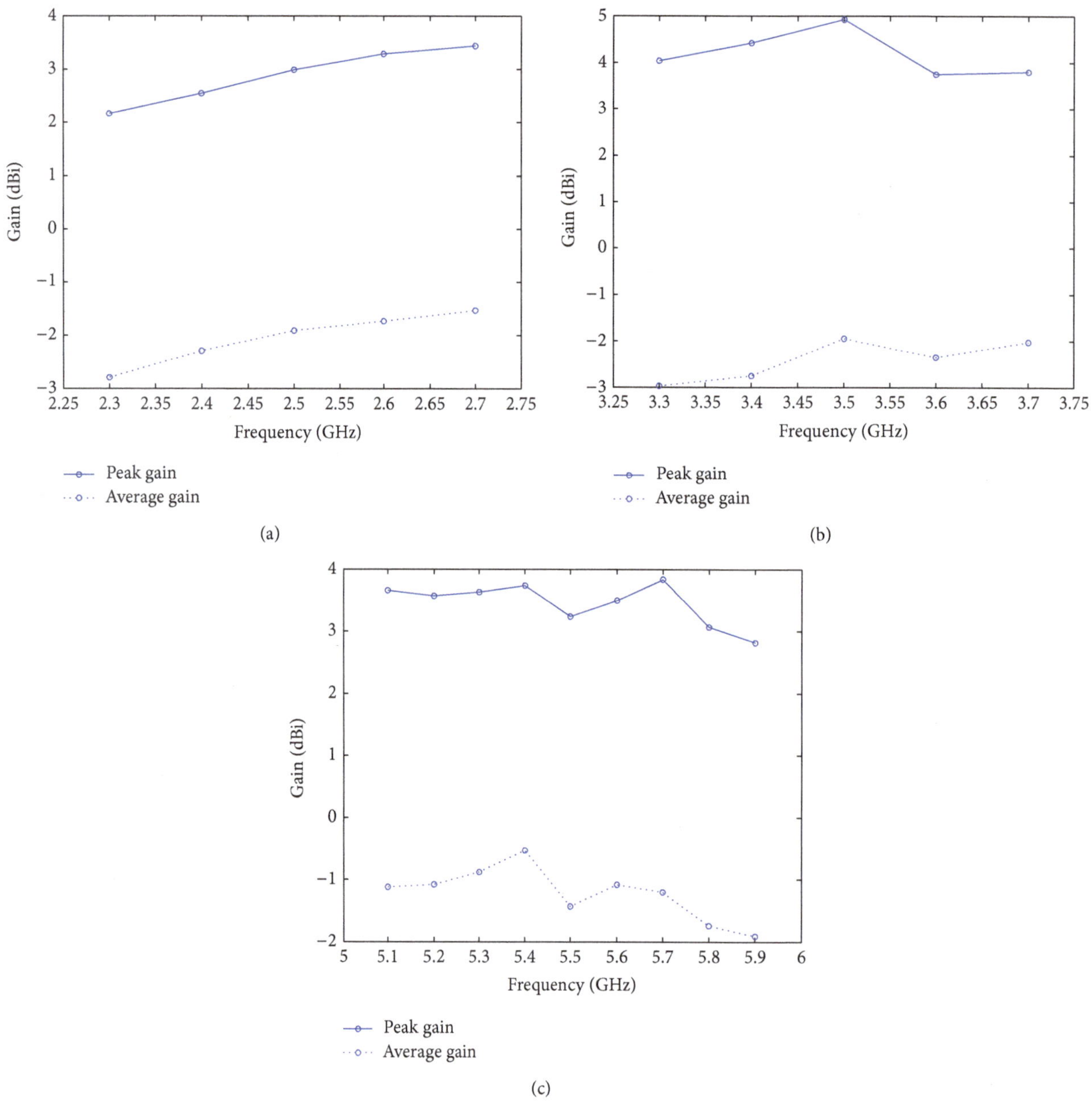

Figure 11: Measured antenna peak and average gains for the following operating frequencies: (a) 2.4/2.5 GHz band, (b) 3.5 GHz band, and (c) 5 GHz band.

Figures 11(a), 11(b), and 11(c) show the 3D measured antenna peak and average gain for the frequencies across the 2.4/2.5, 3.5, and 5 GHz bands. The 2.4/2.5 GHz band had an antenna peak gain level of about 2.17~3.44 dBi (Figure 11(a)), and the 3.5 GHz band about 3.75~4.93 dBi (Figure 11(b)). The measured antenna gain levels were about 3.57~3.74 dBi in the 5.3 GHz band and about 2.82~3.84 dBi in the 5.8 GHz band (Figure 11(c)). The 2.4/2.5 GHz band had an antenna average gain level of about −2.79~ −1.53 dBi (Figure 11(a)) and the 3.5 GHz band about −2.97~ −1.94 dBi (Figure 11(b)). The measured antenna gain levels were about −1.12~ −0.53 dBi in the 5.2 GHz band, and about −1.91~ −1.08 dBi in the 5.8 GHz bands (Figure 11(c)). The 3D antenna gain had a peak value of 2.99 dBi at 2.5 GHz. At 3.5 GHz, the 3D maximum peak gain was 4.93 dBi; at 5.3 GHz, 3.63 dBi; and at 5.8 GHz, 3.07 dBi.

4. Conclusion

A rectangular ring, open-ended planar monopole antenna with two symmetric folded arms and *L* strips for WLAN/WiMAX applications was proposed. By employing these structures, the proposed antenna can yield three different resonances to cover the desired bands while maintaining a simple structure. The results of the parametric studies on the proposed antenna were discussed. In particular, two folded arms and two symmetric *L* strips depend heavily on the characteristics of the proposed antenna in the 2.4/2.5 GHz band and 5 GHz bands, respectively. Furthermore, the studies on surface current distributions of the operating frequencies were discussed. The measured results show that the obtained impedance bandwidths are 36.32% (2.0–2.89 GHz), 21.75% (3.18–3.95 GHz), and about 23.47% (4.7–5.95 GHz), respectively, which are suitable for WLAN and WiMAX applications. In addition, the proposed antenna has good radiation characteristics and gains in the three operating bands, so it can emerge as an excellent candidate for multiband applications in wireless communication.

Acknowledgments

This research was financially supported by the Ministry of Education, Science, and Technology (MEST) and the National Research Foundation of Korea (NRF) through the Human Resource Training Project for Regional Innovation. Likewise, the authors would like to thank the Director S. G. Lee and Jeffrey Park, senior engineer at the R&D Center of Foxconn Corporation in Korea, for the assistance in the measurement of antenna radiation patterns.

References

[1] "WorldWide Interoperability for microwave access forum or WiMAX forum," http://www.wimaxforum.org/.

[2] "IEEE 802.16 workinggroup on broadband wireless access standards," http://grouper.ieee.org/groups/802/16/index.html.

[3] C.-Y. Pan, C.-H. Huang, and T.-S. Horng, "A new printed G-shaped monopole antenna for dual-band wlan applications," *Microwave and Optical Technology Letters*, vol. 45, no. 4, pp. 295–297, 2005.

[4] J. H. Yoon, "Fabrication and measurement of rectangular ring with open-ended CPW-fed monopole antenna for 2.4/5.2-GHz WLAN operation," *Microwave and Optical Technology Letters*, vol. 48, no. 8, pp. 1480–1483, 2006.

[5] Y. Song, Y.-C. Jiao, H. Zhao, Z. Zhang, Z.-B. Weng, and F.-S. Zhang, "Compact printed monopole antenna for multiband WLAN applications," *Microwave and Optical Technology Letters*, vol. 50, no. 2, pp. 365–367, 2008.

[6] Z. Zhang, Y. C. Jiao, Y. Song, T. L. Zhang, S. M. Ning, and F. S. Zhang, "A modified CPW-FED monopole antenna with very small ground for multiband WLAN applications," *Microwave and Optical Technology Letters*, vol. 52, no. 2, pp. 463–466, 2010.

[7] Z.-Y. Liu, Y.-Z. Yin, L.-H. Wen, W.-C. Xiao, Y. Wang, and S.-L. Zuo, "A Y-shaped tri-band monopole antenna with a parasitic M-strip for PCS and WLAN applications," *Journal of Electromagnetic Waves and Applications*, vol. 24, no. 8-9, pp. 1219–1227, 2010.

[8] S.-H. Yeh and K.-L. Wong, "Dual-band F-shaped monopole antenna for 2.4/5.2 GHz WLAN application," in *Proceedings of the IEEE Antennas and Propagation Society International Symposium*, pp. 72–75, San Antonio, Tex, USA, June 2002.

[9] Y.-L. Kuo and K.-L. Wong, "Printed double-T monopole antenna for 2.4/5.2 GHz dual-band WLAN operations," *IEEE Transactions on Antennas and Propagation*, vol. 51, no. 9, pp. 2187–2192, 2003.

[10] W. C. Liu, W. R. Chen, and C. M. Wu, "Printed double S-shaped monopole antenna for wideband and multiband operation of wireless communication," *IEE Proceedings—Microwaves Antennas and Propagation*, vol. 151, pp. 473–476, 2004.

[11] Q. Y. Zhang and Q. X. Chu, "Triple-band dual rectangular ring printed monopole antenna for WLAN/WiMAX applications," *Microwave and Optical Technology Letters*, vol. 51, no. 12, pp. 2845–2848, 2009.

[12] X. S. Ren, Y. Z. Yin, S. F. Zheng, S. L. Zuo, and B. W. Liu, "Triple-band rectangular ring monopole antenna for WLAN/WiMAX applications," *Microwave and Optical Technology Letters*, vol. 53, no. 5, pp. 974–978, 2011.

[13] X. Li, W. Hu, Y. F. Wang, X. W. Shi, and X. T. Gu, "Printed triple band rectangular ring monopole antenna with symmetrical L strips for WLAN/WiMAX applications," *Microwave and Optical Technology Letters*, vol. 54, pp. 1049–1052, 2012.

[14] B. Li, Z. H. Yan, and C. Wang, "Dual rectangular ring with open-ended CPW-fed monopole antenna for WLAN/WiMAX applications," *Progress In Electromagnetics Research Letters*, vol. 25, pp. 101–107, 2011.

[15] Y. F. Wang, B. H. Sun, K. He, R. H. Li, and Y. J. Wang, "A compact tri-band antenna for WLAN/WiMAX applications," *Microwave and Optical Technology Letters*, vol. 53, no. 10, pp. 2371–2375, 2011.

[16] *Ansoft High Frequency Structure Simulator (HFSS) Version 10. 0*, Ansoft Corporation, 2005.

10

A Volume-Surface Integral Equation Solver for Radiation from Microstrip Antenna on Anisotropic Substrate

Yuhuang Ye, Jiade Yuan, and Kaixiong Su

College of Physics and Information Engineering, Fuzhou University, Fuzhou, Fujian 350108, China

Correspondence should be addressed to Jiade Yuan, yjd@fzu.edu.cn

Academic Editor: Tat Yeo

A volume-surface integral equation (VSIE) solver is presented for the calculation of electromagnetic radiation from arbitrary shaped microstrip antenna on anisotropic substrate. The method of moments (MoM) is used to convert the integral equation into a matrix equation, where the equivalent volume current and surface current are expanded into a finite series of SWG and RWG basis function, respectively. A simple strip model is incorporated in the VSIE to simplify the analysis of the probe-fed microstrip antenna. The present approach is sufficiently versatile in handling microstrip antenna with arbitrary shaped anisotropic dielectric substrate. Numerical results indicate the reliability and accuracy of the proposed method.

1. Introduction

During recent years, great interests have been shown in using microstrip antenna deposited on anisotropic substrate since the substrate anisotropy could have important applications on the operation of microstrip antennas [1–4]. With the increasing complexity of geometry and material property, designing these antennas requires more and more dedicated and sophisticated computer-aided-design (CAD) tools to predict the characteristics. The method of moments (MoM) has been proven to be one of the most powerful CAD tools for solving this class of problems. By now, a number of microstrip antennas with anisotropic substrate have been investigated using the MoM-based spectral domain analysis method [1–3]. However, in the MoM-based approaches for the analysis of microstrip antenna, the volume-surface integral equation (VSIE) [5–9] solver is more suitable to analyze complex structure targets and has a number of advantages over the other MoM-based approaches [2, 10], such as the applicability to various inhomogeneous materials, the same simple form regardless of the complexity of the materials and no special treatment for problems with junctions. It should be noted that the VSIE solver has been applied to calculate the electromagnetic scattering by the composite anisotropic material and metal targets in previous work [11, 12].

In this paper, the VSIE solver is extended and applied for the computation of electromagnetic radiation from the microstrip antenna on an anisotropic substrate. A simple strip model is incorporated in the VSIE to simplify the analysis of the probe-fed microstrip antenna. Radiation properties of both the plane and the cylindrical conformal microstrip antenna with uniaxial substrate are calculated using the VSIE solver.

2. Theory

The microstrip antenna is discretized into triangular patches for metallic surfaces and tetrahedral cells for dielectric volume, respectively. The unknown surface currents on the metallic surface and the volume currents in the anisotropic dielectric substrate can be obtained by solving the following coupled volume-surface integral equation [12]:

$$\begin{aligned} \left(\mathbf{E}^i(\mathbf{r}) + \mathbf{E}^s(\mathbf{r})\right)\Big|_{\tan} &= 0, \quad \mathbf{r} \in S, \\ \bar{\bar{\varepsilon}}_r^{\,-1} \cdot \mathbf{D}_v(\mathbf{r}) &= \mathbf{E}^i(\mathbf{r}) + \mathbf{E}^s(\mathbf{r}), \quad \mathbf{r} \in V, \end{aligned} \tag{1}$$

where S and V denote the metallic surface and the dielectric body, respectively. $\mathbf{D}_v(\mathbf{r})$ is the electric flux density in V, while $\mathbf{E}^i(\mathbf{r})$ and $\mathbf{E}^s(\mathbf{r})$ are the electric field due to the applied source and the scattered field from the induced currents on S and in V, respectively. The metallic surface and dielectric body are meshed into triangular patches and tetrahedral cells, and then RWG and SWG basis and test functions are applied to expand the unknown surface current $\mathbf{J}_s(\mathbf{r})$ and

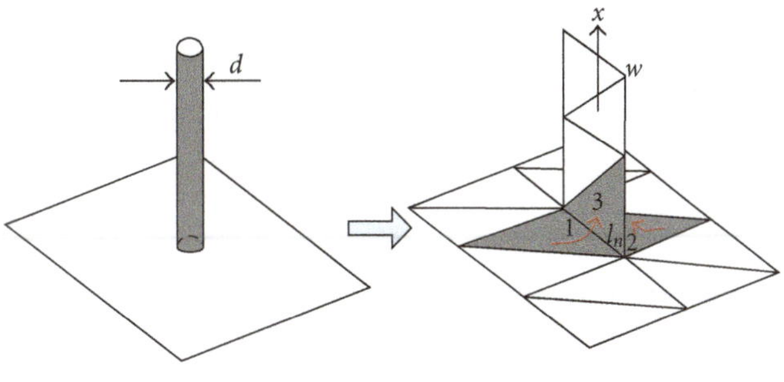

(a) (b)

FIGURE 1: A strip model in the excitation source of antenna. (a) Wire-surface junction. (b) Surface-surface junction model.

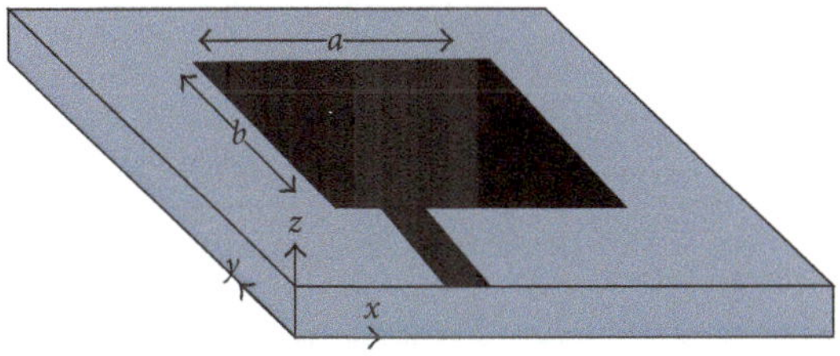

FIGURE 2: Geometrical structure of the planar microstrip antenna.

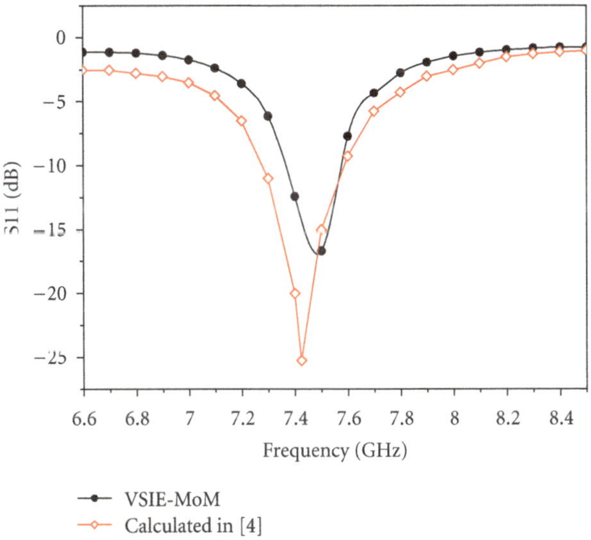

FIGURE 3: S-parameter of the microstrip patch antenna solved by MoM based VSIE solver, and compared with the results calculated in [4].

electric flux density $\mathbf{D}_v(\mathbf{r})$, respectively. $\overline{\overline{\varepsilon}}_r$ is the permittivity tensor of the dielectric material and given as

$$\overline{\overline{\varepsilon}}_r = \begin{bmatrix} \varepsilon_{xx} & 0 & 0 \\ 0 & \varepsilon_{yy} & 0 \\ 0 & 0 & \varepsilon_{zz} \end{bmatrix}. \tag{2}$$

The volume current $\mathbf{J}_v$ is related to the total electric flux density $D_v(\mathbf{r})$ by

$$\mathbf{J}_v = j\omega\overline{\overline{\kappa}}(\mathbf{r}) \cdot \mathbf{D}_v(\mathbf{r}), \tag{3}$$

where $\overline{\overline{\kappa}}(\mathbf{r})$ is the contrast ratio tensor defined as

$$\overline{\overline{\kappa}}(\mathbf{r}) = \left(\overline{\overline{\varepsilon}}_r - \overline{\overline{I}}\right) \cdot \overline{\overline{\varepsilon}}_r^{\,-1}. \tag{4}$$

The hybrid integral equation (1), togther, together with (2)–(4), can be solved by using Galerkin's MoM, and then RWG and SWG basis and test functions are applied to construct the MoM matrix, respectively. The detail derivation of the MoM matrix equation can be found in [11].

A simple equivalent strip model [5] is applied in the treatment of the feeding probe, where the wire-surface junction [13, 14], as shown in Figure 1(a), is changed to a surface-surface junction, as shown in Figure 1(b). The feeding probe of diameter d is replaced by its equivalent strip structure with the width $w = 2d$, as shown in Figure 1. The excitation source locates at the junction, and the $\mathbf{E}^i$ in equation can be established by using the delta-gap voltage model:

$$\mathbf{E}^i = V\delta(l)\mathbf{n}_l, \tag{5}$$

where V is the voltage across the gap and $\mathbf{n}_l$ is normal to the junction l_n.

3. Numerical Results

In this section, two numerical results will be shown. For the first example, We consider a planar microstrip antenna, as shown in Figure 2. The geometric parameters are taken from [4] and are as follows: the dimension of the patch is $a \times b = 12.45\,\text{mm} \times 16\,\text{mm}$, the substrate of the microstrip antenna is $28.1\,\text{mm} \times 32\,\text{mm} \times 0.794\,\text{mm}$, and the permittivity tensor is chosen as $\varepsilon_r = [2.31\ \ 0\ \ 0, 0\ \ 2.31\ \ 0, 0\ \ 0\ \ 2.19]$. (Refer to [4], in Figure 5, when $\theta = 0$.) The planar microstrip antenna is discretized into 1384 tetrahedral cells and 568 triangular patches, respectively, yielding a total of 4091 unknowns.

Figure 3 shows the S-parameter of the microstrip patch antenna solved by the presented method and compared with results in [4]. It is observed that good agreements are observed between the two methods. Moreover, the E-plane radiation pattern of antenna at the frequency of 7.5 GHz is obtained from the MoM and compared with the fine finite-element method- (FEM)- based solution, as shown in Figure 4. Again, good agreements are exhibited. In addition, the CPU time for both the MoM and the FEM-based solution is tested using the same computer, and the test results are 757 seconds and 771 seconds, respectively.

The second example considers a cylindrical conformal microstrip antenna and discusses the effects of the uniaxial substrate on the input impedance. The geometric parameters

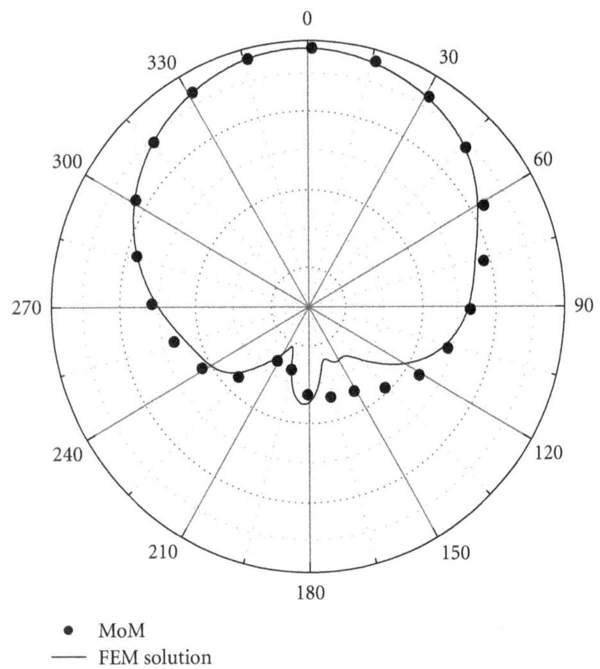

FIGURE 4: E-plane radiation pattern of microstrip antenna at $f = 7.5$ GHz.

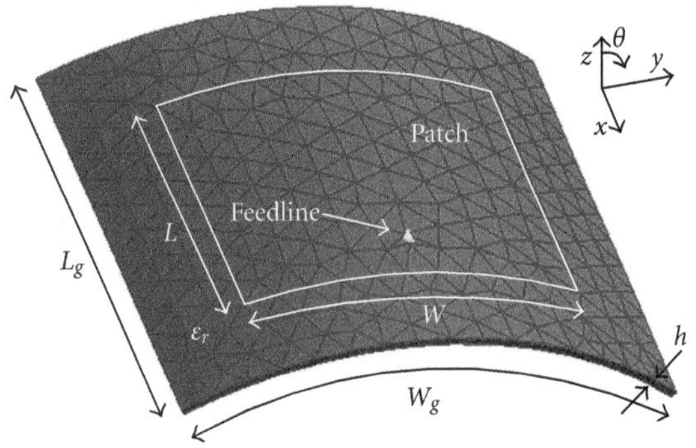

FIGURE 5: Geometry of a conformal microstrip antenna.

are taken from [5], as shown in Figure 5. The cylindrical-rectangular patch has a dimension of $L \times W = 30\,\text{mm} \times 40\,\text{mm}$ with a curvature radius of 50.795 mm, and the cylindrical-rectangular substrate is $L_g \times W_g \times h = 50\,\text{mm} \times 60\,\text{mm} \times 0.795\,\text{mm}$. The feeding point is on the centerline of the curved wide side and is 10 mm from the center of the patch, where the size of the rectangular strip feedline is $2\,\text{mm} \times 0.795\,\text{mm}$. The curved microstrip antenna is modeled with 600 triangular patches and 1340 tetrahedral cells, resulting in 4035 unknowns. The substrate is chosen the uniaxial anisotropic medium and the diagonalized permittivity tensor is taken in the form $\bar{\bar{\varepsilon}}_r = [\varepsilon_1\ 0\ 0,\ 0\ \varepsilon_1\ 0,\ 0\ 0\ \varepsilon_2]$.

Figure 6 shows the input impedance of the antenna versus different permittivity, where ε_1 is 2.32, and ε_2 is chosen as 1.8, 2.32, 2.8, respectively. It can be seen that the permittivity ε_2 has a significant influence on the impedance versus frequency variation, which implies that increasing the ε_2 decreases the resonant frequency of the antenna.

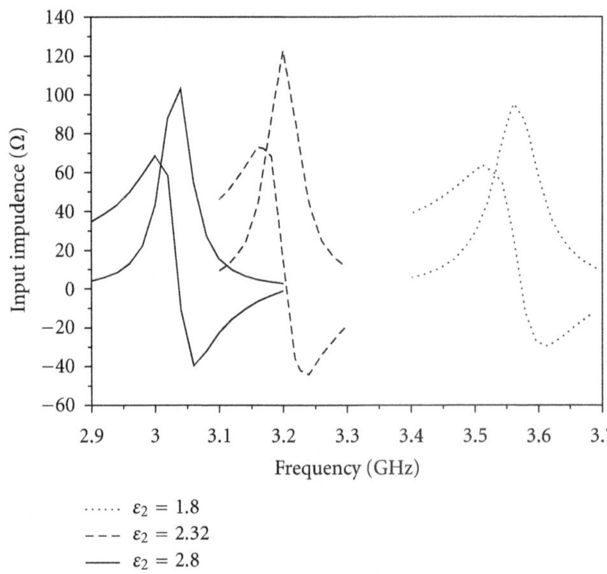

FIGURE 6: Input impedance of the cylindrical conformal microstrip patch antenna ($\varepsilon_1 = 2.32$, $\varepsilon_2 = 1.8, 2.32, 2.8$, resp.).

4. Conclusion

The VSIE solver has been applied to analyze the electromagnetic radiation from the microstrip antenna with arbitrary shaped anisotropic substrate. The targets are modeled using tetrahedral volume elements for substrate and triangle face elements for metal patches. The coupled volume-surface integral equations are derived by introducing tensor permittivity in conventional MoM. A simple strip model is used to simplify the analysis of the probe-fed microstrip antenna. Compared with the conventional MoM-based spectral domain analysis method, the VSIE solver approach offers good flexibility to model arbitrarily shaped microstrip antenna structures while keeping a good accuracy. The presented approach is formulated using the free-space Green's function. This feature makes it easy to apply the MoM-based fast algorithms to reduce the computational complexity of microstrip antenna with arbitrary shaped anisotropic substrate.

Acknowledgments

This work was supported by the Natural Science Foundation of Fujian Province of China (2011J01348) and the Science and Technique Major Program of Fujian Province (2010HZ0004-1).

References

[1] D. M. Pozar, "Radiation and scattering from a microstrip patch on a uniaxial substrat," *IEEE Transactions on Antennas and Propagation*, vol. 35, no. 6, pp. 613–621, 1987.

[2] C. S. Gurel and E. Yazgan, "Characteristics of a circular patch microstrip antenna on uniaxially anisotropic substrate," *IEEE Transactions on Antennas and Propagation*, vol. 52, no. 10, pp. 2532–2537, 2004.

[3] C. Zebiri, M. Lashab, and F. Benabdelaziz, "Rectangular microstrip antenna with uniaxial bi-anisotropic chiral substrate-superstrate," *IET Microwaves, Antennas and Propagation*, vol. 5, no. 1, pp. 17–29, 2011.

[4] A. P. Zhao, J. Juntunen, and A. V. Räisänen, "An efficient FDTD algorithm for the analysis of microstrip patch antennas printed on a general anisotropic dielectric substrate," *IEEE Transactions on Microwave Theory and Techniques*, vol. 47, no. 7, pp. 1142–1146, 1999.

[5] M. He, Q. Chen, Q. Yuan, K. Sawaya, and X. Xu, "A simple strip model in the volume-surface integral equation for analysis of arbitrary probe-fed conformal microstrip antennas," *IEEE Antennas and Wireless Propagation Letters*, vol. 8, pp. 530–533, 2009.

[6] N. Yuan, Tat-Soon Yeo, X. C. Nie, Y. B. Gan, and L. W. Li, "Analysis of probe-fed conformal microstrip antennas on finite grounded substrate," *IEEE Transactions on Antennas and Propagation*, vol. 54, no. 2, pp. 554–562, 2006.

[7] N. Yuan, T. S. Yeo, X. C. Nie, and L. W. Li, "RCS computation of composite conducting-dielectric objects with junctions using the hybrid volume-surface integral equation," *Journal of Electromagnetic Waves and Applications*, vol. 19, no. 1, pp. 19–36, 2005.

[8] X.-C. Nie, N. Yuan, L.-W. Li, Y.-B. Gan, and T. S. Yeo, "A fast volume-surface integral equation solver for scattering from composite conducting-dielectric objects," *IEEE Transactions on Antennas and Propagation*, vol. 53, no. 2, pp. 818–824, 2005.

[9] W. B. Ewe, L. W. Li, and M.-S. Leong, "Fast solution of mixed dielectric/conducting scattering problem using volume-surface adaptive integral method," *IEEE Transactions on Antennas and Propagation*, vol. 52, no. 11, pp. 3071–3077, 2004.

[10] S. M. Rao, T. K. Sarkar, P. Midya, and A. R. Djordevic, "Electromagnetic radiation and scattering from finite conducting and dielectric structures: surface/surface formulation," *IEEE Transactions on Antennas and Propagation*, vol. 39, no. 7, pp. 1034–1037, 1991.

[11] J. Yuan and C. Gu, "A volume-surface integral equation solver for scattering from microstrip antenna on anisotropic substrate," in *Proceedings of the 3rd IEEE International Symposium on Microwave, Antenna, Propagation and EMC Technologies for Wireless Communications (MAPE '09)*, pp. 1083–1085, October 2009.

[12] J. Yuan, Z. Niu, Z. Li, and C. Gu, "Electromagnetic scattering by arbitrarily shaped PEC targets coated with anisotropic media using equivalent dipole-moment method," *Journal of Infrared, Millimeter, and Terahertz Waves*, vol. 31, no. 6, pp. 744–752, 2010.

[13] W. J. Zhao, J. L. W. Li, and L. Hu, "Efficient current-based hybrid analysis of wire antennas mounted on a large realistic aircraft," *IEEE Transactions on Antennas and Propagation*, vol. 58, no. 8, pp. 2666–2672, 2010.

[14] W.-B. Ewe, L.-W. Li, C. S. Chang, and J.-P. Xu, "AIM analysis of scattering and radiation by arbitrary surface-wire configurations," *IEEE Transactions on Antennas and Propagation*, vol. 55, no. 1, pp. 162–166, 2007.

11

Implementation of an Innovative Method to Design Reflectarray Antennas

Abdelhamid Tayebi, Josefa Gómez, Iván González, and Felipe Cátedra

Electromagnetic Computing Group, Computer Science Department, University of Alcalá, Escuela Politécnica, Alcalá de Henares, 28871 Madrid, Spain

Correspondence should be addressed to Abdelhamid Tayebi, hamid.tayebi@uah.es

Academic Editor: Sandra Costanzo

A novel computed aided technique for designing reflectarray antennas is presented. The developed approach automatically generates the geometrical model of reflectarray antennas taking into account some input parameters, such as, the unit cell type and dimensions, frequency, focal length, periodicity, dielectric materials, and desired main beam radiating direction. The characteristics of the reflecting elements are selected considering the spatial phase delay at each unit cell to achieve a progressive phase shift. The implemented procedure also provides the phase characteristic of the unit element, which is rapidly computed by using a parallelized Moment Method (MoM) approach. The MoM is also used to obtain the radiation pattern of the full reflectarray antenna. In order to evaluate the new technique, a dual-interface prototype has been designed and simulated showing high-performance capability.

1. Introduction

Reflectarray antennas [1] have become an attractive alternative to the traditional parabolic reflector antennas due to some advantages, such as, light weight, mechanical robustness, compatibility with active devices, simple manufacturing, low profile, and low cost. Moreover, reflectarray antennas offer the possibility of beam steering, like conventional phased arrays, but eliminate the complexity and losses of the feeding network, showing a higher efficiency. Their main limitation is the narrow-band behavior, which is inherent to the printed antenna structures. To overcome this shortcoming, some proposals based on multilayer configurations [2, 3] and concentric rings [4, 5] have been reported lately. For instance, the design reported in [2] is composed of microstrip stacked patches and shows good performance in dual frequency operation. The reflectarray presented in [3] consists of three stacked layers with rectangular patches and provides dual-polarization operation for space applications. Although the bandwidth is notably increased when using multilayer printed patches, other issues may arise, such as, the manufacturing complexity and higher weight and losses. On the other hand, reflectarrays based on concentric rings are also becoming very popular due to they provide excellent results using one-layer configurations. The schemes proposed in [4] achieve a significant improvement in gain-bandwidth using double square and double cross loop elements. A comparable elementary cell that contains double square and double circular rings is studied in [5], where the phasing characteristic of the single-layer cell is computed showing good results.

Reflectarray antennas are composed of one or more layers of metallic elements and dielectric substrates over a ground plane. They are typically fed by a horn antenna and their operation consists of converting a spherical wavefront into a desired scattered wavefront using a phase shift mechanism. Although it has been demonstrated that reflectarray antennas can be used to generate contoured beams [6], most of them are still designed to radiate a collimate wave front at a certain frequency. To achieve this, according to Figure 1, each

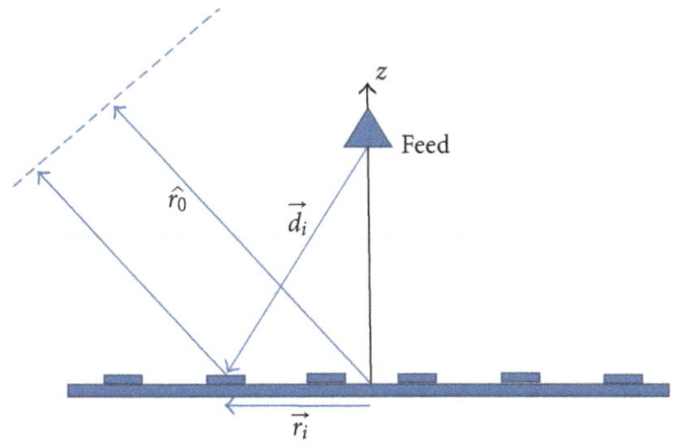

FIGURE 1: Diagram of the reflectarray antenna.

periodic cell must introduce a phase shift that is obtained as follows:

$$\phi_i = k_0(d_i - \vec{r_i} \cdot \hat{r_0}) + 2\pi N; \quad N = 0, 1, 2, \ldots, \tag{1}$$

where k_0 is the propagation constant in vacuum, $\hat{r_0}$ is the unit vector in the desired direction of the main beam, $\vec{r_i}$ is the position vector from the center of the reflectarray plane to the ith radiating element, and d_i is the distance from the feed to the ith element.

Hence, the resonant size, periodicity, orientation, and substrate's properties of every cell must be properly selected to provide the phase shift given by (1).

Due to the radiating elements must reflect the incident electromagnetic field with a progressive phase distribution, the most challenging aspect when designing reflectarray antennas is to find the adequate phase shift to compensate the phase delay caused by the distance from the feed to each unit cell. Thus, the requirement of having a wide database that associates the properties of the cells and their correspondent phase shifts is crucial. In fact, many works proposing different techniques to achieve a wide phase range of the reflection coefficient have been recently published. In the literature, mainly four mechanisms can be found. The first one is the easiest to manufacture and more used [7, 8]. It achieves the phase variation by changing the size of the internal geometry of the unit cells. The second type tries to find the proper phase shift by adjusting the rotation angle of the elements [9]. The third method to provide the progressive phase shift of the reflected wave consists of using variable stub length attached to the radiating elements [10]. Finally, the fourth mechanism consists in using multilayer configurations [2, 3]. The proposed method applies these four techniques, since it is able to find the suitable characteristics of each elementary cell.

Other important aspect is the adequate choice of the radiating element's shape, since the scatter of the incident field strongly depends on its geometry. Hence, an essential goal is to obtain the type of cell that provides a wide reflection coefficient phase range when varying its properties. In fact, it is well-known that a reduced phase range lower than 360° implies a decreasing in directivity. On the other hand, it has been demonstrated [11, 12] that the phase variation depending on any geometrical parameter should be as lineal as possible, providing a smooth slope. This implies that good results are provided only if the 360° range is achieved by varying the considered parameter extremely slowly. In practical situations, the consequences of abrupt variations in the phase are narrower bandwidths and a more complicated manufacturing of the radiation elements due to the minimum tolerances.

Normally, the reflection coefficient phase as a function of the patch size is calculated to determine whether the phase range is wide enough and the curve slope is smooth enough. Thus, a lot of effort is usually required to find a unit cell able to provide the mentioned phase characteristic. The traditional shapes of microstrip patches have been crosses, squares, rectangles, rings, and circles. In this paper, the double square ring has been chosen to prove the efficiency of the developed computer tool.

A new method to automatically design reflectarray antennas is proposed. The properties of the radiating elements, including their resonant dimensions and substrate's characteristics, are computed by means of an efficient process that compensates the phase shift introduced by each elementary cell. Although several works have been recently carried out to investigate the phase characteristics of the cell elements, most of them do not report the reflectarray design once the phase curve has been properly achieved. The contribution of this paper also relies on the final design of the reflectarray, besides implementing the new approach and obtaining the widely used phase curve of the elementary cell. The proposed approach has been included in the computer tool NEWFASANT [13] to validate its performance.

The paper is organized as follows. The procedure to obtain the geometrical model of the reflectarray is described in Section 2, the design of a reduced-scale prototype is stated in Section 3, and finally some conclusions are remarked in Section 4.

2. Design Process

The basic idea of the new technique is the automatic design of reflectarray antennas by varying the parameters of the unit cells. The variation may include the resonant size of the cells, their position, orientation, stub's length, periodicity, number of layers, dielectric materials, and so forth, so that the phase shift given by every element is compensated according to the phase of the reflection coefficient.

First, a database which stores the values of the modified parameters and its correspondent reflection coefficient phase is created at a given frequency and for a determined periodicity. It is important to highlight that the analysis can be conducted by varying one or more parameters. Each entry in the database that associates pairs of parameter values and reflection phases is obtained by analyzing the behavior of a quasi-infinite array of identical elements by applying a full wave method based on the MoM [14], so truncation effects due to the finite size of real reflectarrays are also considered. Despite the big computational effort required to obtain the whole database, it is affordable in time because the MoM code is parallelized. A normal incident plane wave

FIGURE 2: Top view of the reflectarray antenna.

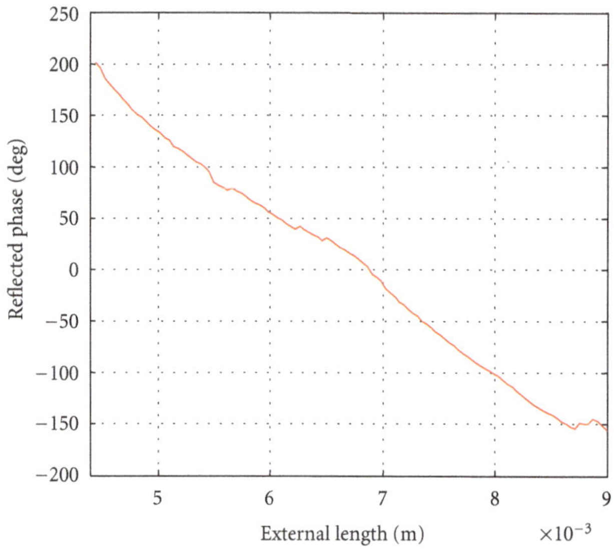

FIGURE 3: Phase characteristic of the unit cell varying the length of the external square ring.

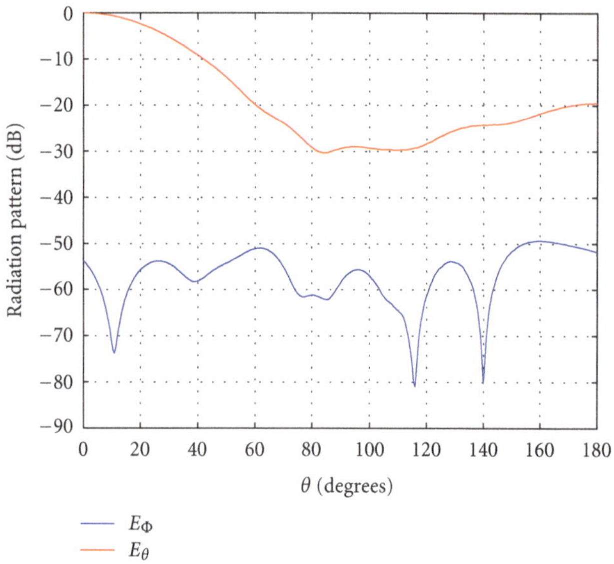

FIGURE 4: Radiation pattern of the feed antenna at 20 GHz.

is considered to compute the reflection coefficient phases. However, due to the fact that the radiation of the real feed antenna has an incidence direction with a certain angle, a further refinement can be done considering some incidence angles of the planar wave. In that case, the tool is able to provide as many databases as incidence angles are indicated.

Then, the phase delay caused by the path difference from the feed point is calculated taking into account the position of each unit cell. Once all phase delays are obtained, the system searches in the database the proper dimensions to compensate the required phase at every element. Hence, each cell can properly adjust their phase delay, introducing no errors and obtaining the maximum directivity. The phase shift needed in each cell when using several incident angles is obtained from one of those databases depending on its position and the feed location. The results provided considering several incident angles are slightly more accurate than only considering normal incidence. The next step consists in creating the geometrical model of the reflectarray, according to the values provided by the database. Finally, the ground plane is correctly located under the cells conforming to the dielectric thicknesses. By default, the reflectarray is created in the XY plane, but it is possible to fit it to curved surfaces. The user is also able to define the shape of the reflectarray, which can be rectangular, square, circular, or elliptical. When the reflectarray has been designed and we have the antenna geometry, the MoM is applied again to obtain a rigorous solution considering the full reflectarray antenna.

The flexibility when designing reflectarray antennas by using the presented tool is enormous, since many combinations of different shapes, dimensions, diverse substrates, incidences, and so forth can be used to design the radiating elements. In addition, the configuration of the unit cell is not constrained to a single layer. In fact, the tool provides a great versatility and allows defining various layers of metallic patches and interfaces of different substrates. Moreover, in order to facilitate the use of the implemented tool, several databases that contain correspondences between dimensions and reflection coefficient phases of many elementary cells (rectangular, circular and square patches, square and circular rings, metallic crosses, patches with holes, etc.) are available.

The module includes three techniques to model the feed antenna. The first one allows importing the radiation pattern of a real horn antenna. The second one allows importing the geometrical model of the feed antenna in several CAD formats, and the third one allows creating the physical model of the feed antenna.

3. Validation of the Implemented Tool

In order to validate the performance of the presented method, the specifications of an existing design recently published [15] have been considered. The reflectarray has

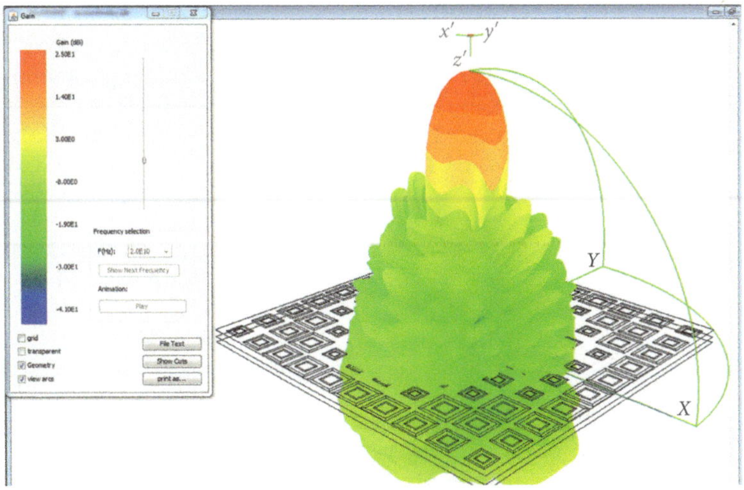

FIGURE 5: Radiation pattern in 3D at 20 GHz.

been redesigned using the propose procedure that adjusts the dimensions of the patches to focus or the main beam in the normal direction at 20 GHz. In our case, it is assumed that the antenna is composed of 100 identical unit cells symmetrically located in a square grid with periodicity of 9 mm. The top view is depicted in Figure 2.

Figure 3 shows the phase variation versus the length of the external square ring that has been automatically provided by our tool, [13]. The complete range of 360° is achieved by varying the length of the external square ring from 4.5 mm to 9 mm. It can be observed that the slope is practically lineal. These results have been achieved using a 2.95 mm substrate of foam under the metallic rings and a 0.85 mm layer of a material whose relative dielectric constant is 2.5 (same multilayer structure than in [15]). This thin layer is located on the cells, acting as a protective surface. On the other hand, the width of both rings and the gap between them have been fixed to 0.3 mm independently of the length variation.

As in [15], the reflectarray antenna is fed by means of a linearly polarized pyramidal horn located at $x = 0$, $y = 0$, $z = 0.0712$ m. The normalized radiation pattern of the fed antenna is depicted in Figure 4 at the central frequency of 20 GHz.

On the other hand, Figure 5 shows the 3D radiation pattern of the reflectarray antenna we have redesigned, where the main beam radiating in the normal direction is clearly depicted. Regarding the polarization purity, Figure 6 shows the cut $\varphi = 0°$ of the normalized radiation pattern at 20 GHz, where it can be observed that the cross-polar component is lower than −30 dB for every θ value. The second side lobes are also quite low, under −20 dB.

Finally, Figure 7 depicts the gain provided by the redesigned reflectarray. Note that the maximum value is 23.5 dBi, which is higher than the value of 20 dBi achieved in [15]. It is important to point out that the analysis of the final reflectarray has been conducted by using the MoM module of NEWFASANT, although it could have been used a different electromagnetic solver. Since reflectarray antennas are electrically very large structures, the multilevel fast

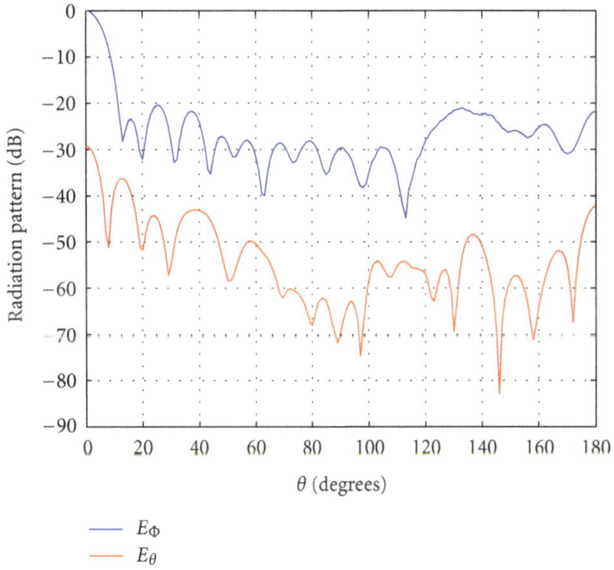

FIGURE 6: Normalized radiation pattern. Cut $\varphi = 0°$, frequency = 20 GHz.

multipole algorithm is used in the full wave analysis to reduce the required memory and time resources.

4. Conclusion

A powerful tool to automatically design reflectarray antennas has been presented. The reported technique is able to generate the geometrical model of the antenna considering some input parameters, such as, the unit cell type, the operating frequency, the focal length, the periodicity, and the desired main beam radiating direction. The tool also computes and provides the phase curve of the unit cell. To evaluate the performance of the new method, a recently published reflectarray has been redesigned. Excellent results

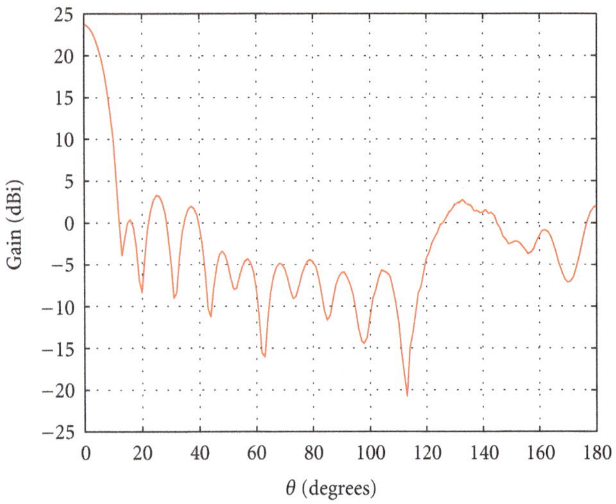

FIGURE 7: Gain. Cut $\varphi = 0°$, frequency = 20 GHz.

in terms of gain and polarization purity have been achieved. The developed tool can be very helpful for researchers.

Acknowledgments

This work has been supported, in part by the Comunidad de Madrid Project S-2009/TIC1485 and by the Castilla-La Mancha Project PPII10-0192-0083, by the Spanish Department of Science, Technology Projects TEC 2010-15706 and CONSOLIDER-INGENIO No CSD-2008-0068.

References

[1] D. Berry, R. Malech, and W. Kennedy, "The reflectarray antenna," *IEEE Transactions on Antennas and Propagation*, vol. 11, no. 6, pp. 645–651, 1963.

[2] J. A. Encinar, "Design of a dual frequency reflectarray using microstrip stacked patches of variable size," *Electronics Letters*, vol. 32, no. 12, pp. 1049–1050, 1996.

[3] G. Zhao and Y.-C. Jiao, "Broadband dual-polarization dual-coverage reflectarray antenna," in *Proceedings of the IEEE 4th International Symposium on Microwave, Antenna, Propagation, and EMC Technologies for Wireless Communications (MAPE '11)*, pp. 102–105, November 2011.

[4] M. R. Chaharmir, J. Shaker, M. Cuhaci, and A. Ittipiboon, "A broadband reflectarray antenna with double square rings as the cell elements," in *Proceedings of the European Conference on Antennas and Propagation (EuCAP '06)*, pp. 1–4, November 2006.

[5] M. E. Bialkowski and K. H. Sayidmarie, "Phasing characteristics of variable size double rings of square or circular shape for design of a single layer microstrip reflectarray," in *Proceedings of the 17th International Conference on Microwaves, Radar and Wireless Communications (MIKON '08)*, pp. 1–4, May 2008.

[6] J. A. Encinar, L. S. Datashvili, J. A. Zornoza et al., "Dual-polarization dual-coverage reflectarray for space applications," *IEEE Transactions on Antennas and Propagation*, vol. 54, no. 10, pp. 2828–2837, 2006.

[7] J. A. Encinar, "Analysis and design of dual frequency reflectarrays using microstrip stacked patches of variable size," in *Proceedings of the 26th European Microwave Conference*, vol. 1, pp. 221–224, September 1996.

[8] J. A. Encinar, "Design of two-layer printed reflectarrays using patches of variable size," *IEEE Transactions on Antennas and Propagation*, vol. 49, no. 10, pp. 1403–1410, 2001.

[9] M. E. Cooley, J. F. Walker, D. G. Gonzalez, Pollon, and G. E, "Novel reflectarray element with variable phase characteristics," *IEE Proceedings*, vol. 144, no. 2, pp. 149–151, 1997.

[10] R. D. Javor, X. D. Wu, and K. Chang, "Design and performance of a microstrip reflectarray antenna," *IEEE Transactions on Antennas and Propagation*, vol. 43, no. 9, pp. 932–939, 1995.

[11] M. E. Bialkowski and K. H. Sayidmarie, "Investigations into phase characteristics of a single-layer reflectarray employing patch or ring elements of variable size," *IEEE Transactions on Antennas and Propagation*, vol. 56, no. 11, pp. 3366–3372, 2008.

[12] P. Krachodnok and R. Wongsan, "Phase-bandwidth enhancement of microstrip patch reflectarray with cross slot loads," in *Proceedings of the Asia-Pacific Microwave Conference (APMC '07)*, pp. 1–4, December 2007.

[13] http://www.fasant.com/.

[14] F. Cátedra, C. Delgado, E. García, and I. González, "Combination of MLFMA and an iterative method to compute large scattering or radiation problems," in *Proceedings of the 4th European Conference on Antennas and Propagation (EuCAP '10)*, Barcelona, Spain, April 2010.

[15] B. Ma, L. Yan, M. Xia, C. Wang, and L. Cui, "Design of reflectarray antenna using double square ring elements," in *Proceedings of the 2nd IEEE International Conference on Microwave Technology and Computational Electromagnetics (ICMTCE '11)*, pp. 213–216, May 2011.

12

Recent Developments of Reflectarray Antennas in Dual-Reflector Configurations

Carolina Tienda,[1] Manuel Arrebola,[2] and José A. Encinar[1]

[1] *Department of Electromagnetism and Circuit Theory, Technical University of Madrid, 30 Complutense Avenue, 28040 Madrid, Spain*
[2] *Department of Electrical Engineering, Group of Signal Theory and Communications, University of Oviedo, Academic Campus, 33203 Gijón (Asturias), Spain*

Correspondence should be addressed to Carolina Tienda, carotienda@gmail.com

Academic Editor: Raphaël Gillard

Recent work on dual-reflector antennas involving reflectarrays is reviewed in this paper. Both dual-reflector antenna with a reflectarray subreflector and dual-reflectarrays antennas with flat or parabolic main reflectarray are considered. First, a general analysis technique for these two configurations is described. Second, results for beam scanning and contoured-beam applications in different frequency bands are shown and discussed. The performance and capabilities of these antennas are shown by describing some practical design cases for radar, satellite communications, and direct broadcast satellite (DBS) applications.

1. Introduction

Dual-reflector antennas are used to reduce the cross-polarization and the overall antenna volume in many applications, particularly in space applications. The use of a reflectarray as sub- or main reflector in a dual-reflector configuration [1] can provide some advantages in the antenna performance and in the manufacturing process. For example, a parabolic reflector with a reflectarray as subreflector has been proposed for compensating the errors on the surface of very large deployable reflectors [2, 3], for beam scanning in a limited angular range [4] and contoured beam applications [5]. This antenna configuration, based on parabolic main reflector and a reflectarray subreflector, offers several advantages compared to a single reflectarray antenna. First of all, it can provide a broad-band performance for large apertures, by using a reflectarray subreflector of small electrical dimensions and a large parabolic reflector. In addition, this dual-reflector configuration can be used to scan or reconfigure the beam by electronically controlling the phase-shift at the elements of a limited-size subreflectarray [4]. These configurations can be used for active or reconfigurable antennas by inserting amplifier modules or phase shifters on the elements of the reflectarray subreflector, in order to achieve high-power transmission or fine beam-pointing adjustment [6].

If the main reflector is replaced by a flat reflectarray, a dual-reflectarray antenna is obtained [7, 8]. This configuration provides phase control on both surfaces, which can be used for different proposals as amplitude and phase synthesis [9], or to reduce the crosspolar radiation and shape the beam. The two-surface control permits to improve the performance of the antenna and extend the applicability of reflectarrays. Moreover this configuration provides advantages for transportable antenna application since the two flat arrays can be easily folded and deployed [10]. In a dual-reflectarray antenna, active control devices can be included in both reflectarrays, being the subreflectarray preferred because of their smaller size. Finally, a dual-reflectarray antenna comprising a parabolic main reflectarray can be used to provide beam shaping in a broad band [11].

These antenna configurations including a controllable reflectarray subreflector can present a great potential in several space applications, such as Synthetic Aperture Radar (SAR), remote sensing, and reconfigurable antennas for direct broadcast satellites (DBSs). Although an SAR antenna

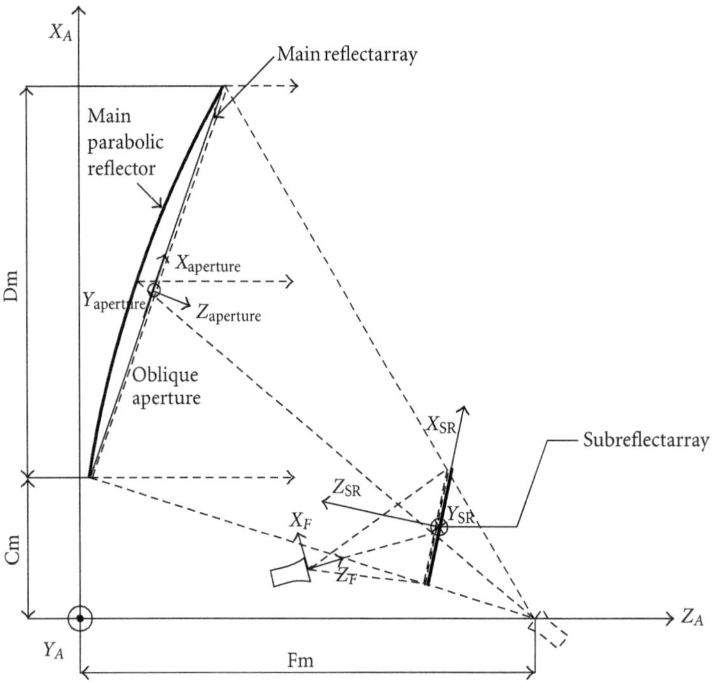

Figure 1: Scheme of dual-reflector antenna with a reflectarray subreflector and with a parabolic main reflector or a flat reflectarray as main reflector.

is usually implemented by active arrays, a reconfigurable reflectarray subreflector can provide beam scanning at a moderate cost and complexity. A second application is in radiometric remote sensing in millimeter and sub-millimeter frequency bands, which requires a beam scanning in a limited range actually implemented by motors [12]. An alternative for electronic beam scanning in this frequency bands can be reflectarrays based on liquid crystals [13–15]. Finally, the reflectarray subreflector can be used to shape the beam in contoured-beam DBS antennas. In this configuration, the beam can be shaped by adjusting the phase-shift distribution on the subreflectarray, while the main parabolic reflector focuses the beam without any bandwidth limitation. If the phase is electronically controlled on the subreflectarray by using MEMS [16, 17] or varactor diodes [18], the beam can be reconfigured to change the coverage during the life-time of the satellite. This aspect is a real need in telecommunications and DBS missions, since the demand in data traffic (TV broadcast, phone, video or data transmission, multimedia direct-to-home, etc....) can change during the 15–20 year lifetime of the satellites.

2. Analysis of Dual-Reflector Antennas Involving Reflectarrays

As shown in Figure 1, the proposed antenna configuration is made up of three components: a primary feed, a flat subreflectarray, and a main reflector that can be either a parabolic classic reflector or a flat or curved reflectarray. Figure 1 shows a dual-offset optics using a horn antenna as primary feed, however a centered configuration can also be analysed. In the analysis of the dual-reflector antenna involving one or two reflectarrays, four coordinate systems are considered. Two of these systems are associated to two components of the antenna: the feed reference system (X_F, Y_F, Z_F), which is centered at the phase centre of the primary feed; and the subreflectarray reference system, defined by (X_{SR}, Y_{SR}, Z_{SR}), whose origin is placed at the centre of the subreflectarray surface. The third coordinate system is used to define the aperture for the computation of the radiated field: the aperture reference system $(X_{aperture}, Y_{aperture}, Z_{aperture})$, which is placed at the centre of the antenna aperture. All these coordinate systems are referred to as a common one, the antenna coordinate system (X_A, Y_A, Z_A), with the Z_A axis on the direction of the beam, as usual in reflector antennas.

This antenna can be analyzed through a modular procedure in four steps. First, the field radiated by the primary feed horn should be evaluated and the incident field must be computed at the centre of each subreflectarray cell. Although, in some cases, a conventional $\cos^q(\theta)$ function can be used to model the field radiated by the feed within a good accuracy [19]. In general, the subreflectarray elements are placed in the Fresnel zone of the primary feed, and therefore the near field of the feed must be considered, which can be obtained through full-wave simulations, measurements of the feed or spherical mode expansion of the field radiated by the feed horn, [20, 21]. Dual linear polarization is considered, vertical and horizontal, which corresponds to X_F and Y_F directions, respectively, according to Figure 1.

In the second step, the subreflectarray is analysed to compute the field reflected by each reflectarray cell. This analysis is carried out element by element through the general technique implemented for multilayer reflectarrays

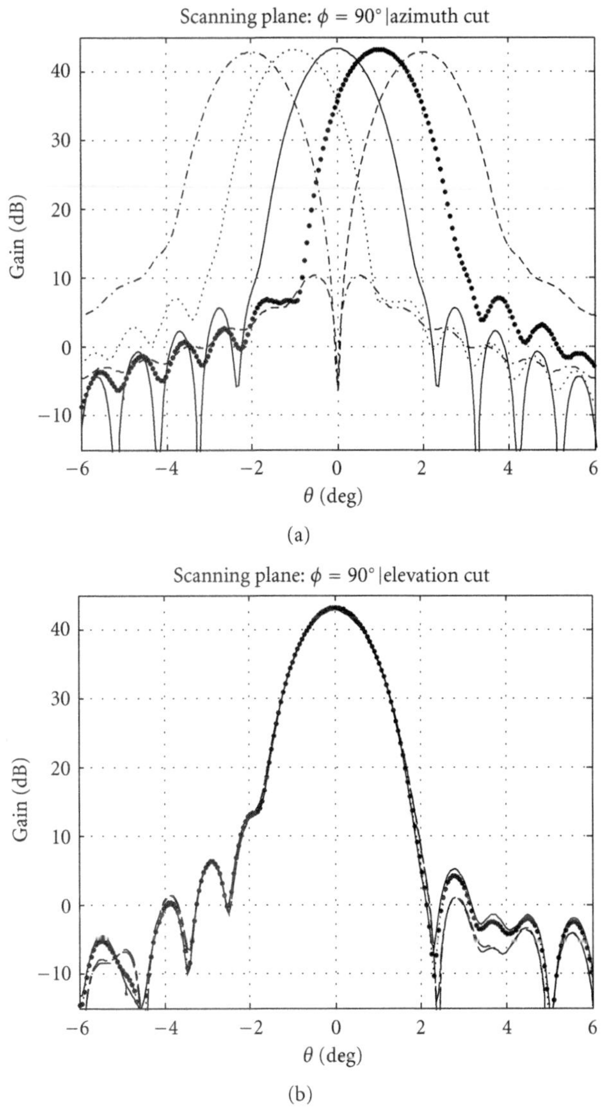

FIGURE 2: Simulated patterns of the dual-reflector antenna at 11.95 GHz for beam scanning in azimuth in the principal planes. (a) Azimuth plane, (b) elevation plane.

[21–23]. The technique is based on Method of Moments in Spectral Domain, considering local periodicity and the real angle of incidence on each periodic cell of the wave coming from the feed, which has been demonstrated to be accurate for reflectarrays in single-reflector configurations. All the field components are taken into account, and the incident and reflected fields on each reflectarray cell, defined by the subindexes (p, q), are related through a reflection matrix $\mathbf{P}^{p,q}$ as follows:

$$E_{\text{ref}}(p,q) = \mathbf{P}^{p,q} \cdot E_{\text{in}}(p,q). \quad (1)$$

The matrix $\mathbf{P}^{p,q}$ depends on the incidence angle and its complex elements relate the Cartesian components of the incident and the reflected fields, considering the cross-polarization produced by printed elements:

$$\mathbf{P}^{p,q} = \begin{pmatrix} \rho_{xx}^{p,q} & \rho_{yx}^{p,q} \\ \rho_{xy}^{p,q} & \rho_{yy}^{p,q} \end{pmatrix}. \quad (2)$$

Third, the analysis of the main reflector is carried out. In case of considering a classic parabolic reflector as main, the analysis will be based on physical optics. Thus, the electric printed currents on the reflector surface are computed, adding the contributions of the field radiated by all the subreflectarray elements. Once the currents are computed, the field is projected on the aperture plane through the Jacobian transformation.

In the case of having a flat reflectarray main reflector, it is analyzed using the same technique as for the subreflectarray, but we have to compute first the incident field on the elements of the main reflectarray. Now, the cells of the main reflectarray are placed in the Fresnel zone of the subreflectarray, but in the far-field region of the subreflectarray cells which are small apertures. Thus, the computation of the incident electric field on the main reflectarray surface can be computed as the superposition of all the contributions from the subreflector elements. In this case, the reflectarray surface is considered as the aperture plane. In the case that the main reflectarray is printed on a curved surface, the same procedure is applied but using a local coordinate system on each element of the main curved reflectarray and analyzing each element, assuming a locally flat periodic environment.

In the fourth step the radiation pattern of the entire antenna is computed starting from the electric field on the antenna aperture (aperture of the main reflector), using an algorithm based in FFT. Because of the analysis method, the radiation patterns are computed taking into account the ohmic losses in the reflectarray, the illumination, and spillover efficiencies.

3. Applications

Dual-reflector antennas involving reflectarrays have been identified as possible solutions for different applications. First, they can be used in beam-scanning applications since the beam pointing can be achieved with a small electronically-reconfigurable subreflectarray. In this applications, the use of a reflectarray as main reflector can be used to improve the folding and deploying performance of the antenna, which is very convenient for transportable antennas. Second, a reflectarray subreflector can be used to synthesize a contoured beam for DBS applications. These beams can also be achieved with a dual-reflectarray antenna using a main parabolic reflectarray to appropriately shape the beam in a broad frequency band and a passive subreflectarray, which can be used to improve the cross-polarization of the antenna. Finally, a reflectarray subreflector has been proposed to compensate the phase error introduced by main parabolic reflector surface because of the manufacturing tolerances [2].

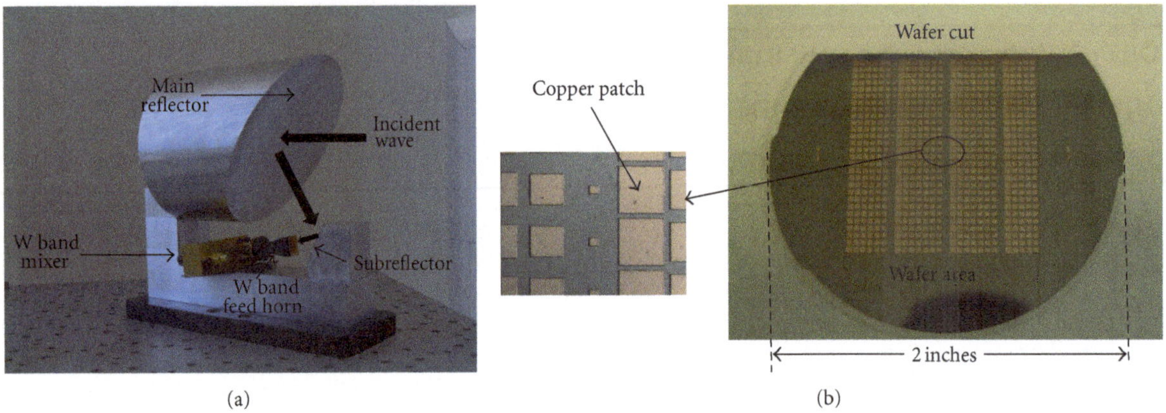

FIGURE 3: Dual reflector antenna at 94 GHz. (a) Photograph and (b) reflectarray subreflector. (From [21] © 2009 IEEE).

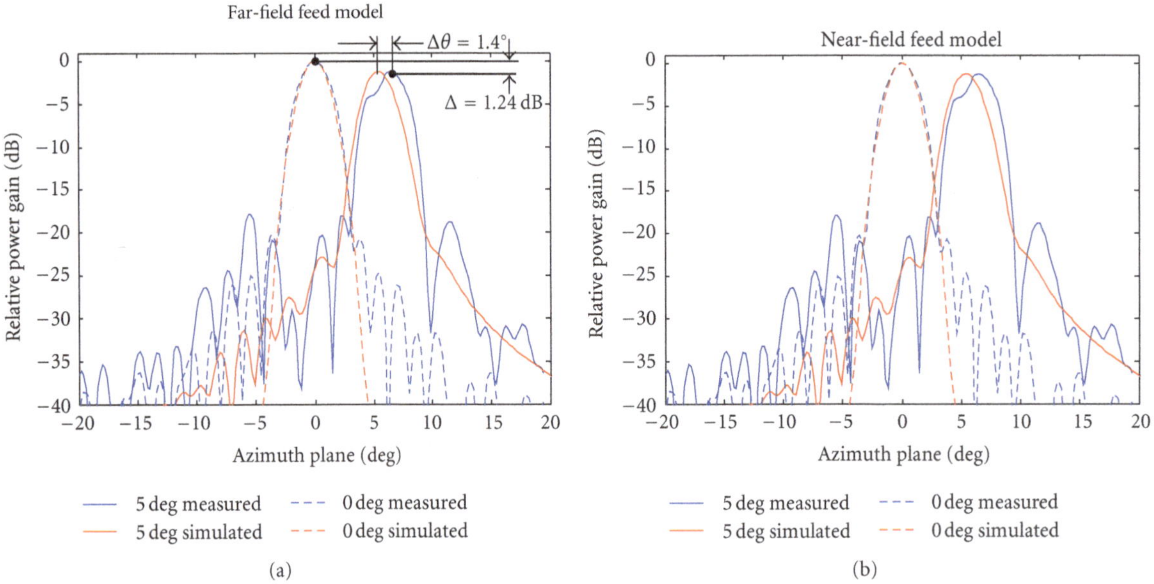

FIGURE 4: Measured and simulated radiation patterns of the dual reflector antenna with reflectarray subreflector in the azimuth plane at 94 GHz, using (a) far-field and (b) near-field models of the feed horn. (From [21] © 2009 IEEE).

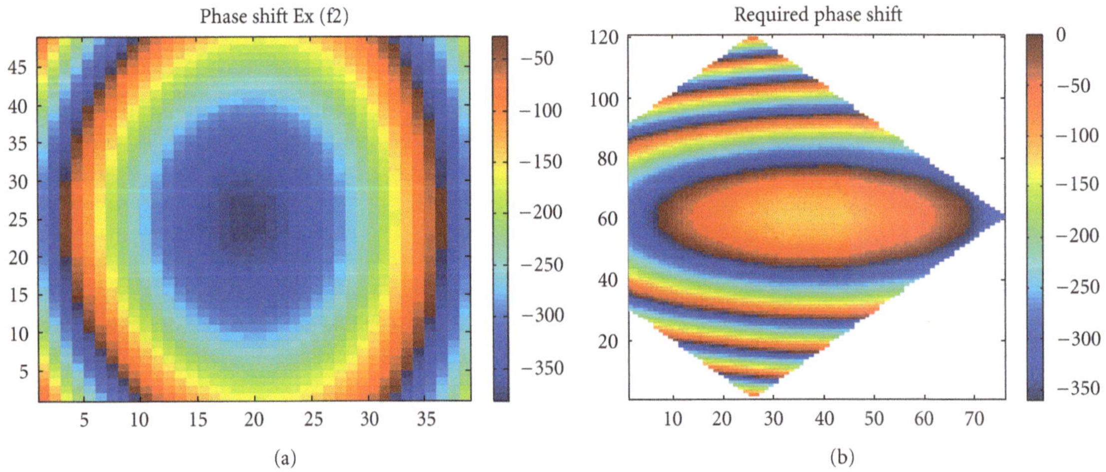

FIGURE 5: Phase distribution to be implemented on the subreflectarray (a) and on the main reflectarray at 12.75 GHz (b).

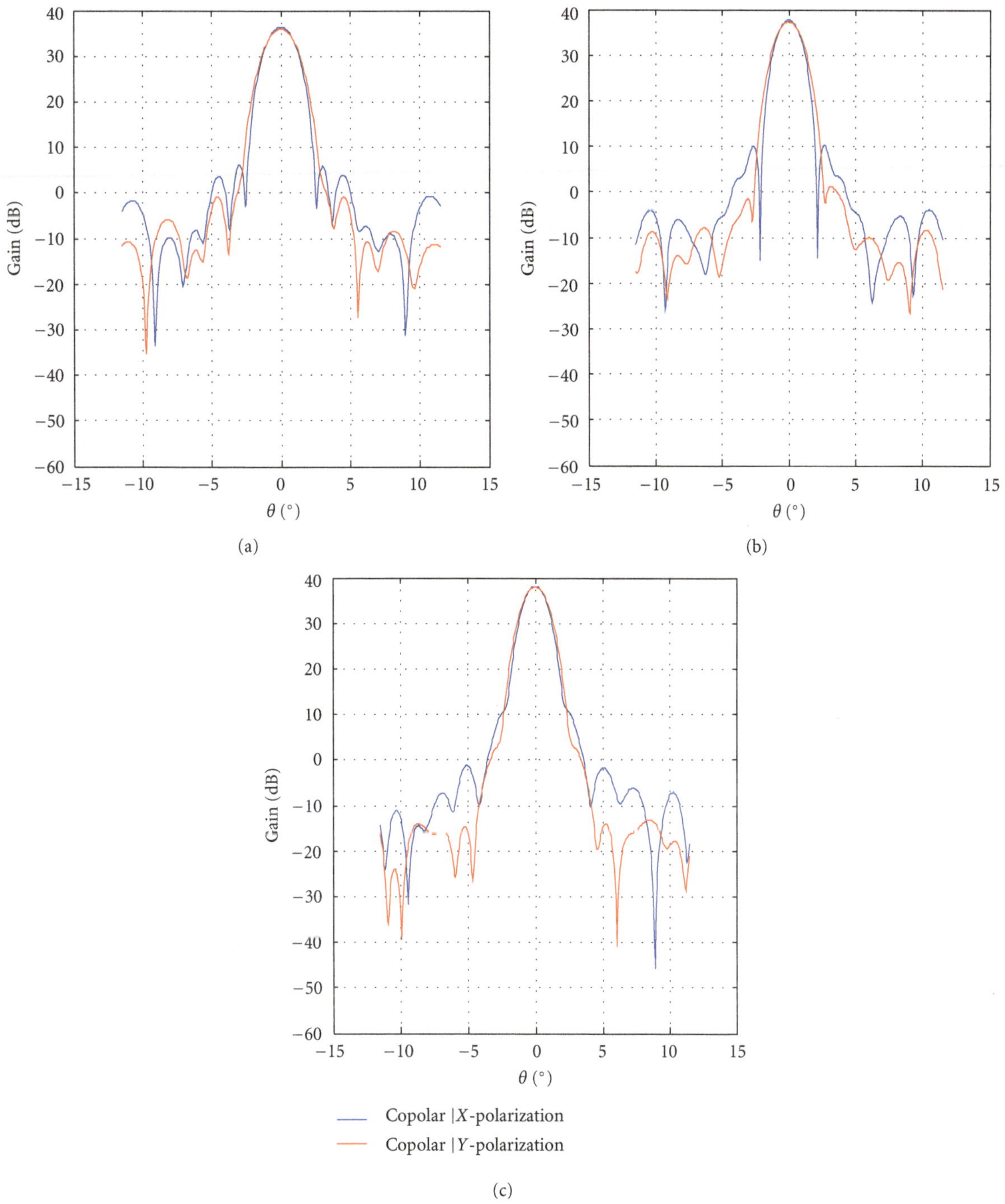

FIGURE 6: Azimuth cut of radiation pattern for X-polarisation at 10.7 GHz (a), 12.75 GHz (b), and 14.25 GHz (c) when using the required phase distribution on the main reflectarray.

3.1. *Beam-Scanning Applications*

3.1.1. Dual-Reflector Antenna with Reflectarray Subreflector. The dual-reflector antenna with a reflectarray subreflector has been proposed as a high-gain beam-scanning antenna by adding controllable phase shifter on the subreflectarray. Thus, the beam pointing can be controlled through a reduced number of controls. In [4], the beam scanning performance of an antenna made of a 1.5 m aperture parabolic reflector and a rectangular subreflectarray (520 mm × 494 mm) at 11.95 GHz was studied in an angular range of ±2° from the antenna boresight. In this example, the feed horn is modeled as a $\cos^q(\theta)$ function, with the q-factor chosen to provide an illumination taper on the subreflectarray around −18 dB, from the centre to the edges. The phase center of the horn is placed at coordinates $(-194, 0, 326)$ mm referring to the subreflectarray coordinate system (X_R, Y_R, Z_R), and the center of the subreflectarray is placed at coordinates $(294, 0, 1174)$ mm

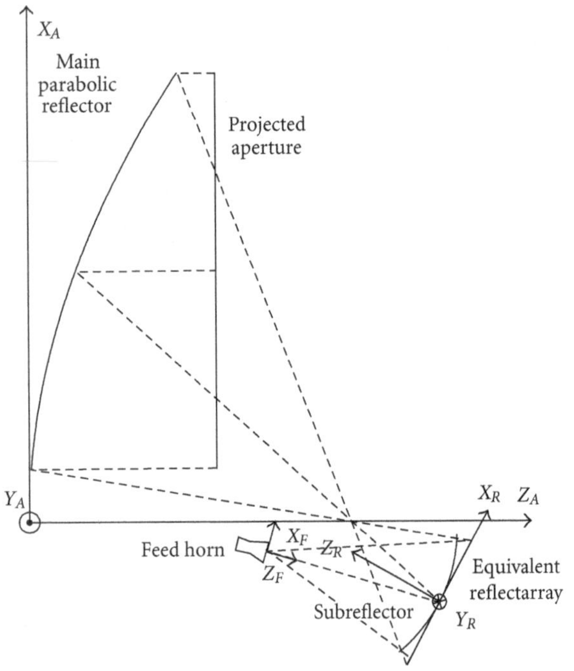

FIGURE 7: Scheme of the subreflectarray-parabolic reflector antenna.

in the antenna reference system (X_A, Y_A, Z_A). In this work, the scanning is achieved by introducing a progressive phase shift along the reflectarray subreflector. This approach permits to deflect the beam in one plane. For example, if a progressive phase is introduced along the Y_R-axis of the subreflectarray, the beam is scanned the azimuth plane ($Y_A Z_A$). As a result, the control circuits are simplified, since the phase control should be the same for all the elements in a column or a row. The radiation patterns of this antenna for the X_A polarization (electric field in X_A direction) in the principal planes are shown in Figure 2, assuming an ideal progressive phase on the subreflectarray for beam scanning in azimuth. The gain has been computed for a beam scanning from $-2°$ to $+2°$, showing a reduction in gain less than 1 dB in all the cases. In the case of ideal phase shifters considered here, the cross-polarization for the overall antenna is slightly increased when the beam is defocused, being in the order of -30 dB below the copolar gain. If a larger beam-scanning angular range is required, the phase control at row or column level is not accurate enough and a synthesis technique must be applied to obtain the phase distribution required on the subreflectarray surface for each beam pointing. A simple synthesis technique for this application is discussed in [24] showing that the angular scanning range can be extended up to $\pm 7°$, with the beam distorted for broader angles because of the spillover.

3.1.2. 94 GHz Dual-Reflector Antenna. The beam deflecting performance of this antenna configuration has been demonstrated in the 94 GHz band, as discussed in [21]. In this work, the dual-reflector antenna is designed, manufactured, and tested to show that it is possible to steer the beam up to $\pm 5°$ from boresight by replacing a flat metal subreflector with a small reflectarray. The antenna demonstrator is made up of an offset parabolic reflector of diameter 120 mm (38λ) and a flat subreflector printed on a 50 mm diameter quartz wafer, see Figure 3. The beam scanning capability of the antenna is demonstrated by replacing a solid metal flat subreflector, which produces a focused beam in the boresight direction, by a passive 28×28 elements reflectarray printed on a 115 μm thick metal-backed quartz wafer. The reflectarray is made of a single layer of varying-size patches, see Figure 3(b), designed to produce a progressive phase distribution in the Y_R direction to deflect the beam 5° in the azimuth plane. The copolar far field radiation patterns of the antenna were measured between $-20°$ and $+20°$ in the azimuth plane at 94 GHz at Queen's University Belfast (QUB). Figure 4 shows the measured radiation pattern using the reflectarray subreflector to tilt the beam 5° and also the results using a flat metallic subreflector to produce a boresight-directed beam. The measured radiation patterns are compared in Figure 4 with the simulated patterns obtained through the technique described in the previous section, in two cases: (a) considering the ideal far-field model and (b) considering the near field of the primary feed horn, see [21]. These results show very good agreement with the measurements when the near-field model is used, which demonstrates the need to use the near field radiated by the feed for an accurate analysis of the antenna. The gain for the boresight beam computed by the analysis tool is 38.7 dBi, with the gain reduced in 1.26 dB for the deflected beam. The predicted reduction in gain is 0.86 dB due to the 5° beam deflection and 0.40 dB because of phase errors and losses in the reflectarray. Some distortion of the tilted beam and a deviation of 1.4° in the beam direction are found because the antenna design was carried by implementing a progressive phase distribution on the reflectarray assuming the far-field model of the feed. However the far-field model of the horn does not provide a realistic estimation of the incident field, neither in amplitude nor in phase, as shown in [21]. In order to reduce the distortion of the deflected beam, the reflectarray subreflector should be designed considering the near field of the feed horn [20].

3.1.3. Dual-Reflectarray Antenna for Bidirectional Satellite Link. A dual-reflectarray antenna can be used for beam-scanning application, [10]. This configuration is very convenient for transportable antennas since the two flat reflectarrays are ease to fold and deploy. In this case, the dual-reflectarray antenna is made up of a passive sub- and main reflectarrays. The main reflectarray can be also implemented with 1-bit electronic control for beam pointing. The antenna is defined to provide a directive beam in receive (10.70–12.75 GHz) and transmit (14.0–14.5 GHz) frequency bands with electronic scanning capabilities within a limited angular range when using 1-bit electronic control elements on the main reflectarray. As discussed in [10], the main reflectarray has pentagonal shape with dimensions of 1500 mm $\times$ 750 mm. This reflectarray comprises 5250 elements distributed in 75 rows and 120 columns. The elements

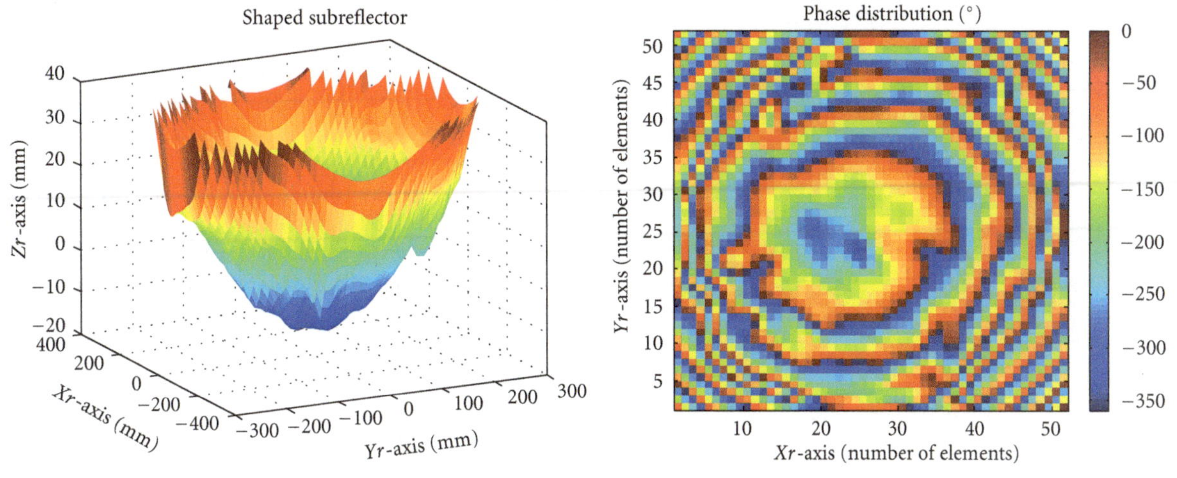

(a) (b)

FIGURE 8: Shaped ellipsoid as subreflector (a) and phase distribution of the reflection coefficient for the equivalent reflectarray at 11.95 GHz (b).

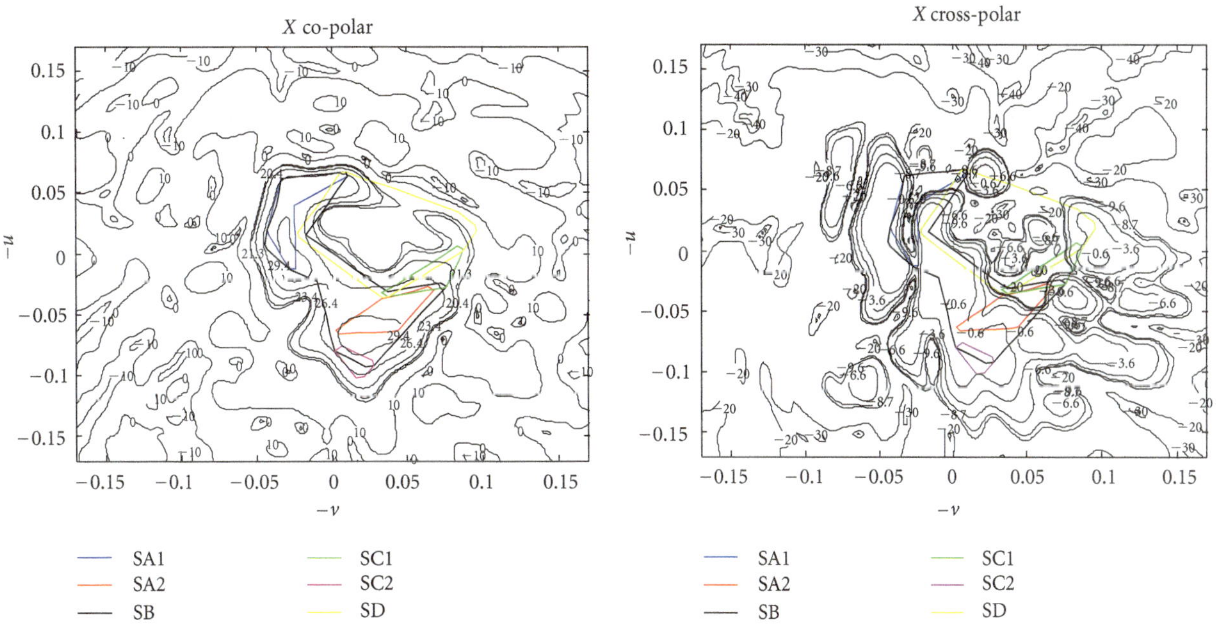

(a) (b)

FIGURE 9: Copolar (a) and cross-polar (b) radiation patterns at 11.95 GHz (X_m-pol.) with the coverage template superimposed.

are arranged in a periodic lattice of 10 mm × 10 mm. The subreflectarray is tilted 27.69° with respect to the main reflectarray, with dimensions of 475 mm × 600 mm and placed with its centre at coordinates $(-473, 0.0, 690)$ mm referring to the main coordinate system. The subreflectarray is rectangular, with 38 × 48 elements in a periodic mesh of 12.5 mm × 12.5 mm periodicity. The phase centre of the feed horn is located at coordinates $(300, 0, 30)$ mm in the main reflectarray coordinate system. The subreflectarray has been designed to provide a very large equivalent focal distance in order to illuminate efficiently the main reflector. A simple model of the horn antenna based on $\cos^q(\theta)$ function is used. The q factor is defined at each frequency to provide an illumination level at the edge of the main reflector between −10 dB and −15 dB.

Figure 5(a) shows the phase distribution on the subreflectarray. Figure 5(b) shows the required phase shift for the main reflectarray to obtain a pencil beam pointing at 40° from the z-axis. The radiation patterns at 10.7 GHz, 12.75 GHz (Rx), and 14.25 GHz (Tx) are shown in Figure 6.

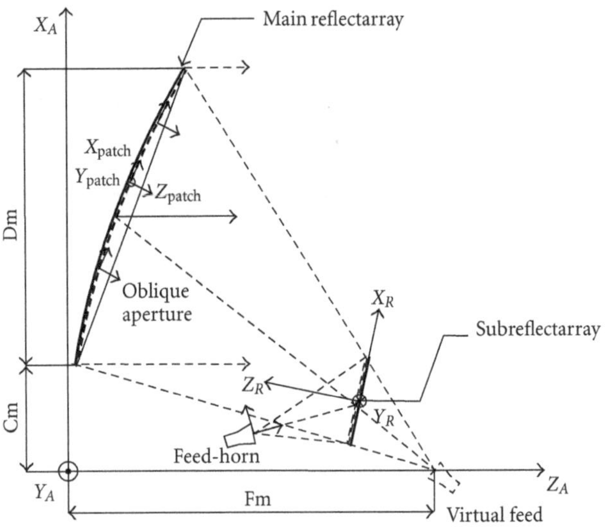

FIGURE 10: Scheme of a dual-reflectarray structure analysis.

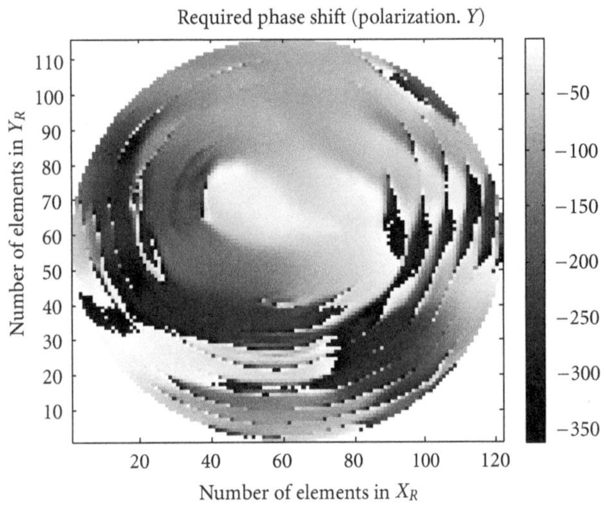

FIGURE 11: Required phase-shift for a 1.5 m parabolic reflectarray at 11.95 GHz, $F/D = 1.2$ [25].

3.2. Contoured-Beam Antenna for DBS Applications

3.2.1. Contoured-Beam Antenna with a Shaped Subreflectarray. A dual-reflector antenna based on a reflectarray sub-reflector and a main parabolic reflector has been proposed to obtain contoured beams for DBS applications [5]. The classic solution in this application consists on single or dual reflector antenna based on shaped reflectors. This solution is expensive and time consuming since the reflector surface shaping depends on the coverage footprint, and the manufacture requires specific moulds which are different for each mission. In the case of a dual-reflector antenna with a reflectarray sub-reflector, the beam shaping can be achieved by the phase-shift introduced on the reflectarray surface. In this work, the coverage specified for a South American mission on Amazonas satellite in Ku-band transmit band (11.95 GHz) is selected. The geometry of the antenna is derived from a dual-offset Gregorian optics with a main aperture of 1.5 m diameter, see Figure 7. In this antenna, the ellipsoidal subreflector (0.5 m diameter) is replaced by a reflectarray with the appropriate phasing. As shown in Figure 7, the equivalent subreflectarray is placed on the $X_R Y_R$ plane, being its centre placed at $(-308, 0, 1596)$ mm in the antenna coordinate system. The subreflectarray is made up of 51×51 elements arranged in regular mesh with period 11 mm $\times$ 11 mm. As in previous sections, the feed horn is modeled as a $\cos^q(\theta)$ function which provides an illumination taper at edges of -22 dB in order to reduce the spillover.

In order to synthesize the subreflectarray phases, the shape of a subreflector that fulfils the coverage requirements is achieved and used as the starting point in the process to obtain the phase distribution of the reflection coefficient of the equivalent reflectarray, see Figure 8(a). The phase distribution shown in Figure 8(b) is the same for the two linear polarizations of the antenna, since it has been derived from the distances from the shaped ellipsoid to the equivalent reflectarray plane. The radiation pattern for X_A polarization of the antenna is plotted in Figure 9, being the results similar for the orthogonal polarization. Co- and cross-polar components have been calculated considering the elements of the reflectarray as ideal phase shifters and no losses (ideal reflectarray elements), see Figures 9(a) and 9(b). The shape of the footprint obtained shows some distortion. However, the performance is near to the coverage required mask; see Figure 9(a). This phase distribution can be a suitable starting point for a further optimization process.

3.2.2. Parabolic Reflectarray in Dual-Reflector Configuration. The South American coverage antenna shown in the previous section can also be implemented with a dual-reflectarray antenna using a main parabolic reflectarray and a flat sub-reflectarray. The contoured beam was previously synthesized in a single offset parabolic reflectarray as discussed in [25]. The general scheme of this antenna is shown in Figure 10.

The antenna projected aperture has a diameter of 1.5 m, resulting an elliptic parabolic reflectarray of axes 1551 mm $\times$ 1500 mm. The clearance of the antenna is 200 mm and the focal distance is 1800 mm. The subreflector is a flat reflectarray with axes 598 mm $\times$ 559 mm. In principle, the antenna optics is defined to compact the overall volume. In this case, the subreflectarray behaves as a metal plane introducing a constant phase distribution, equivalent to a flat metal plate, in order to be able to compare the results with those previously reported for a single offset parabolic reflectarray [25]. Although the degree of freedom in the phase distribution on the subreflectarray can be used to reduce the cross-polarization of the antenna. The phase distribution on the main reflectarray surface required to obtain the South American coverage is shown in Figure 11. To implement this phase distribution, a unit cell with two stacked rectangular patches has been used as a reflectarray element. The patches are arranged in a periodic lattice of 13 mm$\times$13 mm and they are assumed to be printed on a 125-micron Kapton film bonded to a quartz-fabric composite

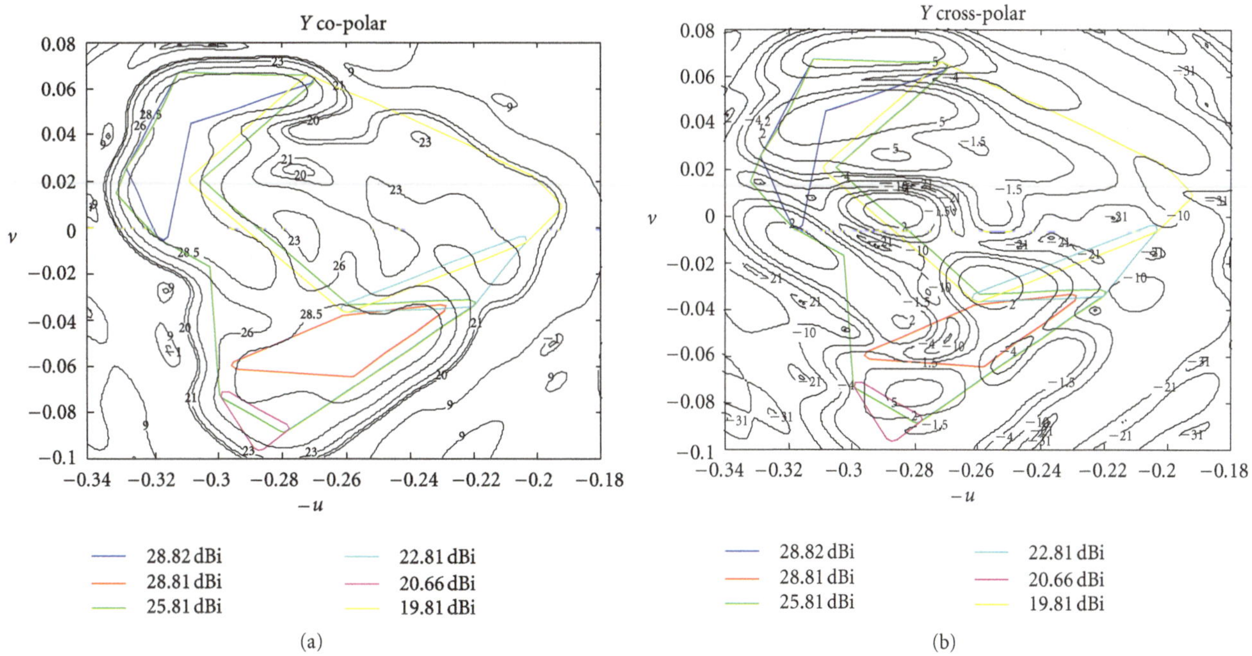

FIGURE 12: Radiation patterns obtained by MoM for a two-layer reflectarray in Tx band, 11.95 GHz. (a) Copolar component. (b) Crosspolar component.

layer and separated by quartz-honeycomb separators with 3.5 mm thickness.

The radiation pattern was computed at three frequencies: 11.7, 11.95, and 12.2 GHz, corresponding to the central and extreme frequencies of the transmit band (TX). Accurate predictions of both copolar and cross-polar radiation patterns at central frequency are provided in Figure 12, showing that the contoured patterns are close to the requirements, which are also shown in the figures as coloured polygons. Similar results were obtained for the extreme frequencies of the band. Two small holes at the border of the 28.81 dBi contour are found as a result of ohmic losses and phase errors produced by the approximate design technique, in which the pattern synthesis is applied on the planar aperture and the required phase distribution is obtained by geometrical optics. The contoured patterns can be improved by a suitable pattern synthesis taking into account the curved surface.

The analysis technique provides an accurate prediction of the cross-polarization pattern performance, since it takes into account the cross-polarization produced by the printed patches and by the offset reflector. The cross-polarization shown in Figure 8(b) can be reduced by synthesizing a proper phase distribution on the subreflectarray surface.

4. Conclusions

A review of recent work on dual-reflector antennas involving reflectarrays has been presented, considering a flat reflectarray subreflector and parabolic reflector, flat reflectarray, or parabolic reflectarray as main reflectors. A modular technique for the analysis of dual-offset antennas involving one or two reflectarrays has been described. The technique has been applied to analyze several antenna configurations for beam scanning and contoured beam for DBS applications.

Three examples of beam scanning dual-reflector antennas involving reflectarrays have been presented. First, a high-gain beam-scanning antenna based on a flat subreflectarray and a main parabolic reflector has been studied. The beam scanning can be realized by adding controllable phase shifters on the subreflectarray. The beam scanning performance of an antenna in a reduced angular range of $\pm 2°$ has been presented. Second, a dual-offset antenna with a reflectarray sub-reflector working at 94 GHz has been designed to provide electronic beam scanning in a limited angular range ($\pm 5°$). This antenna has been manufactured and tested, demonstrating the capability of this configuration for beam scanning in submillimeter-wave frequencies. The results show that a near-field model of the feed horn is necessary for an accurate analysis and design of the antenna. Third, a dual-reflectarray antenna with electronic beam steering has been presented for portable systems to provide Ku-band satellite links with automatic pointing capabilities. The antenna configuration comprises a passive subreflectarray and a main reflectarray with 1-bit electronic control to provide a pencil beam steerable in a $\pm 5°$ angular range.

Dual-reflector antennas with one or two reflectarrays have also been proposed for contoured-beam applications. Two antennas providing South American coverage have been presented. A first solution consists of a dual-reflector dual-offset antenna with a 56 cm subreflectarray and a 1.5 m main parabolic reflector. In this case the coverage is achieved by synthesizing the phase distribution on the subreflectarray surface. The same coverage is also obtained by a dual-reflectarray antenna with 1.5 m parabolic reflectarray. In this antenna the beam shaping is achieved by the main reflectarray, while sub reflectarray can be used to reduce the cross-polar radiation of the entire antenna.

All the configurations and applications presented in the paper show the versatility that a dual-offset antenna involving reflectarrays may offer, as beam scanning, beam shaping, or reduction of cross-polarization.

Acknowledgments

This work has been supported in part by the "Ministerio de Ciencia e Innovación" (Spanish Government) under projects CONSOLIDER-INGENIO CSD2008-00068 (TERASENSE) and TEC2010-17567 and also by the European Space Agency (ESA).

References

[1] B. Khayatian, Y. Rahmat-Samii, and J. Huang, "Radiation characteristics of reflectarray antennas: methodology and applicatios to dual configurations," in *Proceedings of the European Conference on Antennas and Propagation (EuCAP '06)*, Nice, France, November 2006.

[2] S. Xu, H. Rajagopalan, Y. Rahmat-Samii, and W. A. Imbriale, "A novel reflector surface distortion compensating technique using a sub-reflectarray," in *Proceedings of the IEEE Antennas and Propagation Society International Symposium (AP-S '07)*, pp. 5315–5318, Honolulu, Hawaii, USA, June 2007.

[3] S. Xu, Y. Rahmat-Samii, and W. A. Imbriale, "Subreflectarrays for reflector surface distortion compensation," *IEEE Transactions on Antennas and Propagation*, vol. 57, no. 2, pp. 364–372, 2009.

[4] M. Arrebola, L. de Haro, and J. A. Encinar, "Analysis of dual-reflector antennas with a reflectarray as subreflector," *IEEE Antennas and Propagation Magazine*, vol. 50, no. 6, pp. 39–51, 2008.

[5] M. Arrebola, L. De Haro, J. A. Encinar, and L. F. De La Fuente, "Contoured-beam Gregorian antenna with a reflectarray as subreflector," in *Proceedings of the 2nd European Conference on Antennas and Propagation (EuCAP '07)*, Edinburgh, UK, November 2007.

[6] W. Huang and J. R. Lesh, "Reflectarray elements on sub-reflector for fine beam pointing," JPL New Technology Report 41562, 2004.

[7] W. Menzel and D. Kessler, "A folded reflectarray antenna for 2D scanning," in *German Microwave Conference*, pp. 1–14, March 2009.

[8] C. Tienda, M. Arrebola, J. A. Encinar, and G. Toso, "Analysis of a dual-reflect array antenna," in *Proceedings of the IET on Microwaves, Antennas & Propagation*, vol. 5, no. 13, pp. 1636–1645, October 2011.

[9] R. Leberer and W. Menzel, "A dual planar reflectarray with synthesized phase and amplitude distribution," *IEEE Transactions on Antennas and Propagation*, vol. 53, no. 11, pp. 3534–3539, 2005.

[10] C. Tienda, J. A. Encinar, S. Montori, R. Vincenti Gatti, M. Arrebola, and R. Sorrentino, "Dual-reflectarray antenna for bidirectional satellite links in Ku-band," in *Proceedings of the 5th European Conference on Antennas and Propagation (EUCAP '11)*, pp. 1404–1407, Rome, Italy, April 2011.

[11] Y. Álvarez, F. Las-Heras, and M. R. Pino, "Reconstruction of equivalent currents distribution over arbitrary three-dimensional surfaces based on integral equation algorithms," *IEEE Transactions on Antennas and Propagation*, vol. 55, no. 12, pp. 3460–3468, 2007.

[12] R. J. Martin and D. H. Martin, "Quasi-optical antennas for radiometric remote-sensing," *Electronics and Communication Engineering Journal*, vol. 8, no. 1, pp. 37–48, 1996.

[13] A. Möessinger, R. Marin, S. Mueller, J. Freese, and R. Jakoby, "Electronically reconfigurable reflectarrays with nematic liquid crystals," *Electronics Letters*, vol. 42, no. 16, pp. 899–900, 2006.

[14] W. Hu, M. Y. Ismail, R. Cahill et al., "Liquid-crystal-based reflectarray antenna with electronically switchable monopulse patterns," *Electronics Letters*, vol. 43, no. 14, pp. 899–900, 2007.

[15] A. Moessinger, R. Marin, D. Eicher, R. Jakoby, and H. Schlaak, "Liquid crystal reflectarray with electronic 2D-reconfiguration capability," in *Proceedings of the 29th ESA Antenna Workshop on Multiple Beams and Reconfigurable Antennas*, pp. 67–70, April 2007.

[16] L. Marcaccioli, B. Mencagli, R. Vincenti Gatti et al., "Beam steering MEMS mm-wave reflectarrays," in *MEMSWAVE International Symposium*, Orvieto, Italy, June 2006.

[17] H. Rajagopalan, Y. Rahmat-Samii, and W. A. Imbriale, "RF MEMS actuated reconfigurable reflectarray patch-slot element," *IEEE Transactions on Antennas and Propagation*, vol. 56, no. 12, pp. 3689–3699, 2008.

[18] S. V. Hum, M. Okoniewski, and R. J. Davies, "Realizing an electronically tunable reflectarray using varactor diode-tuned elements," *IEEE Microwave and Wireless Components Letters*, vol. 15, no. 6, pp. 422–424, 2005.

[19] M. Arrebola, J. A. Encinar, and M. Barba, "Multifed printed reflectarray with three simultaneous shaped beams for LMDS central station antenna," *IEEE Transactions on Antennas and Propagation*, vol. 56, no. 6, pp. 1518–1527, 2008.

[20] M. Arrebola, Y. Álvarez, J. A. Encinar, and F. Las-Heras, "Accurate analysis of printed reflectarrays considering the near field of the primary feed," *IET Microwaves, Antennas and Propagation*, vol. 3, no. 2, pp. 187–194, 2009.

[21] W. Hu, M. Arrebola, R. Cahill et al., "94 GHz dual-reflector antenna with reflectarray subreflector," *IEEE Transactions on Antennas and Propagation*, vol. 57, no. 10, part 2, pp. 3043–3050, 2009.

[22] C. Wan and J. A. Encinar, "Efficient computation of generalized scattering matrix for analyzing multilayered periodic structures," *IEEE Transactions on Antennas and Propagation*, vol. 43, no. 11, pp. 1233–1242, 1995.

[23] J. A. Encinar, "Design of two-layer printed reflectarrays using patches of variable size," *IEEE Transactions on Antennas and Propagation*, vol. 49, no. 10, pp. 1403–1410, 2001.

[24] M. Arrebola, E. Carrasco, and J. A. Encinar, "Beam scanning antenna using a reflectarray as sub-reflector," *ACES Journal*, vol. 26, no. 6, 2011.

[25] J. A. Encinar, M. Arrebola, and G. Toso, "A parabolic reflectarray for a bandwidth improved contoured beam coverage," in *Proceedings of the 2nd European Conference on Antennas and Propagation (EuCAP '07)*, EICC, November 2007.

13

Comparison of Mutual Coupling Phenomena in Subwavelength Ridged Circular Apertures and Half-Wavelength Dipole Antenna Arrays

Jong-Eon Park, Ki Young Kim, and Jae-Won Song

School of Electrical Engineering and Computer Science, Kyungpook National University, Daegu 702-701, Republic of Korea

Correspondence should be addressed to Jong-Eon Park, parkje00@naver.com

Academic Editor: Zhongxiang Q. Shen

The resonant microwave transmission characteristics of several coupled subwavelength ridged circular aperture arrangements in a thin metallic film are investigated using the three-dimensional finite-difference time-domain (3D FDTD) method. Simple equivalent circuits represented by the self and mutual conductances that have been quantitatively extracted for each resonance condition, including the mutual coupling effects, are also devised with the help of virtual magnetic current elements. Furthermore, a duality is identified between the ridged circular apertures and conventional half-wavelength dipole arrays based on comparing the respective resonance conditions.

1. Introduction

Bethe's theory of the diffraction of plane electromagnetic waves during transmission through a circular hole in an infinitesimally thin perfect conducting screen states that the power transmitted through the subwavelength hole is very restricted, being proportional to $(a/\lambda)^4$ [1], where a is the radius of the circular hole, and λ is the wavelength. Moreover, the transmission efficiency is exponentially decreased when the thickness of the screen is considered [2]. As such, this poor transmission efficiency has been perceived as a fundamental restriction in dealing with electromagnetic transmission through a small hole. Therefore, several cousin structures have been studied to overcome this limitation. Harrington analyzed the concept of an electrically small aperture in a conducting screen backed by a conducting body [3], referred to as aperture-body resonance (ABR), where a subwavelength hole loaded with a nearby conductor considerably increases the power throughput when compared with a hole without a nearby scatterer. Inspired by ABR, Leviatan investigated the electromagnetic coupling phenomena between two half-space regions separated by two slot-perforated parallel conducting planes [4]. In this case, when varying the distance between the centers of each slot, a resonant transmission phenomenon is observed, occurring at resonance spacings that approach multiples of $\lambda/2$. Similarly, the problem of electromagnetic transmission through two slits in parallel plates can be solved with the help of equivalent circuit representation [5]. Here, when the gap between the two plates is close enough, meaning much less than the wavelength of the incident wave, the resulting maximum transmitted power becomes constant, that is, λ/π, irrespective of the slit width and incident angle. Meanwhile, in the case of a thick conducting screen, the resonance transmission has been investigated by altering the thickness of the conducting screen [6]. In this situation, different resonance phenomena are observed when the resonance cavity is built along the longitudinal direction inside the slot region. Alternatively, the addition of two ridges in an electrically small circular aperture also considerably enhances the transmitted power [7–9]. The resonance power transmission of ridged circular apertures as a function of the distance between them was also recently explored [10], experimentally verified [11], compared with dipole arrangements [12], and investigated up to its infinite periodic structure [13]. Furthermore, the recent interest in surface plasmons (SPs), followed by Ebbesen's seminal discovery of extraordinary optical transmission (EOT) [14], have led to optical transmission via bowtie, H-, and C-shaped apertures perforated in metallic plates [15–21]. This resonance mechanism is also very similar to the ridged circular apertures used in the microwave region. Extensive analyses of ridged circular apertures (or H-shaped apertures) have already been performed. Here, the resonant frequency of a ridged circular aperture can be adjusted by modifying the width of each ridge and the gap between the ridges without rescaling the host circular shape, while the resonant frequency of a slot antenna, which is one of the basic aperture antenna structures, depends on the width of the slot. Notwithstanding, if the thickness of the metal in the optical regime is considered, the resonant transmission mechanism is not exactly the same for a perfect electric conductor in the microwave region.

However, despite the extensive previous work on ridged structures, an in-depth study of multiple subwavelength resonant apertures is still required and fascinating as well. Once the structure of a single ridged aperture is determined, the arrangement of the multiple apertures becomes an important issue in terms of the potential manipulation of the electromagnetic waves, meaning that the transmission and/or reflection at a particular spectrum of interest can be tailored. Accordingly, this study checks the transmission as a function of the center-to-center distance for vertically and horizontally arranged structures. Inspired by our previous observations [12], the distances for both the minimum and maximum transmitted power are examined for different arrangements and then compared with the spacing distances in conventional dipole arrays. Moreover, as discussed in [12], an equivalent circuit representation is obtained, which is helpful to understand the resonant transmission mechanism. Furthermore, the radiation conductance variation and susceptance variation of a single circular aperture are discussed in the case of a ridge-loaded and ridge-removed aperture. As a result, the self-conductance and mutual conductances for both the maximum and minimum transmitted power in the case of the ridged circular apertures are found to be very similar to the self-resistance and mutual resistances for both the maximum and minimum gains of dipole antenna arrays, allowing the establishment of a duality between them. The usage of a resonant ridged aperture has already been proposed in many applications, including nanoscale lithography [22], single-molecule fluorescence measurement [23], and optical data storage [24–27], and so on [28–30], and this study may be a stepping stone from the single resonant aperture to two-dimensional array of the aperture, for example, a frequency selective surface (FSS) [31], a significant electromagnetic structure with many potential applications.

2. Resonant Transmission for Various Coupling Cases

This study presents the problem of electromagnetic wave scattering using subwavelength ridged circular apertures perforated in an infinitesimally thin metallic screen (here,

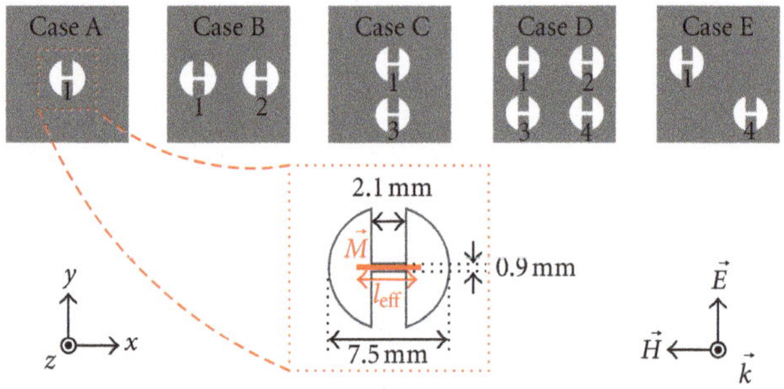

FIGURE 1: Ridged circular apertures perforated in an infinitesimally thin metal screen; (case A: single aperture; case B: collinear; case C: side-by-side; case D: 2-by-2; case E: parallel-in-echelon).

a perfect electric conductor). Five different aperture arrangements, shown in Figure 1, are considered to investigate the mutual coupling phenomena and related physics. The aperture dimensions were arbitrarily chosen to allow the structures to resonate in the X-band (8–12 GHz); the diameter of each circular aperture was 7.5 mm, the ridge width was 2.1 mm, and the gap between the two ridges was 0.9 mm. The plane wave was assumed to be normally incident to the screen from the back side, that is, along the $+z$-direction, while the electric field was linearly polarized along the y-direction to cross the ridge. If the polarization of the electric field had been aligned with the ridge, that is, in the x-direction, this would have restricted the electromagnetic wave transmission [7]. The three-dimensional finite-difference time-domain (3D FDTD) method [32] was used to analyze the structures. Perfectly matched layer absorbing boundary conditions (PML ABCs) were set up at all six ($\pm x$, $\pm y$, and $\pm z$) surrounding boundaries. The structure was numerically solved using a simulation volume of 16.8 mm × 16.8 mm × 9.9 mm (x-, y-, and z-axes in order) including the PML region. Plus, 25 air buffer cells were used for both the incident and transmitted regions in the z-axis. The unit time step for each update of the Maxwell equations was set at 2.60×10^{-13} sec, which was determined using the stability criteria and Courant factor of the FDTD algorithm.

From the numerical analysis of the FDTD simulation, a transmission cross-section (TCS) was obtained to calculate the transmission efficiency of the small ridged circular apertures. The definition of a TCS is the ratio of the transmitted power P_{trans} to the incident power density P_{inc} as follows:

$$\text{TCS}\ [\text{m}^2] = \frac{P_{\text{trans}}\ [\text{W}]}{P_{\text{inc}}\ [\text{W/m}^2]} = \frac{0.5 \times \text{Re}\left[\iint \left\{\vec{E}_t \times \vec{H}_t^*\right\} \cdot d\vec{s}\right]}{\left|\vec{E}_{\text{inc}}\right|^2 / (2\eta_0)}, \tag{1}$$

where $\vec{E}_t$ and $\vec{H}_t$ are the transverse electric and magnetic fields of the aperture, respectively, the asterisk denotes the complex conjugate, and $d\vec{s}$ is the unit cell area of the aperture. The incident power density is given by $P_{\text{inc}} = |\vec{E}_{\text{inc}}|^2/2\eta_0$, where $\vec{E}_{\text{inc}}$ is the electric field of the incident plane wave with a Gaussian form, and η_0 (= 120π) is the intrinsic impedance of the free space. Using this definition,

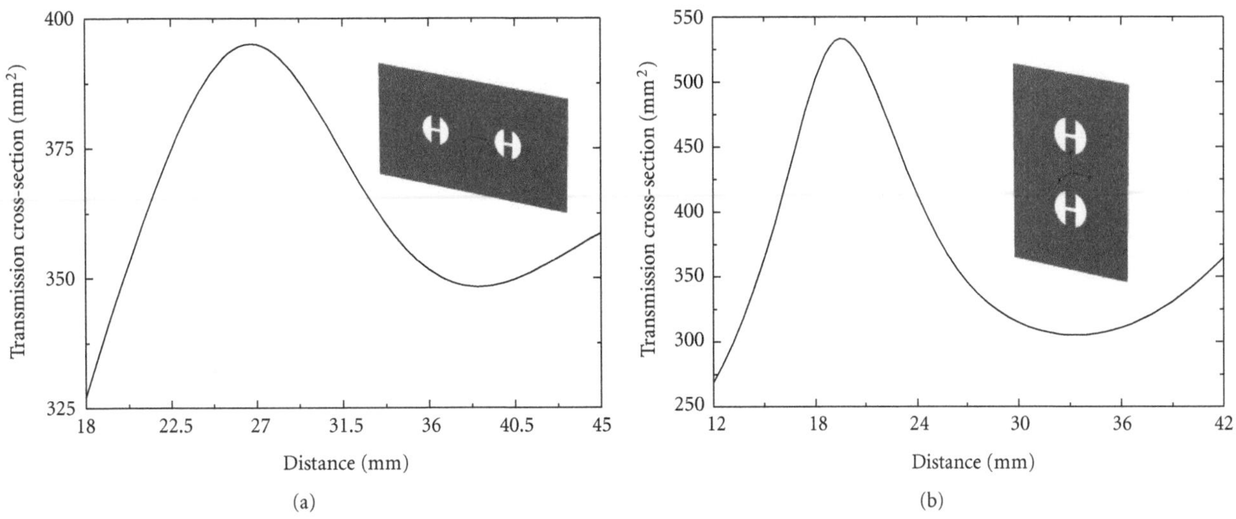

FIGURE 2: TCS versus center-to-center distance for (a) case B (collinear case) and (b) case C (side-by-side case).

the TCS for the single ridged circular aperture, shown as case A in Figure 1, was determined as 180.84 mm^2 at the resonant frequency of 11.02 GHz.

Prior to investigating the TCS variations for the multiple aperture cases, the virtual magnetic current was defined. When the plane wave was incident on the aperture, the y-component of the electric field E_y was concentrated on the gap between the two ridges (ignoring a small amount of E_x), thereby producing an equivalent magnetic current of $\vec{M} = \vec{E} \times \vec{n}$, depicted as the short red line in Figure 1. Here, $\vec{E}$ is the induced electric field in the gap between the ridges, and $\vec{n}$ is the normal vector on the xy plane. As will be examined later, the physics of a half-wavelength dipole array also apply to arrays of an equivalent magnetic current $\vec{M}$, which is useful for understanding the coupling mechanisms in coupled ridged circular aperture structures [12]. The classification of the five cases in Figure 1 is based on the arrangement of the equivalent magnetic current elements $\vec{M}$, allowing a convenient comparison of the duality of the current structures and dipole arrays.

Figure 2 shows the TCS variations at the resonant-frequency of 11.02 GHz as a function of the distance between the two coupled resonant structures, cases B and C, respectively. The distances were measured between the centers of the two resonant apertures. The distances for the maximum and minimum TCSs for cases B and C are listed in Table 1. In terms of the relative arrangements of the magnetic dipole antenna elements, the distances for the maximum (minimum) TCSs were very close to the cases of half-wavelength dipole antennas [33, 34], where the maximum (minimum) gain ratios for two dipole antennas in collinear and side-by-side arrangements are achieved when the dipoles are approximately 1.0λ (1.48λ) and 0.7λ (1.2λ) away from each other, respectively. Here, the two terms collinear and side-by-side reflect the dipole arrangements [34]. Similarly, these two terms can also be applied to the equivalent magnetic current elements $\vec{M}$. As such, the collinear and side-by-side cases correspond to cases B and C in Figure 1, respectively.

TABLE 1: Center-to-center distance for maximum and minimum TCS cases (cases B and C).

	Distance for case B	Distance for case C
Maximum TCS case	(i) 26.70 mm = 0.981 λ	(ii) 19.50 mm = 0.716 λ
Minimum TCS case	(iii) 38.55 mm = 1.416 λ	(iv) 33.60 mm = 1.234 λ

TABLE 2: Total TCS and TCS per unit aperture (at resonant frequency of 11.02 GHz).

Case	Total TCS (mm^2)	TCS per unit aperture (mm^2)	Distances for each case (Table 1)
A	180.84	180.84	—
B^{max}	395.08	197.54	(i)
B^{min}	348.32	174.16	(iii)
C^{max}	533.35	266.68	(ii)
C^{min}	305.15	152.58	(iv)
D^{max}	1116.07	279.02	(i) and (ii)
D^{min}	615.62	153.91	(iii) and (iv)
E^{max}	341.72	170.86	(i) and (ii)
E^{min}	380.25	190.13	(iii) and (iv)

Table 2 shows the total TCS and TCS per unit aperture for cases A to E at the resonant frequency of 11.02 GHz. The superscripts "max" and "min" for each case represent the maximum and minimum TCS for cases B and C, respectively. For cases D and E, the distance for cases B and C was applied without sweeping the distance. The distances applied are summarized in the last column of Table 2. The TCS-enhancing effect of case C^{max} (C^{min}) was greater (smaller) than that of case B^{max} (B^{min}), as seen in the third column of Table 2, implying that the mutual coupling in case C was

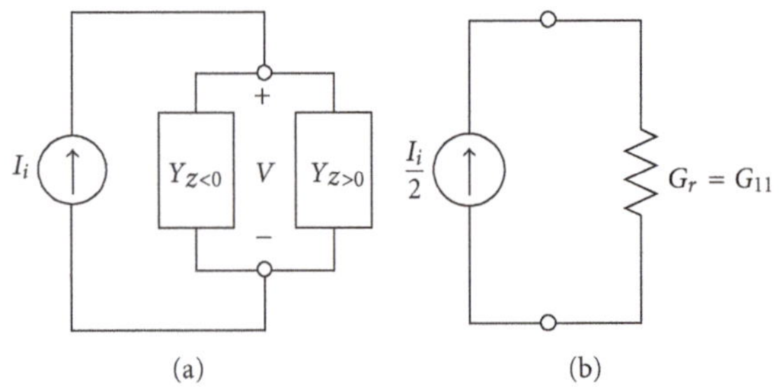

FIGURE 3: Equivalent circuit representation of single aperture for (a) general case and (b) its simplification from viewpoint of resonant transmission.

stronger than that in case B. This stronger coupling in the side-by-side case is very similar to the side-by-side case for dipole arrays. In addition, the resonant frequencies for cases B, C, and D remained the same as that for the single aperture case, that is, 11.02 GHz, as each aperture was resonant at the same frequency, and the distance between the two (or four) apertures did not produce any resonant frequency shift. However, the resonant frequency for case E did shift a little due to a slight difference between the resonant frequency for each aperture and that for the maximum TCS of the diagonal center-to-center distance [12]. The mutual coupling associated with the enhanced TCS will be further discussed later.

3. Equivalent Circuit Representations with Self and Mutual Conductances

3.1. Equivalent Circuit and Radiation Admittance for Single Aperture. Based on the numerical results in Table 2 and radiation conductance concept, an equivalent circuit was constructed. Figure 3(a) shows the equivalent circuit for a single aperture under general circumstances. The equivalent current source can be represented as $I_i = 2H_0 \cdot l_{\text{eff}}$ [8], where H_0 is the incident magnetic field, and l_{eff} is the effective physical length of $\vec{M}$, as marked in Figure 1. $Y_{z<0}$ and $Y_{z>0}$ are the total admittances for the half-space regions of $z < 0$ and $z > 0$, respectively, which become identical due to the symmetry of the structure. In general, each admittance in Figure 3(a) is expressed as $Y_r = G_r + jB_r$, where G_r and B_r are the radiating conductance and susceptance for the aperture, respectively. At the resonant frequency of 11.02 GHz, the imaginary part $|B_r|$ became smaller than the real part $|G_r|$, as shown in Figure 4, implying that the most of the incident power was transmitted into the opposite region, while a small amount of power was stored near the aperture. Thus, a simplified equivalent circuit with only the conductance, as shown in Figure 3(b), was used to represent the resonant transmission to the $z > 0$ region. In contrast, at an off-resonance, the transmitted power was significantly reduced, implying that the conductance was decreased and the susceptance became more dominant than the conductance. Indeed, this represents that the power was not transmitted to the opposite region but rather remained near the aperture. This was also observed from

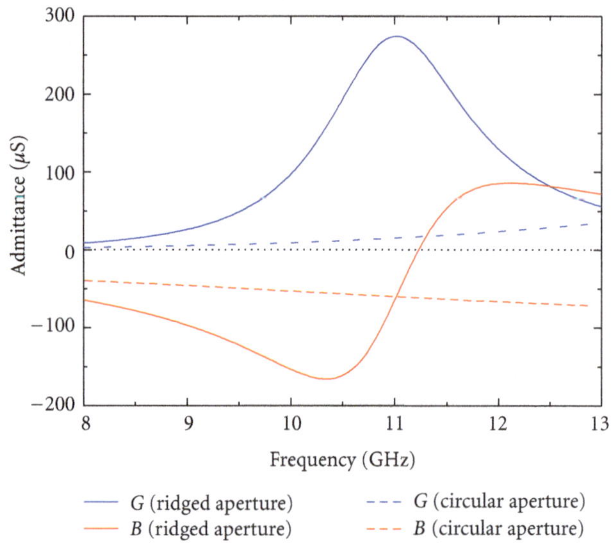

FIGURE 4: Admittance (conductance G and susceptance B) for ridged circular aperture and original host circular aperture with diameter of 7.5 mm.

the admittance (conductance and susceptance) at an off-resonant frequency (i.e., except around 11.02 GHz) and the admittance over the whole frequency region for the original circular aperture without any ridges, shown in Figure 4. Derivations of the admittance for the ridged and original host circular apertures were conducted using the analytic expression in [35] and the 3D FDTD numerical results, as shown later in (4). Note that the discrepancy in the frequency points for the maximum radiation conductance and zero susceptance of the coupled subwavelength aperture was not that abnormal, since a small frequency difference has also been observed between the maximum radiation resistance and zero reactance of a dipole [34]. Thus, the duality of both types of structure allows them to be considered as related to each other, which will be discussed later. However, if the thickness of the structure is included in the analysis (as in [12]), this small frequency difference between the maximum radiation conductance and zero susceptance of the structure will not hold any more and could be larger than that of the present structure.

3.2. Self-Conductance and Effective Length of Magnetic Dipole Antenna Element. As shown in Figure 1, the magnetic dipole antenna element $\vec{M}$, which is the same as the equivalent magnetic current or virtual magnetic current in the previous section, yet expressed differently in this case since this terminology is more convenient for comparison with conventional dipole antennas, is considered to be virtually generated between the two ridges, since a strong electric field is induced in this region. As the length of $\vec{M}$ is very short compared to the wavelength of the resonant frequency, that is, $l_{\text{eff}} \ll \lambda$, this element is considered to be the same as the dual structure of a Hertzian dipole, for which the radiation resistance is given by $R_r = 80\pi^2 \cdot (l_{\text{H}}/\lambda)^2$ [36], where l_{H} is the physical length of the Hertzian dipole. Thus,

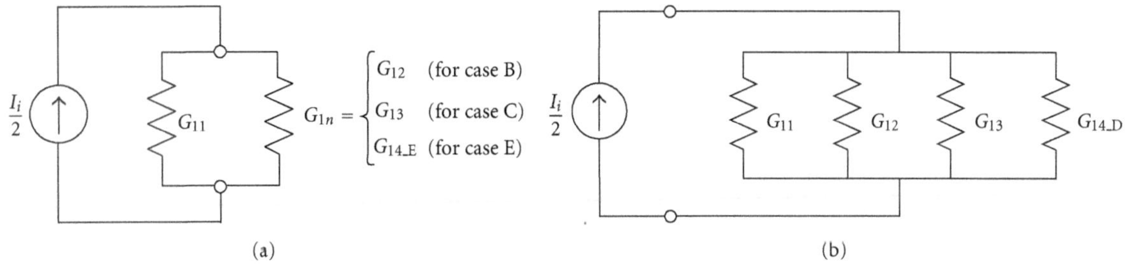

FIGURE 5: Equivalent circuit representation for (a) cases B, C, and E and (b) case D.

the radiation conductance of $\vec{M}$ for the ridged aperture is written as follows based on the duality of a Hertzian dipole,

$$G_r = \frac{4}{\eta_0^2} R_r = \frac{8\pi}{3\eta_0}\left(\frac{l_{\mathrm{H}}}{\lambda}\right)^2, \quad (2)$$

which is already well known as Booker's relation [37]. This can also be referred to as the self-conductance of a single aperture, that is, $G_{11} = G_r$ (Figure 3(b)), together with $l_{\mathrm{eff}} = l_{\mathrm{H}}$. The subscript 11 of G_{11} represents not only the self-radiating conductance but also aperture 1 of case A in Figure 1. The transmitted power P_{trans} for G_{11} with the sourcing current $I_i/2$, as shown in Figure 3(b), is then given as

$$P_{\mathrm{trans}} = \frac{|I_i|^2}{4G_{11}} = \frac{|2H_0 \cdot l_{\mathrm{eff}}|^2}{4 \cdot (8\pi/3\eta_0)(l_{\mathrm{eff}}/\lambda)^2} = \frac{1}{2}\eta_0 |H_0|^2 \cdot \frac{3\lambda^2}{4\pi}. \quad (3)$$

If this transmitted power is then divided by the incident power density, that is, $P_{\mathrm{inc}} = \eta_0|H_0|^2/2$, the TCS is $3\lambda^2/(4\pi)\,\mathrm{m}^2$, which agrees well with the numerically obtained TCS for case A and is exactly the same as the maximum TCS for general 3D resonant apertures, as predicted by the coupled-mode theory [38]. From (2), G_{11} with the proportional coefficient g_{11} can be rewritten as $G_{11} = (g_{11}/\eta_0) \cdot (l_{\mathrm{eff}}/\lambda)^2$, where $g_{11} = 8\pi/3 = 8.220$. Meanwhile, G_{11} can also be alternatively expressed as in [35], that is,

$$G_{11} = \mathrm{Re}\{Y\} = \mathrm{Re}\left\{\frac{P^*}{|V|^2}\right\} = \mathrm{Re}\left[\frac{\left\{\left(\vec{E} \times \vec{H}^*\right) \cdot d\vec{S}\right\}^*}{|V|^2}\right], \quad (4)$$

where P is the radiated power from the aperture, V is the voltage across the two ridges, as depicted in Figure 3(a), and $d\vec{S}$ is the surface vector normal to the aperture. P^* and $|V|$ were easily obtained from the electromagnetic fields of the aperture at $z = 0$ via the 3D FDTD simulations, from which the self-conductance was found to be $G_{11} = 274.51\,\mu$S. From (4) and the previous relation of G_{11}, $l_{\mathrm{eff}} = 3.058$ mm was also obtained, which was much shorter than the wavelength of the resonant frequency at 11.02 GHz, that is, 27.223 mm, while slightly longer than the ridge width of 2.1 mm due to the strong confinement of the electric field in the gap between the two ridges with its fringing around the gap.

TABLE 3: Self and mutual conductances of coupled ridged circular apertures.

	Self and mutual conductances	Cases
G_{11}	274.51 μS	A
$G_{12}^{\max}$	−23.18 μS	$\mathrm{B}^{\max}$
$G_{13}^{\max}$	−88.36 μS	$\mathrm{C}^{\max}$
$G_{14\_D}^{\max}/G_{14\_E}^{\max}$	14.94/16.08 μS	$\mathrm{D}^{\max}/\mathrm{E}^{\max}$
$G_{12}^{\min}$	10.50 μS	$\mathrm{B}^{\min}$
$G_{13}^{\min}$	50.79 μS	$\mathrm{C}^{\min}$
$G_{14\_D}^{\min}/G_{14\_E}^{\min}$	−13.31/−13.38 μS	$\mathrm{D}^{\min}/\mathrm{E}^{\min}$

Consequently, the proposition to obtain self-conductance utilizing a Hertzian dipole is clearly convincing.

3.3. Equivalent Circuits and Mutual Conductances for Coupled Structures. For the maximum TCS in case B (case $\mathrm{B}^{\max}$ in Table 2), the mutual conductance $G_{12}^{\max}$ between apertures 1 and 2 together with G_{11} were considered for a proper equivalent circuit representation. Similar to obtaining G_{11}, $G_{12}^{\max}$ was also simply expressed as $G_{12}^{\max} = (g_{12}^{\max}/\eta_0) \cdot (l_{\mathrm{eff}}/\lambda)^2$, where $g_{12}^{\max}$ is the proportional coefficient for $G_{12}^{\max}$. As with the notation for the self-radiating conductance G_{11}, the subscript 12 of $G_{12}^{\max}$ represents the mutual coupling effect between apertures 1 and 2, as depicted in case B in Figure 1. Subscripts 13 and 14 hereinafter are also termed in the same way. The corresponding equivalent circuit was established by replacing G_{11} in Figure 3(b) with a parallel connection of G_{11} and $G_{12}^{\max}$, as shown in Figure 5(a).The total conductance for case $\mathrm{B}^{\max}$ is then expressed as

$$G_{\mathrm{B}}^{\max} = G_{11} + G_{12}^{\max} = (g_{11} + g_{12}^{\max})\frac{1}{\eta_0}\left(\frac{l_{\mathrm{eff}}}{\lambda}\right)^2. \quad (5)$$

Recalling the total TCS for case $\mathrm{B}^{\max}$ in Table 2, that is, 395.08 mm^2 = $0.533\lambda^2\,\mathrm{m}^2$, $G_{12}^{\max} = -23.18\,\mu$S was obtained using $P_{\mathrm{trans}} = |I_i|^2/(4G_{\mathrm{B}}^{\max})$ (similar to (3)) with (5). The mutual conductance between apertures 1 and 3 for case $\mathrm{C}^{\max}$ was also obtained in the same way, that is, $G_{13}^{\max} = -88.36\,\mu$S (Figure 5(a) also holds for this case). Similarly, the mutual conductance for cases $\mathrm{B}^{\min}$ and $\mathrm{C}^{\min}$ was obtained, and $G_{12}^{\min}$ and $G_{13}^{\min}$ are listed in Table 3, which summarizes the self-conductance G_{11} and mutual conductances G_{1n} ($n = 2, 3,$ and 4) for each case.

Meanwhile, the total conductance for case D (both maximum and minimum TCS cases are applied) is written as

$$G_D = G_{11} + G_{12} + G_{13} + G_{14} = (g_{11} + g_{12} + g_{13} + g_{14}) \frac{1}{\eta_0} \left(\frac{l_{\text{eff}}}{\lambda} \right)^2, \quad (6)$$

where g_{14} is also the proportional coefficient for the mutual conductance G_{14}. The parallel-in-echelon elements, that is, apertures 1 and 4, and G_{14} were also obtained, since G_{11}, G_{12}, and G_{13} were already known. The corresponding equivalent circuit for case D was represented as shown in Figure 5(b). In a similar way to cases B and C, from the total TCS for case D^{max} (D^{min}) in Table 2, $G_{14\_D}^{max} = 14.94\,\mu S$ ($G_{14\_D}^{min} = -13.31\,\mu S$) was obtained (see Table 3), which was much smaller than G_{12} and G_{13}, yet a different sign. In contrast, G_{14_E} was obtained from the TCS for case E in Table 2 and (5) using g_{14} instead of g_{12}^{max}, as listed in Table 3. Both $G_{14\_D}^{max}$ and $G_{14\_E}^{max}$ (or $G_{14\_D}^{min}$ and $G_{14\_E}^{min}$) were almost the same. Hence, the results confirmed the assumption of an equivalent magnetic current element $\vec{M}$ for subwavelength ridged circular apertures, and the very simple equivalent circuit representations were quite reasonable. Furthermore, the similar quantities for $G_{14\_D}^{max}$ and $G_{14\_E}^{max}$ (or $G_{14\_D}^{min}$ and $G_{14\_E}^{min}$) showed that the mutual conductance between two apertures can be obtained as long as the self-conductance of a single element and the total TCS for the two apertures are known, that is, without solving cases B to E totally and even when the two apertures are arbitrarily located.

3.4. Duality between Coupled Ridged Apertures and Multiple Dipole Antennas. From the previous procedures for obtaining the self-conductance and mutual conductances, the relations between the total conductances and the TCSs can be explained. Clearly, the summation of G_{11} and G_{12} for case B was larger than the summation of G_{11} and G_{13} for case C, while the TCS for case B was much lower than the TCS for case C. Thus, the TCS was found to be inversely proportional to the summation of the self and mutual conductances, as confirmed by the results in Tables 2 and 3. While this has already been discussed by the present authors according to the quantitative relations [12], it should be noted that a negative mutual conductance enhanced the TCS. As for a dual structure, the self and mutual resistances of half-wavelength dipole antenna arrays are listed in Table 4. Here too, the gain ratio of the structure is also inversely proportional to the summation of the self and mutual resistances [33], where the gain ratio was obtained as a function of the distance and arrangement of the two dipole elements, as with the present structures. More specifically, the gain ratio was maximized (minimized) when the mutual resistance was minimized (maximized) at a center-to-center distance of 1.0λ (1.48λ) and 0.7λ (1.2λ) in the collinear and side-by-side arrangements, respectively, and the relations were very similar to the distances satisfying the inverse proportionality between the TCSs and the summation of the conductances, as mentioned earlier. When extending

TABLE 4: Self and mutual resistances of half-wavelength dipole antennas.

Self and mutual resistances		Distance for each case (Table 1)
R_{11}	73.130 Ω	—
R_{12}^{max}	−4.173 Ω	(i)
R_{13}^{max}	−24.309 Ω	(ii)
R_{14}^{max}	8.822 Ω	(i) & (ii)
R_{12}^{min}	1.573 Ω	(iii)
R_{13}^{min}	14.926 Ω	(iv)
R_{14}^{min}	−3.809 Ω	(iii) & (iv)

TABLE 5: Duality between half-wavelength dipole antenna array and ridged circular apertures.

Dipole antenna array [33]	Coupled ridged circular apertures (this work)
Gain ratio	Total TCS
Self resistance	Self conductance
Mutual resistance	Mutual conductance

the inverse proportionalities, G_{14}^{max} (G_{14}^{min}) had a minimal effect on the TCS in case E^{max} (E^{min}), as shown in the TCS per unit aperture column in Table 2, in contrast to case B^{max} or C^{max} (B^{min} or C^{min}). Plus, the small positive (negative) value of G_{14}^{max} (G_{14}^{min}) was quite similar to the mutual resistance R_{14}^{max} (R_{14}^{min}), shown in Table 4, in the parallel-in-echelon case of a half-wavelength dipole antenna. After all, by comparing the self and mutual conductances with the self and mutual resistances as listed in Tables 3 and 4, the negative or positive signs of them are exactly the same, and the quantities' proportions among them are quite similar. Thus, despite the physical dissimilarities of the two resonant structures, the key physical quantities regarding the resonance behaviors interacting with incident electromagnetic waves were closely correlated due to the duality, as pointed out in Table 5.

4. Conclusion

The resonant power transmissions of various single and multiple subwavelength ridged circular aperture arrangements in a thin conducting screen were quantitatively investigated using equivalent circuits via the TCS with an admittance concept. An equivalent magnetic current element was introduced to effectively investigate the mutual coupling effect. As a result, for each case, the TCS for the ridged apertures was found to be inversely proportional to the summation of the conductances. Furthermore, the gain ratios in half-wavelength dipole antenna arrangements for both the maximum and minimum cases corresponded to the maximum and minimum TCSs for the current structures, respectively. Therefore, these observations identified the duality between coupled subwavelength ridged circular apertures and electric dipole antenna arrays. The current results may be useful for analyzing and/or synthesizing pairs of subwavelength metallic structures for such applications as optical data storage, high resolution near-field microscopy, and FSS.

Acknowledgments

The work of J.-E. Park and J.-W. Song was supported in part by the 2nd stage of Brain Korea 21 (BK21) at Kyungpook National University. The work of J.-E. Park was supported in part by the Korea Student Aid Foundation (KOSAF) Grant funded by the Korean government (MEST) (no. S2-2009-000-00793-1).

References

[1] H. A. Bethe, "Theory of diffraction by small holes," *Physical Review*, vol. 66, pp. 163–182, 1944.

[2] A. Roberts, "Electromagnetic theory of diffraction by a circular aperture in a thick, perfectly conducting screen," *Journal of the Optical Society of America A*, vol. 4, no. 10, pp. 1970–1983, 1987.

[3] R. F. Harrington, "Resonant behavior of a small aperture backed by a conducting body," *IEEE Transactions on Antennas and Propagation*, vol. 30, no. 2, pp. 205–212, 1982.

[4] Y. Leviatan, "Electromagnetic coupling between two half-space regions separated by two slot-perforated parallel conducting screens," *IEEE Transactions on Microwave Theory and Techniques*, vol. 36, no. 1, pp. 44–52, 1988.

[5] J.-I. Lee, J.-H. Ko, and Y.-K. Cho, "Electromagnetic transmission through slits in two adjacent conducting parallel-plates," *The Journal of Korean Institute of Electromagnetic Engineering and Science*, vol. 20, no. 1, pp. 83–90, 2009.

[6] Y. K. Cho, K. W. Kim, J. H. Ko, and J. I. Lee, "Transmission through a narrow slot in a thick conducting screen," *IEEE Transactions on Antennas and Propagation*, vol. 57, no. 3, pp. 813–816, 2009.

[7] J.-E. Park, J. Yeo, J.-I. Lee, J.-W. Ko, and Y.-K. Cho, "Resonant transmission of an electrically small aperture with a ridge," *Journal of Electromagnetic Waves and Applications*, vol. 23, no. 14-15, pp. 1981–1990, 2009.

[8] J. E. Park and Y. K. Cho, "Comparison of transmission resonance phenomena through small coupling apertures between two kinds of transmission resonance structures," in *Proceedings of the 20th URSI International Symposium on Electromagnetic Theory (EMTS '10)*, pp. 744–747, Berlin, Germany, August 2010.

[9] J.-H. Ko, Y.-K. Cho, J. Yeo, and J.-I. Lee, "A Study on the resonant transmission through a ridge-loaded small circular aperture," *The Journal of Korean Institute of Electromagnetic Engineering and Science*, vol. 22, no. 6, pp. 654–660, 2011.

[10] J. Yeo, J. H. Ko, J. E. Park, J. I. Lee, and Y. K. Cho, "Analysis of resonant transmission characteristics of two sub-wavelength apertures with a ridge located in parallel," in *Proceedings of the IEEE Antennas and Propagation on Symposium*, Toronto, Canada, July 2010.

[11] J.-I. Lee, Y.-K. Cho, J.-H. Ko, and J. Yeo, "Resonant transmission through a pair of ridge-loaded circular sub-wavelength apertures," *Progress In Electromagnetics Research M*, vol. 24, pp. 113–126, 2012.

[12] J.-E. Park, K. Y. Kim, and J.-W. Song, "Resonant power transmission through coupled subwavelength ridged circular apertures," *Journal of Electromagnetic Waves and Applications*, vol. 26, no. 4, pp. 423–435, 2012.

[13] J.-H. Ko, J.-I. Lee, and Y.-K. Cho, "Electromagnetic scattering by a two-dimensional periodic array of small resonant apertures," *The Journal of Korean Institute of Electromagnetic Engineering and Science*, vol. 22, no. 3, pp. 320–326, 2011.

[14] T. W. Ebbesen, H. J. Lezec, H. F. Ghaemi, T. Thio, and P. A. Wolff, "Extraordinary optical transmission through sub-wavelength hole arrays," *Nature*, vol. 391, no. 6668, pp. 667–669, 1998.

[15] E. X. Jin and X. Xu, "Finite-difference time-domain studies on optical transmission through planar nano-apertures in a metal film," *Japanese Journal of Applied Physics*, vol. 43, no. 1, pp. 407–417, 2004.

[16] K. Şendur, W. Challener, and C. Peng, "Ridge waveguide as a near field aperture for high density data storage," *Journal of Applied Physics*, vol. 96, no. 5, pp. 2743–2752, 2004.

[17] E. X. Jin and X. Xu, "Plasmonic effects in near-field optical transmission enhancement through a single bowtie-shaped aperture," *Applied Physics B*, vol. 84, no. 1-2, pp. 3–9, 2006.

[18] H. Guo, T. P. Meyrath, T. Zentgraf et al., "Optical resonances of bowtie slot antennas and their geometry and material dependence," *Optics Express*, vol. 16, no. 11, pp. 7756–7766, 2008.

[19] L. Wang and X. Xu, "High transmission nanoscale bowtie-shaped aperture probe for near-field optical imaging," *Applied Physics Letters*, vol. 90, no. 26, Article ID 261105, 2007.

[20] T. Grosjean, M. Mivelle, F. I. Baida, G. W. Burr, and U. C. Fischer, "Diabolo nanoantenna for enhancing and confining the magnetic optical field," *Nano Letters*, vol. 11, no. 3, pp. 1009–1013, 2011.

[21] B. Lee, I. M. Lee, S. Kim, D. H. Oh, and L. Hesselink, "Review on subwavelength confinement of light with plasmonics," *Journal of Modern Optics*, vol. 57, no. 16, pp. 1479–1497, 2010.

[22] L. Wang, S. M. Uppuluri, E. X. Jin, and X. Xu, "Nanolithography using high transmission nanoscale bowtie apertures," *Nano Letters*, vol. 6, no. 3, pp. 361–364, 2006.

[23] A. Kinkhabwala, Z. Yu, S. Fan, Y. Avlasevich, K. Müllen, and W. E. Moerner, "Large single-molecule fluorescence enhancements produced by a bowtie nanoantenna," *Nature Photonics*, vol. 3, no. 11, pp. 654–657, 2009.

[24] S. Park and J. W. Hahn, "Plasmonic data storage medium with metallic nano-aperture array embedded in dielectric material," *Optics Express*, vol. 17, no. 22, pp. 20203–20210, 2009.

[25] H. G. Frey, S. Witt, K. Felderer, and R. Guckenberger, "High-resolution imaging of single fluorescent molecules with the optical near-field of a metal tip," *Physical Review Letters*, vol. 93, no. 20, pp. 1–200801, 2004.

[26] M. Mansuripur, A. R. Zakharian, A. Lesuffleur et al., "Plasmonic nano-structures for optical data storage," *Optics Express*, vol. 17, no. 16, pp. 14001–14014, 2009.

[27] P. Zijlstra, J. W. M. Chon, and M. Gu, "Five-dimensional optical recording mediated by surface plasmons in gold nanorods," *Nature*, vol. 459, no. 7245, pp. 410–413, 2009.

[28] Z. Rao, L. Hesselink, and J. S. Harris, "High transmission through ridge nano-apertures on vertical-cavity surface-emitting lasers," *Optics Express*, vol. 15, no. 16, pp. 10427–10438, 2007.

[29] C. E. Talley, J. B. Jackson, C. Oubre et al., "Surface-enhanced Raman scattering from individual Au nanoparticles and nanoparticle dimer substrates," *Nano Letters*, vol. 5, no. 8, pp. 1569–1574, 2005.

[30] L. Tang, D. A. B. Miller, A. K. Okyay et al., "C-shaped nanoaperture-enhanced germanium photodetector," *Optics Letters*, vol. 31, no. 10, pp. 1519–1521, 2006.

[31] B. A. Munk, *Finite Antenna Arrays and FSS*, John Wiley & Sons, Hoboken, NJ, USA, 2003.

[32] A. Taflove and S. C. Hagness, *Computational Electrodynamics: The Finite-Difference Time-Domain Method*, Artech House, Norwood, Mass, USA, 3rd edition, 2005.

[33] P. S. Carter, "Circuit relations in radiating systems and applications to antenna problems," *Proceedings of the Institute of Radio Engineers*, vol. 20, no. 6, pp. 1004–1041, 1932.

[34] C. A. Balanis, *Antenna Theory*, Wiley-Interscience, New Jersey, NJ, USA, 3rd edition, 2005.

[35] R. F. Harrington, *Time-Harmonic Electromagnetic Fields*, chapter 4, McGraw-Hill, New York, NY, USA, 1961.

[36] U. Inan and A. Inan, *Electromagnetic Waves*, Prentice Hall, New Jersey, NJ, USA, 2000.

[37] H. G. Booker, "Slot aerials and their relation to complementary wire aerials (Babinet's principle)," *Journal of the Institution of Electrical Engineers A*, vol. 93, no. 4, pp. 620–626, 1946.

[38] L. Verslegers, Z. Yu, P. B. Cartysse, and S. Fan, "Temporal coupled- mode theory for resonant apertures," *Journal of the Optical Society of America B*, vol. 27, no. 10, pp. 1947–1956, 2010.

14

Wearable Quarter-Wave Folded Microstrip Antenna for Passive UHF RFID Applications

Thomas Kaufmann,[1,2] Damith C. Ranasinghe,[2] Ming Zhou,[2] and Christophe Fumeaux[1]

[1] *School of Electrical and Electronic Engineering, The University of Adelaide, Adelaide, SA 5005, Australia*
[2] *The Auto-ID Laboratory, The University of Adelaide, Adelaide, SA 5005, Australia*

Correspondence should be addressed to Thomas Kaufmann; thomaska@eleceng.adelaide.edu.au

Academic Editor: Charles Bunting

A wearable low-profile inset-fed quarter-wave folded microstrip patch antenna for noninvasive activity monitoring of elderly is presented. The proposed antenna is embedded with a sensor-enabled passive radio-frequency identification (RFID) tag operating in the ultra-high frequency (UHF) industrial-scientific-medical (ISM) band around 900 MHz. The device exhibits a low and narrow profile based on a planar folded quarter-wave length patch structure and is integrated on a flexible substrate to maximise comfort to the wearer. An extended ground plane made from silver fabric successfully minimises the impact of the human body on the antenna performance. Measurements on a prototype demonstrate a reflection coefficient (S_{11}) of −30 dB at resonance and a −10 dB bandwidth from 920 MHz to 926 MHz. Simulation results predict a maximum gain of 2.8 dBi. This is confirmed by tag measurements where a 4-meter read range is achieved using a transmit power of 30 dBm, for the case where the passive wearable tag antenna is mounted on a body in a practical setting. This represents an almost 40% increase in read range over an existing dipole antenna placed over a 10 mm isolator layer on a human subject.

1. Introduction

The miniaturisation of sensors and wireless systems is generating an explosive growth in body-centric wireless computing applications based on wearable electronic devices, especially in healthcare [1–6]. Consequently, the need for lightweight, low-profile, low cost, and wearable antennas has also grown rapidly during recent years. In this paper we present a successful design of such a wearable antenna for an emerging class of sensor-enabled passive radio-frequency identification (RFID) tags for use in healthcare applications.

The wireless identification and sensing platform (WISP) [7] is a passive RFID-based platform suitable for wireless wearable applications. In the present case, it integrates an accelerometer and a passive RFID tag into a batteryless system with the capability of being integrated into clothing. The intended application is the monitoring of high falls risk activities for elderly in hospitals and residential care facilities [1, 2]. Typical WISP devices are currently built using a dipole antenna on a FR4 substrate and as a result exhibit poor performance when worn in close proximity to a human body. In this case, the read range of the WISP is less than 2.5 m once attached to clothing worn by a person. This read range is inadequate for monitoring an elderly subject in a care facility. Furthermore, the thin dipole antenna on the WISP is fragile and can pose a significant discomfort to the wearer of the device while possible breakages can cause injury. Thus, a new antenna design is proposed to replace the dipole structure.

A few researchers have investigated wearable RFID tag antennas in the ultra-high frequency (UHF) range [8–11]. Most designs have been based on half-wave length dimensions, which correspond to approximately 16 cm in length at 923 MHz. In order to further reduce the physical dimensions, approaches including tapering in a bow-tie [9], meandering, [10], and meandering with a superstrate [11] have been considered. Further size reductions are achieved in different geometries using, for example, Planar Inverted F Antennas (PIFAs) [12], metamaterial superstrates [13], and folded shorted patches [14]. These designs usually use coaxial probe feeding mechanisms. However, in the context of wearable antennas, coaxial line fed antennas are (i) not easy to fabricate;

(ii) fragile since wire connections are likely to fail or break in a wearable application using a flexible substrate; and (iii) impractical because of the difficulties of feeding the balanced two-port WISP tag circuitry.

Currently, a major development for WISP antenna is focused on applying paper and Inkjet printing to lower the fabrication cost of standard designs while retaining acceptable performance [8]. In this paper we present a different geometry of a low profile antenna meeting stringent requirements to support wearability in WISP-based applications. This is achieved by applying a microstrip inset-feeding method in a quarter-wave folded patch antenna on a flexible substrate. The increased performance of our design improves the read range for a body worn WISP. In particular, the proposed approach delivers improved performance and low profile without sacrificing efficiency. It is also worth noting that the tag does not require high permittivity materials, which helps in satisfying the substrate flexibility requirement and reduces the cost of production. Furthermore, we also demonstrate a design procedure for integrating the proposed antenna design to the 2-port WISP chip without an external balun and thus greatly simplifying the production of the antenna and reducing cost, while enhancing wearability. To the best of our knowledge, this is the first wearable antenna design for a WISP.

The remaining sections of the paper are organised as follows Section 2 presents a brief overview of RFID technology, with the aim of introducing the performance metrics relevant for the analysis; Section 3 describes the antenna requirements and the proposed design with a parametric study of the antenna to justify the dimension selection and demonstrates the ability of the antenna to be tuned to a different input impedance; Section 4 describes the simulation studies of the antenna performance in the vicinity of a wearer; Section 5 presents experimental results validating the simulation results as well as practical measurements of tag antenna performance; and finally, Section 6 concludes this paper.

2. RFID Primer

Radio frequency identification (RFID) is a wireless technology capable of contactless and automatic identification of individual objects or people [1] without line of sight requirements.

2.1. RFID Systems. RFID systems can be categorised into three primary components: tags, readers, and back-end systems. Figure 1 demonstrates the basic structure of a RFID system.

2.1.1. Tags. Tages also called labels, contain a microchip controller that stores unique identifying information of the object or person to which the tag is attached and an antenna for communicating the information via radio waves [9]. When a tag is placed in a radio frequency (RF) field, it reflects back the information to the reader (modulated backscatter). Tags can be categorised into three types in terms of powering. Semi-passive tags have an on-board power source and may have on-board sensors. Active tags have an on-board power supply

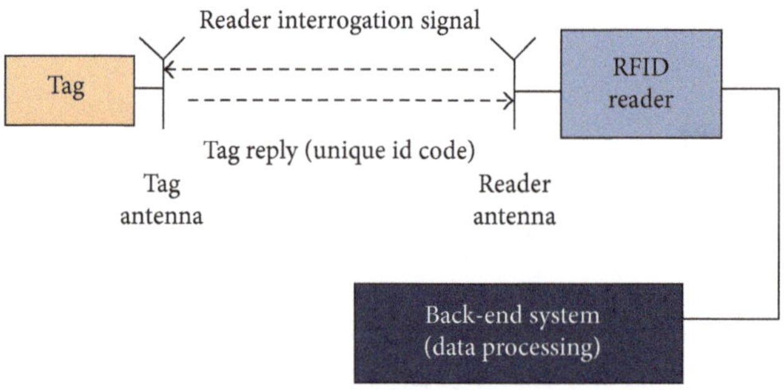

Figure 1: Basic RFID system block.

and also include an active receiver and a distinct active transmitter and may talk with each other and form a network. Passive tags do not have their own power supply and therefore have infinite life time as they obtain power from the incident electromagnetic (EM) radiation provided by the reading system.

2.1.2. RFID Readers. Readers are devices that communicate with the individual tags to interrogate and power them. The readers are usually distributed in fixed locations chosen to maximise the coverage of a given area. Modern RFID readers are capable of reading multiple tags (several hundred) at the same time.

2.1.3. Back-End Systems. A network of computers is connected to the readers to collect, process, and store data gathered in the interrogation process.

2.2. Tag Antenna Performance Metrics. While key characteristic measures exist for measuring the performance of antennas (such as radiation efficiency), there are a number of additional performance measures that must be investigated to ensure performance of an antenna design in a practical RFID application. In this paper we have used the following metrics: (i) received signal strength indicator (RSSI), (ii) read rate, and (iii) read range as described below.

2.2.1. Receive Signal Strength Indicator. The RSSI indicates the power level of the backscattered tag response detected by the reader. Therefore, a larger RSSI value is indicative of a stronger received signal from the WISP. The power of an RF signal received by a monostatic reader antenna can be given as [15]

$$\mathrm{RSSI} = P_t G_t^2 G_{\mathrm{path}}^2 K. \tag{1}$$

In this equation, P_t is the output power of the reader, G_t is the gain of the reader antenna, K is the WISP backscatter gain, and G_{path} is the path gain of the deterministic multipath channel,

$$G_{\mathrm{path}} = \left(\frac{\lambda}{4\pi d_o}\right)^2 |H|^2, \tag{2}$$

where d_o is the direct path length and H is the channel response due to multipath. As visible in (1) and (2), the RSSI

TABLE 1: Antenna requirements and specifications.

Operational frequency band	920–926 MHz ISM band in Australia (regulated by Australian Communications and Media Authority).
Impedance	The WISP has a measured input impedance of $78 + 16j\ \Omega$ after an on-chip matching network. A good impedance matching is crucial for maximum performance.
Read range	The tag should have a suitable read range to properly track the human. The current dipole has a read range of 2.5 m; it is desired to increase this range. It is also desirable to have a low directivity, that is, semidirectional pattern, for good angular coverage.
Materials	The antenna must be flexible so that it can be subject to bending and can be worn by a human.
Reliability	The tag should not be easily detuned because of its location on or near the human body.

is inversely proportional to the quadruple power of distance and can be used as a measure to compare the relative performance of tags operating at various reader transmit power levels and reading distances.

2.2.2. Read Rate. Upon interrogation by a reader, an RFID tag replies a unique identification code such as an electronic product code or EPC. The tag read rate is the number of successful tag reads per second as obtained by the RFID reader over the entire frequency hopping spectrum, for example, 920 MHz–926 MHz ISM band in Australia.

2.2.3. Read Range. The read range is the maximum line of sight distance from a reader antenna to the WISP tag for which successful interrogation can be achieved. It can be defined both in a real environment or a reflection free environment such as an anechoic chamber.

Based on the Friis transmission equation (3) [7], the theoretical maximum read range can be expressed in dB according to

$$P_r = P_t - 20\log\left(\frac{4\pi d_o}{\lambda}\right) + G_t + G_r - L_p. \qquad (3)$$

In (3), the free space wavelength is $\lambda = 325$ mm at the centre frequency of 923 MHz; the power transmitted by the reader is $P_t = 1$, W = 30 dBm, and G_t and G_r are the gain of the reader antenna and the tag antenna, respectively. The power required to turn on the WISP chip is $P_r = -9.5$ dBm [7] and L_p is the polarisation loss. Evaluating the theoretically possible range sets an upper bound on the expected values for read range measurements. These computations aid in the design of the overall system.

3. Antenna Design

The antenna design aims at converting the WISP to a wearable WISP (W^2ISP). The term WISP generally refers to a completely passive RFID tag supporting EPC Class 1 Generation II RFID protocol [7]. It is a sensor-enabled and programmable tag realisation that has been demonstrated for applications in human activity recognition [1, 2] when worn over the chest (above the sternum). However, one key issue remains the wearability of the tag [1].

An essential requirement for a wearable RFID-based monitoring system is to ensure the comfort of the subject to which the tag is attached, for example, patients in acute care or elderly in residential care. This is especially important for elderly patients in a prone position during sleep. In this case, the centre chest position has been identified as most appropriate, as an object placed on the sternum will likely have a minimal negative impact on the wearer's comfort. Hence, the size of the antenna must be restricted into a space of 35 mm × 90 mm × 5 mm and the construction materials must be flexible and soft. The design requirements for the antenna to develop a wearable and passive human activity monitoring device are summarised in Table 1.

3.1. Antenna Structure. Figure 2 shows the antenna structure. In order to limit the area occupied by the antenna to the sternum of a human body, the ground plane is limited to a width of 35 mm (0.11λ) and a length of 83 mm (0.255λ). Given the very restrictive constraints from Table 1, we have chosen to design a folded inset-fed microstrip patch structure with effective length of 0.25λ ($L = 79$ mm). Folding the patch can be interpreted as the introduction of a metallic wall symmetry in the centre of a microstrip antenna. This effectively reduced the length in half compared to the original dipole design.

In the design of the folded microstrip antenna, the choice of substrate is important in terms of antenna performance. The radiation efficiency of a patch antenna depends strongly on the electrical properties of the substrate. A thicker substrate with lower dielectric constant has better efficiency, increased bandwidth, and antenna gain at the cost of increased size [16]. The substrate selected in the present design is a highly flexible Cuming Microwave C-Foam PR-4, which is a closed-cell low loss flexible foam. The relative permittivity is $\varepsilon_r = 1.06$ and the loss tangent is $\tan\delta = 0.0001$. The height is chosen as $h = 3.2$ mm as a compromise between wearability and performance. The exceptional flexibility and softness of this material are of paramount importance considering the comfort and safety of the antenna wearers.

Due to the small width of the antenna, the input impedance is very high at the first edge of the resonator on the feed side. This can be addressed using a rather long inset feed. The inset-feed point (y_0) is 13 mm away from the shortened edge, which denotes the effective center of the patch. Using such a low-permittivity substrate, a relatively wide microstrip line of 8 mm would be required to obtain 78 Ohms characteristic impedance. This would be too wide to efficiently feed the antenna which has a specified maximum width of 30 mm. Furthermore, slots with a nonneglectable width have to be incorporated to allow for the inset feed. This is not practical,

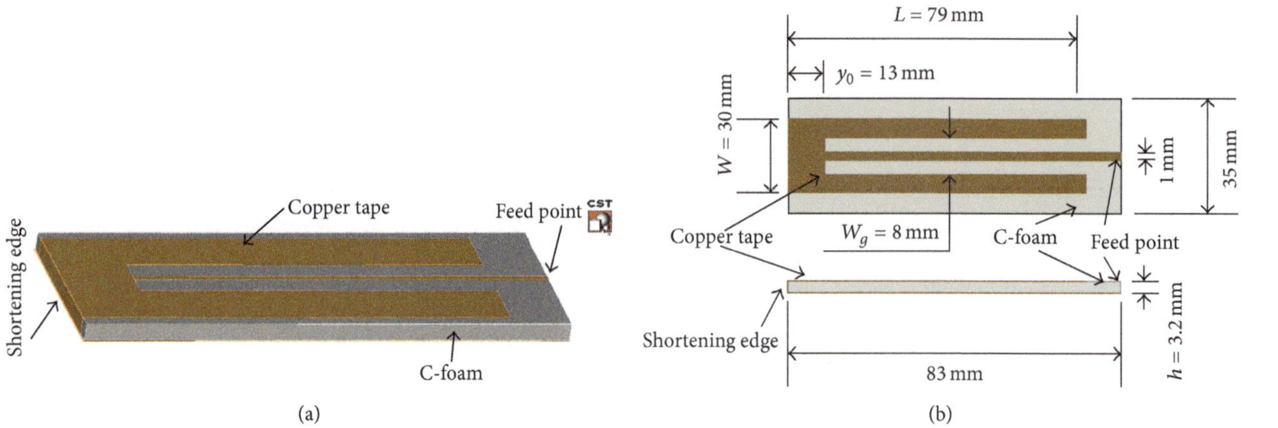

Figure 2: Structure of quarter-wave folded inset-fed rectangular patch antenna placed on silver fabric ground plane in three-dimensional view (a) and with dimensions (b).

as the effective width of the radiating elements would be severely reduced, yielding a degraded antenna performance. Instead, in an optimisation process, the width of the inset-feed line is reduced to 1 mm. Considering the electric field distribution, the overall inset width, including the slots and feeding strip, is selected as $W_g = 8$ mm. The resulting radiating elements have a width of 11 mm on each side.

By using a discrete voltage port rather than a microstrip line wave port in the simulations, we have also been able to show that the thin inset feed allowed the antenna to be fed using a balanced structure that can be directly connected to the two-port WISP chip output (see Section 4.4). The proposed antenna is modelled and simulated using CST Microwave Studio 2012 using a 78 Ω discrete port to provide an approximate match to the input impedance of the unbalanced WISP chip.

The physical dimensions are intentionally chosen to be integer millimetre values to ensure that the antenna remains practicable and simpler to fabricate. The conductor used to construct the antenna is copper foil (0.035 mm thick). A nonconductive adhesive attached to the foil allows a reliable construction of the antenna.

The power gain is defined as the product of the directivity and the radiation efficiency of the antenna. A directivity of 5–8 dB can be expected from the cavity model [16] for a rectangular microstrip patch. However, the actual values are expected to be severely affected by the limited size of the antenna ground plane as well as the close proximity of a human body as discussed later in Section 4. Therefore, to decouple the antenna from the effects of the human body while providing an extended ground plane, a silver fabric ground plane is added to the realized design. The RIPSTOP silver fabric sheet with dimensions of 200 × 200 mm is selected for this purpose, because of its low sheet resistivity of less than 0.25 Ω/sq.

3.2. Study of the Critical Design Parameters. This section considers the impact of various parameters on the resonance frequency and reflection coefficient (S_{11}) in the operation band. These considerations are important, especially for an RFID

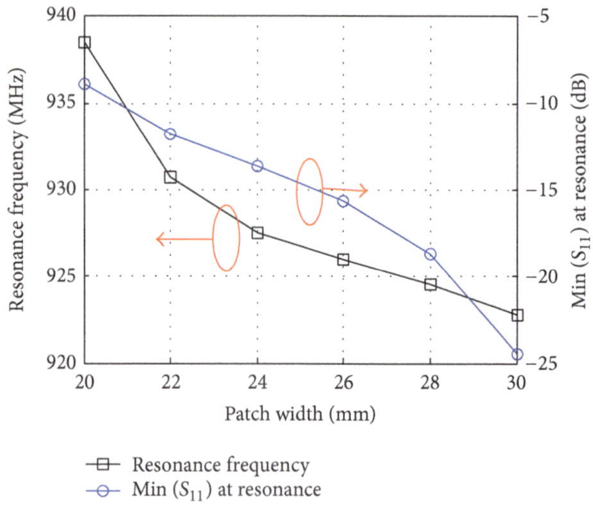

Figure 3: Simulated antenna resonant frequency and minimum reflection coefficient within a specific bandwidth versus variation of patch width W.

tag antenna designer to (i) select appropriate physical dimensions to optimise the performance of the design and (ii) to tune the final antenna once connected to the WISP. Our main focus is to achieve a high impedance match and to increase the gain of the antenna.

3.2.1. Width of the Radiation Patch. The width of the antenna patch (W) controls the input impedance at the first edge of the patch and the radiation pattern. Figure 3 shows that with increasing patch width W, the antenna tends to resonate at lower frequencies and has a better impedance match (lower reflection coefficient) to the target input impedance for a fixed inset point. Hence, in order to maximise the match to the WISP's input impedance, the patch width is chosen to be the largest value permitted within the dimension restrictions (see Table 1).

3.2.2. Inset-Feed Position. Figure 4 shows the reflection coefficient according to the change in feed position. The antenna feed input impedance can be lowered by shifting the inset-feed position y_0 towards the folded edge. The minimum value

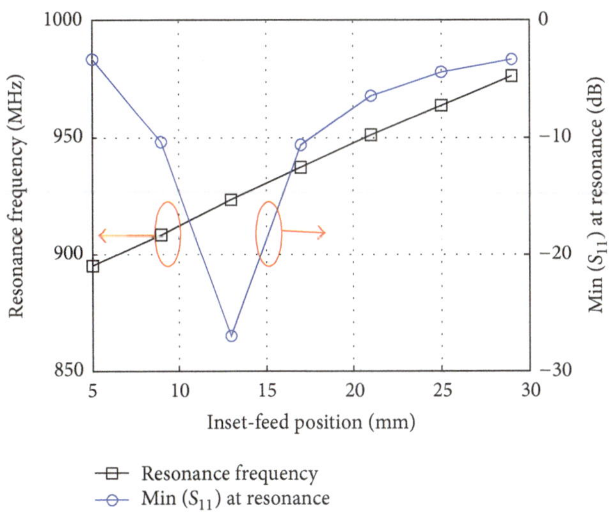

FIGURE 4: Simulated antenna resonant frequency and minimum reflection coefficient within the specified bandwidth versus variation of feed position y_0.

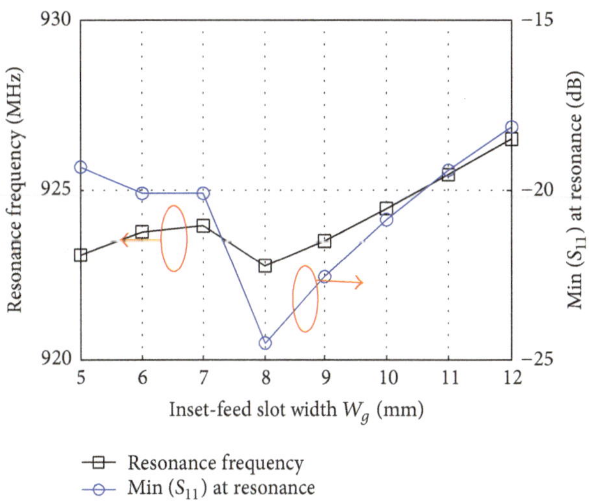

FIGURE 5: Simulated antenna resonant frequency and minimum reflection coefficient within the specified bandwidth versus variation of gap width.

of reflection coefficient varies accordingly and attains a minimum at a specific distance $y_0 = 13$ mm at the centre frequency of the operation bandwidth, at 923 MHz. Figure 4 also illustrates that the resonant frequency increases when the feed position is shifted away from the shorting edge of the patch.

3.2.3. Inset-Feed Slot Width. The reflection coefficient and the corresponding resonant frequency are also dependent on the width of the overall inset-feed width W_g. The fringing field distribution along the feed line can influence the radiating pattern as additional, orthogonally placed effective radiators. If the inset-feed slot is too wide, the antenna will radiate from these positions. If the width is too small, the fields of the travelling wave in the transmission line will couple into the microstrip resonator. It can be observed from the field distribution that the antenna starts to radiate through the inset-feed slots when they become larger than 8 mm. This is manifested in Figure 5 as discontinuity in the curves.

4. Impact of Human Body

For the intended application of the device in human activity recognition, the antenna is by nature placed near a human body. Thus, the effects of human tissue in close proximity, such as skin or fat tissue, must be included in our considerations. In simulation models, the antenna is placed at distances of $d_h = 3$ mm, 10 mm, and 20 mm from the human body model consisting of a 1 mm thick skin layer and 5 mm thick fat tissue layer. The dielectric constants chosen in this design are $\varepsilon_r = 31.2$ for skin and $\varepsilon_r = 4.60$ for fat, while the conductivities are $\sigma = 0.585$ S/m for skin and $\sigma = 8.01$ S/m for fat.

For this investigation of the body effects, the standalone antenna, that is, without an extended ground plane, is modelled first. Afterwards, a 200 × 200 mm layer of silver fabric is placed below the antenna to extend the ground plane and reduce the impact of the human body. For each situation, the reflection coefficient and radiation patterns are compared.

4.1. Standalone Antenna without Silver Fabric Ground Plane. The reflection coefficient of the standalone antenna without the silver fabric is shown in Figure 6 in free space and in the vicinity of the human body model. A severe detuning can be observed, with a widening of the curve resonant dip suggesting large losses, when the human body is placed close to the body model. Figure 7 shows the radiation pattern of the simulated antenna in free space and at a distance of 3 mm to the tissue at 923 MHz. In free space, due to the very limited size of the ground plane, a gain of 1.3 dBi is predicted from simulations. In the vicinity of the body, the antenna performance is significantly degraded and the maximum gain is decreased to −2.1 dBi as a consequence of the severe coupling into the human tissue.

4.2. Elimination of Human Body Attenuation Using a Silver Fabric Ground Plane. To minimise the attenuation caused by the human body as well as to increase both the gain and efficiency of the antenna, a large ground plane is added to this design. For this purpose, a sheet of flexible RIPSTOP silver fabric is attached to the bottom copper plane of the antenna. The use of silver fabric facilitates the integration of the antenna into clothing while maintaining the flexibility of the original design.

Figure 8 demonstrates the effectiveness of the sheet of silver fabric in eliminating the detuning and attenuation caused by a human body. The S_{11} characteristics experience no noticeable variation when the antenna is placed in proximity of the human body model, and the overall performance is highly insensitive to the distance d_h from the tissue model. The same statement can be made about the simulated radiation characteristics as shown in Figure 9, where the gain patterns are almost identical in both simulated xy- and xz-principal planes.

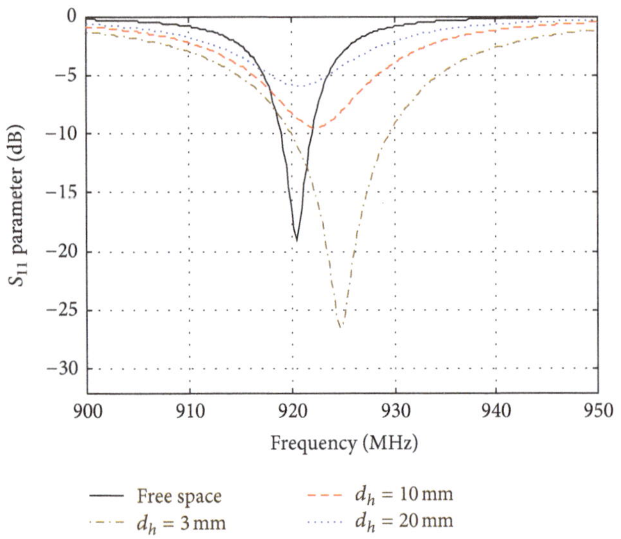

FIGURE 6: Simulated antenna (without silver fabric) reflection coefficient in free space at various distances from the human body.

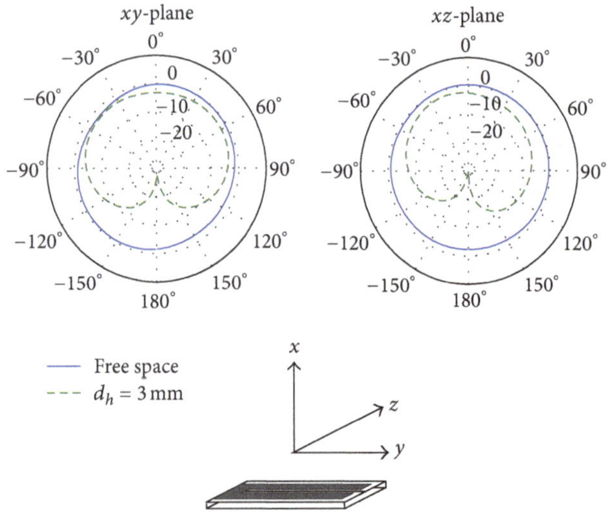

FIGURE 7: Simulated standalone antenna gain (without silver fabric) at a frequency of 923 MHz.

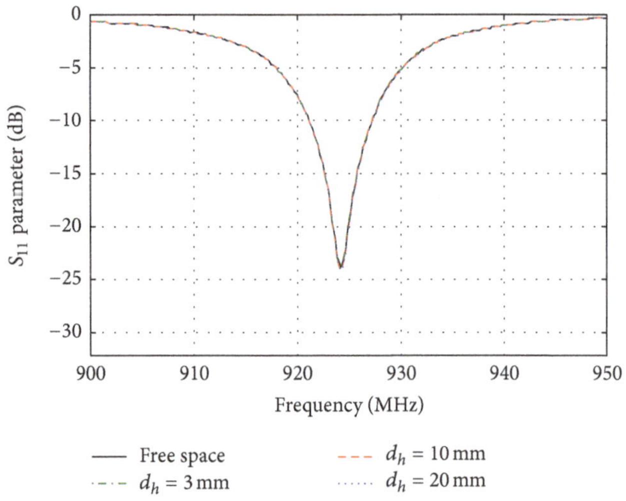

FIGURE 8: Simulated antenna (with silver fabric) reflection coefficient in free space and for various distances to the human body.

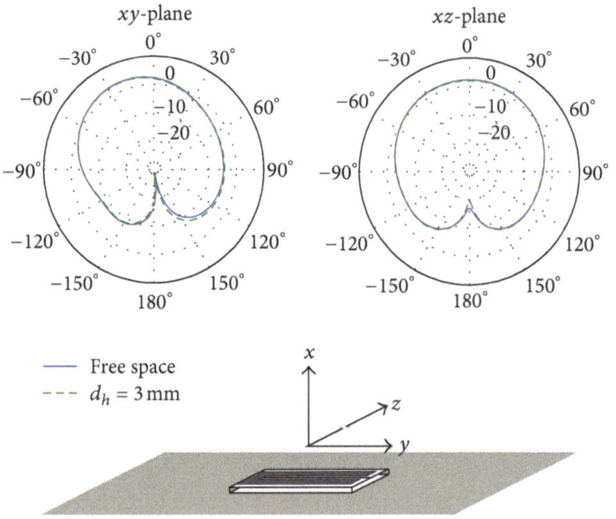

FIGURE 9: Simulated gain of the final design (with silver fabric) at a frequency of 923 MHz.

4.3. Final Design. Based on the parametric study in Section 3.2 and the results presented in this section, the antenna structure in Figure 1 is fabricated with a 200 × 200 mm silver fabric ground plane. The free space simulation results in Figure 8 indicate that the antenna resonates at f_r = 924.23 MHz and has an impedance bandwidth of 6.6 MHz defined for $|S_{11}| < -10$ dB. The predicted maximum gain is 2.8 dBi as shown in Figure 9, with an efficiency of around 60%, where the losses can be mainly attributed to the imperfect conductivity of the silver fabric.

4.4. Feeding the Antenna with a WISP. The connection from the two ports of the WISP chip to the antenna is realised using parallel metallic wires. This is illustrated in Figure 10 and functions as narrow band balun. The effect of the exact geometry of this arrangement on the resonant frequency and minimum reflection coefficient is investigated through a parametric study. The two parameters of interest are the radius r and length l of the balanced paired-wire transmission line. The simulated range for the radius is $r = [0.025, 0.05, 0.1, 0.25, 0.5]$ mm and the wire length is $l = [0, 2.5, 5, \dots, 17.5, 20]$ mm.

Figure 11 shows variations of the resonance frequency as a function of the wire length l. It can be observed that longer wires cause a detuning towards lower frequencies. This effect is more pronounced with thinner wires. To further illustrate the effect, the maximum reflection coefficient within the specified frequency band is plotted in Figure 12 and indicates a change in the bandwidth. The results again show that for wire lengths longer than l = 5 mm, the connection method is unsuitable, and even more so for thin wires. Figure 12 also shows that an optimum length of l = 5 mm yields an acceptable return loss for all radii. Therefore, this wire length l = 5 mm and a radius of 0.1 mm are selected to connect the antenna to the WISP chip.

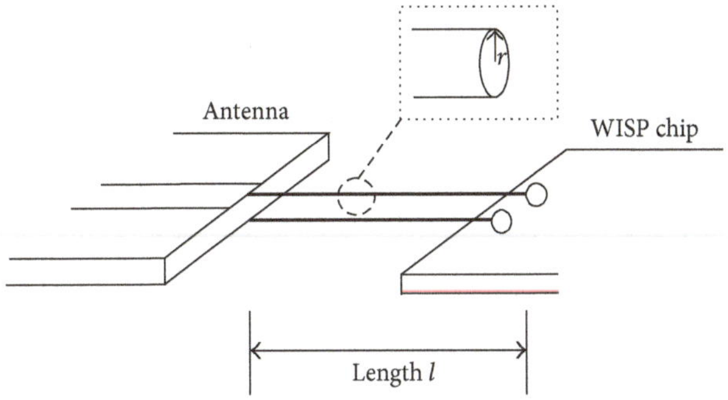

FIGURE 10: Connection between the antenna and the WISP chip.

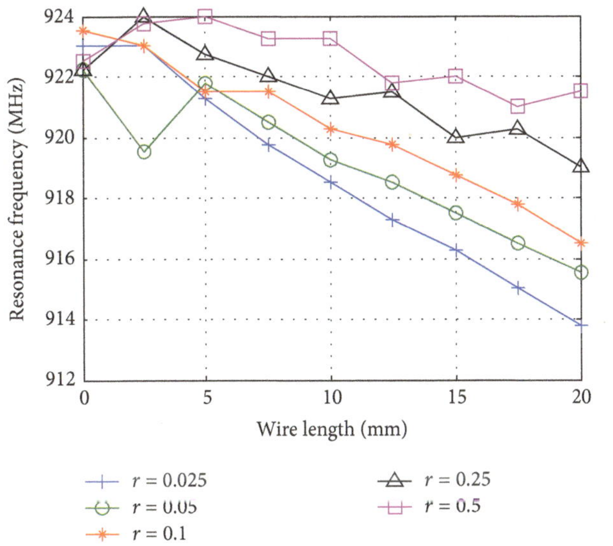

FIGURE 11: Simulated resonance frequency for different wire lengths l and radius r.

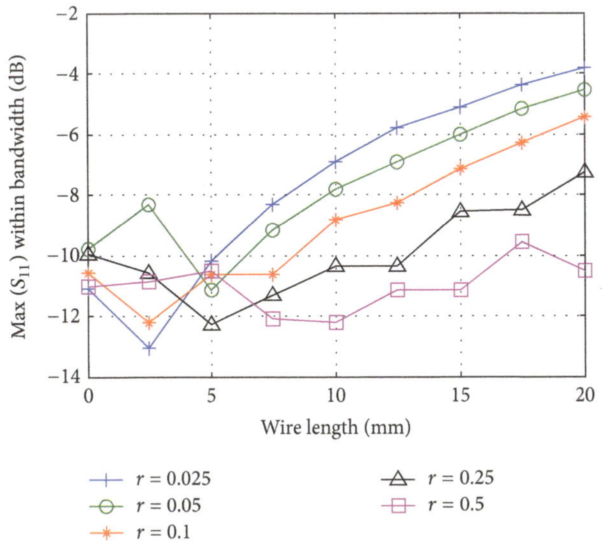

FIGURE 12: Simulated maximum reflection coefficient within the specified bandwidth from 920 to 926 MHz for different wire lengths l and radius r.

FIGURE 13: Fabricated antenna with SMA connector for impedance measurements.

5. Experimental Results

This section describes the measured results obtained from a manufactured antenna prototype and compares them with the predictions from simulations presented in Section 4. The experiments include measurement of the antenna input impedance, reflection coefficient, RSSI, read rate, and read range tests. The performance testing of the WISP [2] has been carried out first in an anechoic box and then worn on a human subject in a practical operating environment (laboratory).

5.1. Antenna Reflection Coefficient. Two similar antennas have been fabricated and the measurements of reflection coefficient were carried out by connecting the fabricated antenna to a vector network analyser using a 50 Ω SMA connector, as shown in Figure 13. The validity of measuring the reflection coefficient for determining the resonance frequency and bandwidth using a connector of slightly different impedance has been confirmed in separate simulations. During the measurement phase, the antenna was tuned by progressively adding copper tape such that the inset-feed point was eventually placed at y_0 = 14 mm from the shorting edge. This tuning step was necessary to meet resonance at 923 MHz, that is, at the centre frequency in the 920–926 MHz band. Figure 14 shows the measured reflection coefficient of the two antennas and −10 dB bandwidths are listed in Table 2.

Comparing the two manufactured antennas, *Antenna 1* (see Table 2) shows a better reflection coefficient and has been chosen for connection to the WISP chip. In the original version, the WISP chip is connected to a dipole. So, the original dipole was removed as illustrated in Figure 15. The WISP was then soldered onto the folded microstrip patch antenna using two short wires. The device with WISP chip connected to the proposed folded patch antenna is shown in Figure 16. The silver fabric was attached using conductive epoxy.

5.2. Wearable WISP Tag Performance in Anechoic Box. The next set of experiments compares the performance of the W^2ISP, that is, the wearable WISP with the inset-fed microstrip patch structure, to the performance of the original WISP including a half-wave dipole structure. In the setup for these

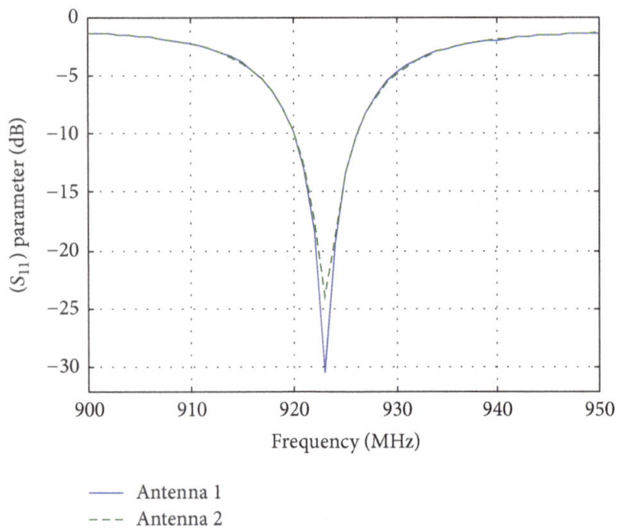

FIGURE 14: Measured reflection coefficient of fabricated antennas with silver fabric ground plane.

TABLE 2: Measured −10 dB bandwidth of fabricated antennas.

Antenna No.	f_r (MHz)	−10 dB BW (MHz)	−10 dB Band (MHz)
Antenna 1	923.07	6.07	920.04–926.11
Antenna 2	923.11	6.10	920.07–926.17

measurements, a circular right-hand circularly polarised (RHCP) reader antenna with a gain of $G_t = 6$ dBi has been used. Considering that the tag antennas are horizontally polarised, a polarisation loss factor of $L_p = 3$ dB has to be taken into account in the power budget calculation. Using the simulated gain, the calculated maximum read ranges in free space based on the Friis transmission equation (3) are $r = 4.7$ m for the folded microstrip antenna and $r = 4.4$ m for the dipole antenna. The estimated read range for the dipole antenna assumes no coupling onto the human body and the performance can potentially be severely degraded due to losses in the human body. In a practical complex indoor multipath environment, these read ranges can be further reduced in worst case scenarios.

An Impinj Speedway R420 RFID reader was used for obtaining the RFID performance metrics. The frequency range was set to Australian UHF ISM (industrial-scientific-medical) band used for RFID (920 MHz to 926 MHz). The WISP was activated by the interrogation signal received from the Impinj reader [1].

For comparison, all of the measurements of the dipole and the proposed folded microstrip antenna have been performed under identical environmental conditions, that is, placed in an anechoic box at the same distance from the reader, as shown in Figure 17. This ensures the consistency of the results since both devices are affected by the same environmental factors, such as reflections and interference from other equipments operating in the same band.

The first measured performance indicator is the RSSI, described by (1). RSSI values were measured at distances from

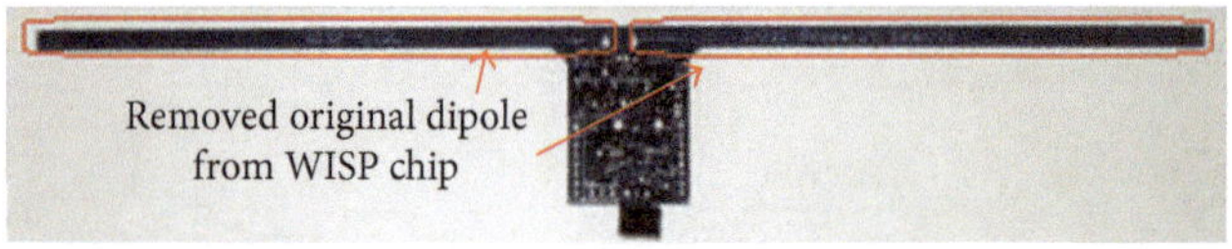

FIGURE 15: The original dipole is cut off from the WISP and the new folded microstrip antenna is connected to the balanced port in the centre. The length of the original dipole is 145 mm.

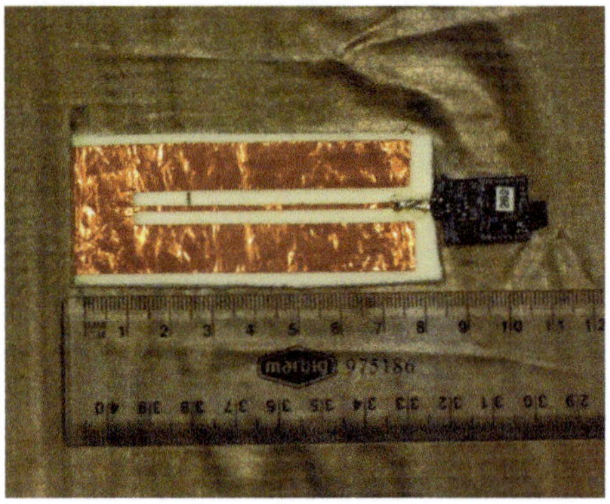

FIGURE 16: WISP connected to the proposed antenna. The silver fabric is attached using conductive epoxy.

0.5 meter to 2 meters for both the dipole antenna and the wearable folded microstrip patch antenna without a human body in its vicinity. Comparison of the RSSI values displayed in Figure 18 clearly scales with the higher maximum gain of the wearable antenna design and the results indicate that the proposed wearable antenna receives on average 1 dB stronger signals than the dipole. These results correspond well with the expected results, as the simulations predict a gain of 2.1 dB for the dipole versus a gain of 2.8 dB for the wearable folded microstrip antenna. It is emphasized that this slight gain-related improvement of the performance is achieved in free space configuration and is only a small contribution to overall more significant improvement achieved when worn on a body. In addition, the main beam of the wearable microstrip antenna is pointed slightly upwards with its maximum at approximately 30 degrees elevation (see Figure 11). This is a desirable characteristic for the human activity monitoring application since the reader antennas are likely to be mounted on the ceiling.

Figure 19 shows the read rates (see Section 2.2) of the two antennas at different transmitted power levels. Again, the tests have been performed using the experimental arrangement shown in Figure 17, while using different transmit power levels at a distance of 1 meter. The proposed wearable antenna consistently provides better read rates at all power levels. The minimum reader transmit power at which the wearable WISP could be read is at 19 dBm, while the dipole-based WISP stopped being activated below 22 dBm. The W^2ISP is successfully read by the reader using a lower signal power and consequently a longer read range is expected.

Using the experimental setup in Figure 17, the performance of the tag in different directions has been measured to ensure satisfactory tag read performance. The read rate

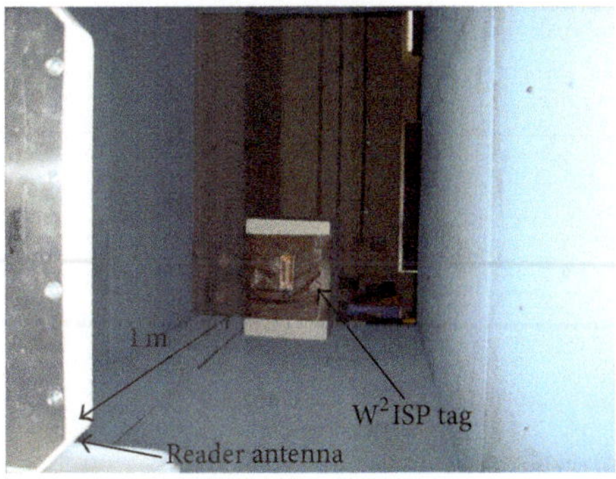

FIGURE 17: Anechoic box experimental arrangement.

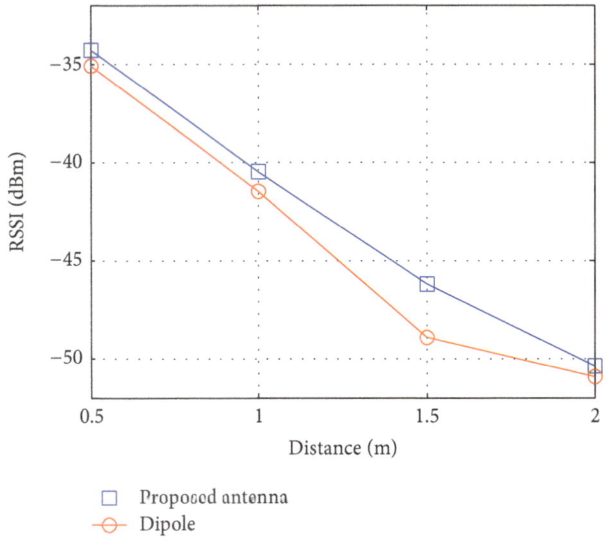

FIGURE 18: Measured RSSI versus distance in the anechoic box without a human body in the vicinity.

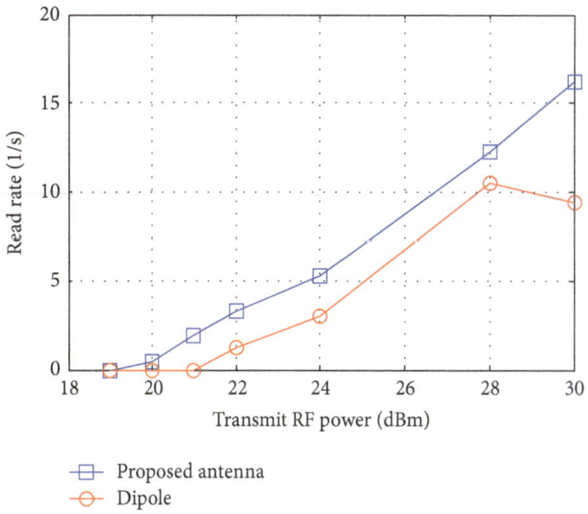

FIGURE 19: Measured read rates versus transmitted power at a fixed distance of 1 meter in the anechoic box without a human body close by.

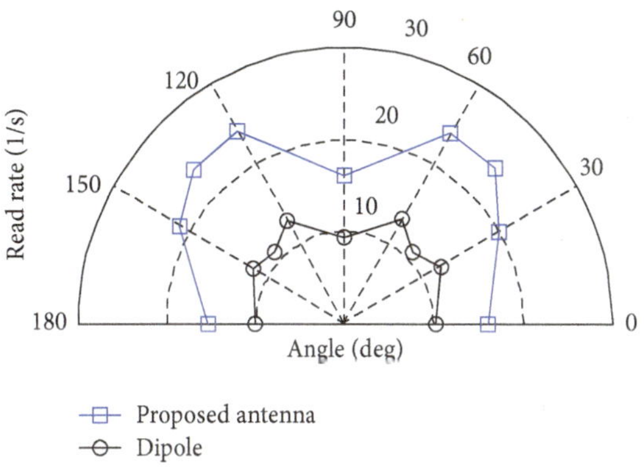

FIGURE 20: Measured read rates versus antenna facing angle in the anechoic box.

shown in Figure 20 is a good indicator of the gain of the antenna. When varying the orientation of the antenna with respect to the reader in ±90 degree angular range, the proposed folded microstrip antenna achieves consistently higher read rates than the original dipole. The slight reduction in read rate at an angle of 0 degree can be explained through multipath in the imperfectly anechoic measurement environment.

5.3. Wearable WISP Tag Performance on a Human Body. Measurements were performed to compare the performance of the wearable WISP when worn over clothing by a human subject, including the isolating silver fabric extended ground as shown in Figure 21, and compared to the performance of the original dipole. In order to reduce the detuning of the dipole, a polystyrene isolator with a thickness of 10 mm is used as shown in Figure 22.

The read range measurements were performed under real-world conditions in a laboratory room at different distances using a transmit power of $P_t = 1$, W = 30 dBm. It can be observed from Figure 23 that the read range performance of the proposed wearable antenna is significantly improved from the original dipole-based WISP. The proposed antenna meets all the requirements outlined in Table 1. The original WISP with the dipole antenna fails to power the WISP beyond 2.5 m. This degraded performance compared to the 4.4 m estimated from the Friis transmission equation (3) is expected because of the losses arising from coupling of the dipole antenna into the human body. The WISP with the new wearable folded microstrip patch antenna continues to be read up to a maximum read range of 4 m. These results are in good agreement with the theoretical maximum read range of 4.7 m calculated with the Friis transmission equation (3).

6. Conclusion

In this paper, a new implementation of a wearable, flexible antenna for the RFID sensor platform WISP has been proposed. The specifications require the whole system to be small and comfortable enough to be worn over a long period of time in the chest area for human activity recognition, for example, fall detection in elderly care.

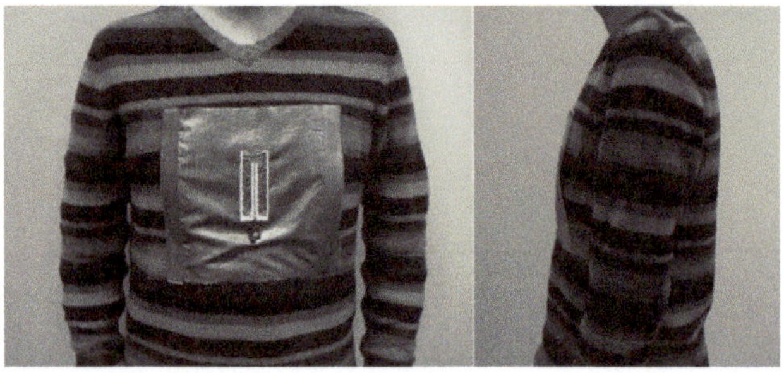

FIGURE 21: Proposed folded microstrip antenna worn on human body.

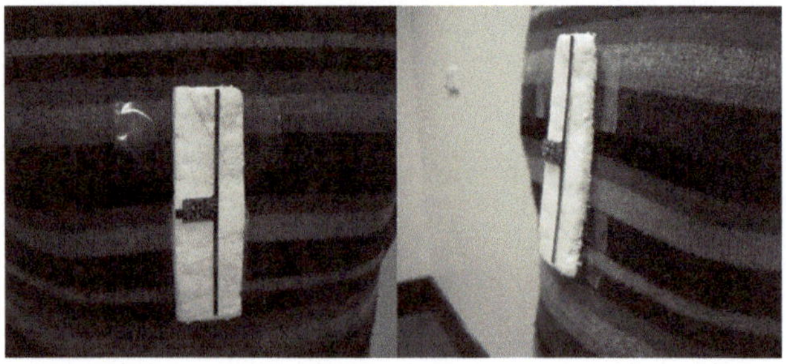

FIGURE 22: Dipole antenna worn on human body. The thickness of the polystyrene is about 10 mm, which leaves enough space from human body to the WISP tag.

The main features of the new antenna are its low-profile, its flexibility, and a reliable read range. The simulated and experimental results demonstrate that the proposed design satisfies all the W^{2}ISP system requirements. The following features can be concluded for this antenna design.

(1) The folded microstrip implementation allows to reduce the size of the antenna to only half that of the original dipole design.

(2) An inset feed instead of a coaxial line feed can simplify the manufacturing process and improve the ruggedness of the antenna, especially when using a flexible substrate.

(3) The tag is low profile and flexible, and therefore, it is suitable for wearable applications.

(4) With the sheet of silver fabric, the detuning and attenuation arising from the human body can be minimised while the gain of the antenna is improved.

(5) Comparing to the original dipole, the read rate of the antenna is increased while the size is reduced. The 4-meter read range achieved satisfies current experimental requirements of the WISP application [1, 2].

Based on this design, further studies will be conducted with the goal of improving the usability in a real-world environment.

Acknowledgments

This research was supported by The Hospital Research Foundation (THRF) and the Australian Research Council (DP130104614). C. Fumeaux acknowledges the support by the Australian Research Council DP120100661. The authors would like to thank Professor P.H. Cole at the University of Adelaide for his support and guidance and Dr. J. R Smith and A. P. Sample at the University of Washington for their support of this research project.

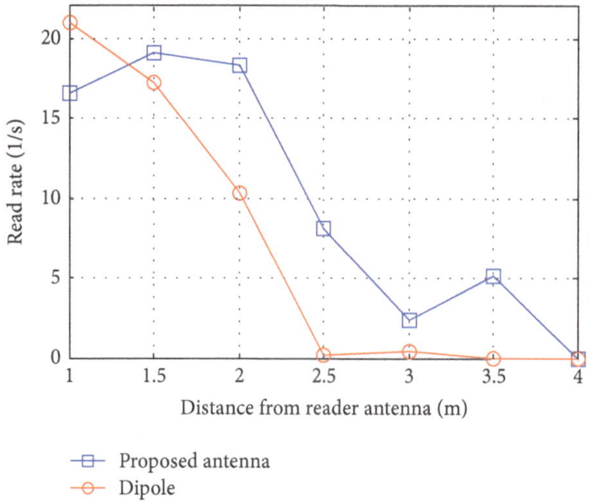

FIGURE 23: Measured read rates in the laboratory room with the antenna being worn on a human body.

References

[1] D. C. Ranasinghe, R. L. S. Torres, A. P. Sample, J. R. Smith, K. Hill, and R. Visvanathan, "Towards falls prevention: a wearable wireless and battery-less sensing and automatic identification tag for real time monitoring of human movements," in *Proceedings of the 34th Annual International IEEE EMBS Conference*, 2012.

[2] R. Visvanathan, D. C. Ranasinghe, and R. L. S. Torres, "Framework for preventing falls in acute hospitals using passive sensor enabled radio frequency identification technology," in *Proceedings of the 34th Annual International IEEE EMBS Conference*, 2012.

[3] B. H. Calhoun, J. Lach, J. Stankovic et al., "Body sensor networks: a holistic approach from silicon to users," *Proceedings of the IEEE*, vol. 100, no. 1, pp. 91–106, 2012.

[4] M. Kangas, A. Konttila, I. Winblad, and T. Jämsä, "Determination of simple thresholds for accelerometry-based parameters for fall detection," in *Proceedings of the 29th Annual International Conference of IEEE-EMBS, Engineering in Medicine and Biology Society (EMBC '07)*, pp. 1367–1370, Lyon, France, August 2007.

[5] A. K. Bourke, J. V. O'Brien, and G. M. Lyons, "Evaluation of a threshold-based tri-axial accelerometer fall detection algorithm," *Gait and Posture*, vol. 26, no. 2, pp. 194–199, 2007.

[6] T. E. Lockhart, A. T. Barth, X. Zhang, R. Songra, E. Abdel-Rahman, and J. Lach, "Portable, non-invasive fall risk assessment in end stage renal disease patients on hemodialysis," in *Proceedings of the 1st Wireless Health Conference (WH '10)*, pp. 84–93, October 2010.

[7] A. P. Sample, D. J. Yeager, P. S. Powledge, A. V. Mamishev, and J. R. Smith, "Design of an RFID-based battery-free programmable

sensing platform," *IEEE Transactions on Instrumentation and Measurement*, vol. 57, no. 11, pp. 2608–2615, 2008.

[8] M. Marroncelli, D. Trinchero, V. Lakafosis, and M. M. Tentzeris, "Concealable, low-cost paper-printed antennas for WISP-based RFIDs," in *Proceedings of the 5th IEEE International Conference on RFID (RFID '11)*, pp. 6–10, April 2011.

[9] A. Rida, L. Yang, R. Vyas, and M. M. Tentzeris, "Conductive inkjet-printed antennas on flexible low-cost paper-based substrates for RFID and WSN applications," *IEEE Antennas and Propagation Magazine*, vol. 51, no. 3, pp. 13–23, 2009.

[10] K. V. S. Rao, P. V. Nikitin, and S. F. Lam, "Antenna design for UHF RFID tags: a review and a practical application," *IEEE Transactions on Antennas and Propagation*, vol. 53, no. 12, pp. 3870–3876, 2005.

[11] B. A. Sani, M. Rajab, R. Foster, and Y. Hao, "Antennas and propagation of implanted RFIDs for pervasive healthcare applications," *Proceedings of the IEEE*, vol. 98, no. 9, 2010.

[12] L. Mo and C. Qin, "Tunable compact UHF RFID metal tag based on cpw open stub feed PIFA antenna," *International Journal of Antennas and Propagation*, vol. 2012, Article ID 167658, 8 pages, 2012.

[13] H. Attia, M. M. Bait-Suwailam, and O. M. Ramahi, "Enhanced gain planar inverted-F antenna with metamaterial superstrate for UMTS applications," in *Proceedings of the Progress in Electromagnetics Research Symposium (PIERS '10)*, vol. 6, pp. 585–588, July 2010.

[14] R. Li, G. DeJean, M. M. Tentzeris, and J. Laskar, "Development and analysis of a folded shorted-patch antenna with reduced size," *IEEE Transactions on Antennas and Propagation*, vol. 52, no. 2, pp. 555–562, 2004.

[15] P. V. Nikitin, R. Martinez, S. Ramamurthy, H. Leland, G. Spiess, and K. V. S. Rao, "Phase based spatial identification of UHF RFID tags," in *Proceedings of the 4th Annual IEEE International Conference on RFID (RFID '10)*, pp. 102–109, April 2010.

[16] C. A. Balanis, *Antenna Theory: Analysis and Design*, John Wiley & Sons, New York, NY, USA, 2005.

15

CPW-Fed Fractal Slot Antenna for UWB Application

Hou Zhang, Hai-Yang Xu, Bo Tian, and Xian-Feng Zeng

Radar Engineering Department, Missile Institute of Air Force Engineering University, Shaanxi Province Sanyuan 713800, China

Correspondence should be addressed to Hai-Yang Xu, xuhaiyang.8888@163.com

Academic Editor: Ahmed A. Kishk

A novel printed ultrawideband (UWB) CPW-fed Koch fractal slot antenna is proposed. The UWB bandwidth is largely enhanced by the three-iteration Koch fractal slot whose extrusive angle is 900 instead of 600 to obtain better convoluted shape and self-similarity. The −10 dB impedance bandwidth is 3400 MHz (1270–4670 MHz) or about 114% in comparison with 1290 MHz (1620–2910 MHz) or about 57% of the referenced square slot antenna. Details of the proposed antenna design and performance are presented and discussed.

1. Introduction

Owing to the progress in wireless communication systems and the increase in their applications, the ultrawideband (UWB) antenna has become a key component in developing the UWB technique [1]. Conventional antennas are designed to radiate in a relatively narrow range of frequencies. As opposed to traditional narrowband antennas, the UWB antennas can transmit and receive electromagnetic waves in a wide range. The UWB antennas are used to transmit the pulses with minimum distortion in the shape of the pulses. What is more, the demand for portable, low profile, and compact wireless systems has increased research interests in miniature antennas [2]. As an important member of the UWB family, the slot antenna has the merits of wideband, low cost, light weight, easy fabrication and so on [3–5]. It can be easily integrated with microwave integrated circuits. What is more, they have the merits of low radiation loss, less dispersion too. The coplanar-waveguide- (CPW-) fed printed slot antennas have the simplest structure of a single metallic layer [3–7]. It can be easily integrated with monolithic microwave-integrated circuits. Some attempts have been made to enhance the bandwidth among various slot antennas, including the use of taper CPW [3], loaded strips [4], inset L-strip tuning stub [5], and U-shaped tunning stub [6]. Reference [7] used a widened tuning stub to enhance the bandwidth of the square slot antenna to 60% bandwidth.

The term fractal, which means broken or irregular fragments, was originally coined by Mandelbrot to describe a family of complex shapes that possess an inherent self-similarity in their geometrical structure. Fractal antennas, due to the property of self-similarity and the ability of filling the space, have the self-loaded characteristic to allow for smaller, multiband and broadband antenna design [8–14]. So they can enhance the bandwidth without any additional matching element.

In this paper, a promising bandwidth-enhancement technique is applied in the CPW-fed slot antenna in [15]. The modified Koch fractal is applied in the slot of the proposed antenna. The slot is achieved by three Koch iterations. And then through adjusting the gap between the tuning stub and the ground plane, the impedance bandwidth can enhance largely.

2. Antenna Design and Performance

The geometry of the proposed antenna is shown in Figure 1. The antenna is printed on the FR-4 substrate with the thickness of $h = 1.6$ mm and relative permittivity of $\varepsilon_r = 4.6$. The ground is chosen to be square, whose side length is $G = 72$ mm. The antenna is fed by a 50 Ω CPW feed line, which has a metal strip of width $w_f = 6.37$ mm and a gap of distance $g = 0.5$ mm. A widened tuning stub of length $L_3 = 22.5$ mm and width $w = 36$ mm is connected to the

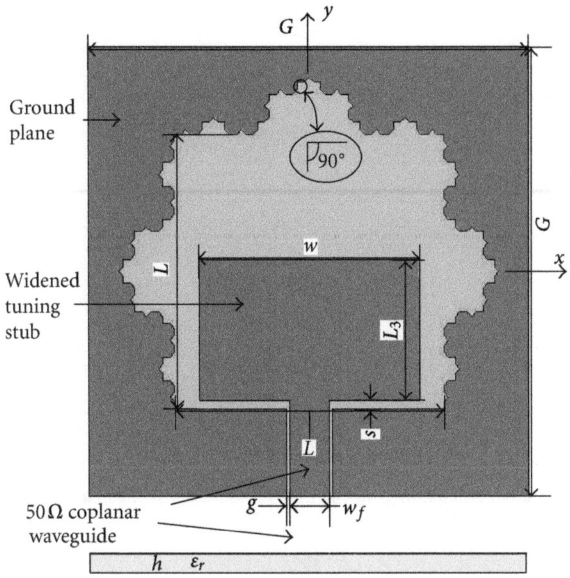

FIGURE 1: Geometry of the proposed antenna.

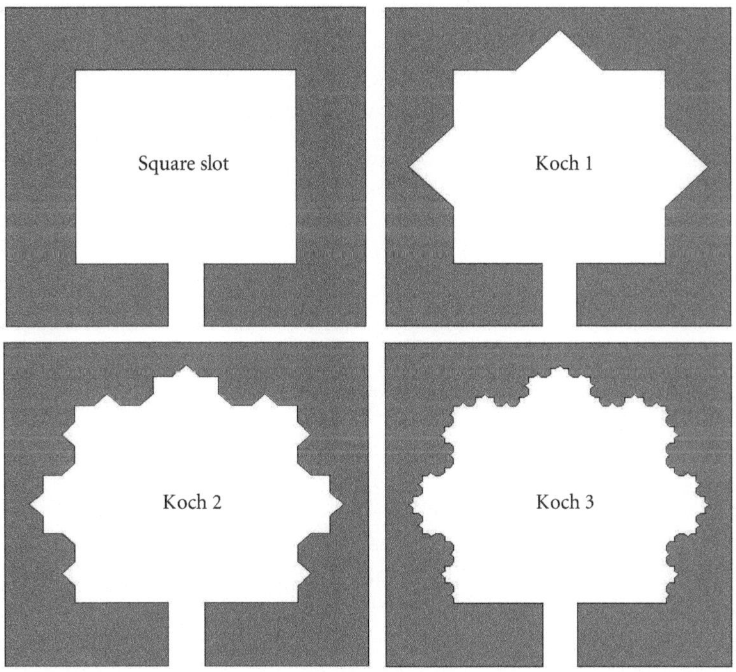

FIGURE 2: Geometries of the square slot and the Koch island for different iterations.

end of the CPW feed line. Unlike the tuning stub in [12], that of the proposed antenna is symmetrical with respect to the feed line and printed in the same layer with the metal ground. This design can obtain more symmetrical radiation patterns, especially in the high-frequency section. The spacing between the tuning stub and edge of the ground plane is $S = 1.4$ mm. The Koch fractal slot is achieved by 3 iterations based on the square slot of $L = 44$ mm.

Figure 2 gives the geometry of square slot and three modified Koch slots with different iterations. Unlike the Koch fractal forms in [13, 14], the extrusive angle of fractal is 90° instead of 60° so as to obtain better convoluted shape in the slot and better self-similarity in the corners. Three sides of the square slot are replaced by the 90°-angle Koch islands. Higher iterations of the modified Koch curves are constructed by subtracting smaller and smaller right-angled isosceles triangles from the former structure, as shown in Koch1–3 of Figure 2.

At higher iteration k, the area of the slot increases. Let A_k be the area of the slot at Koch k (let the area of the square slot

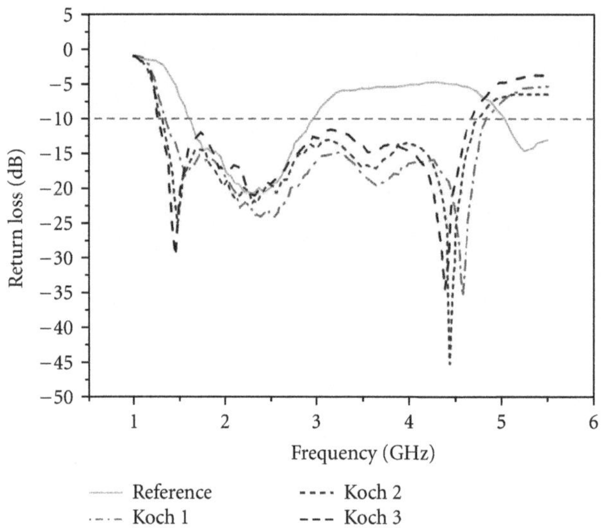

FIGURE 3: Comparison of the measured return loss.

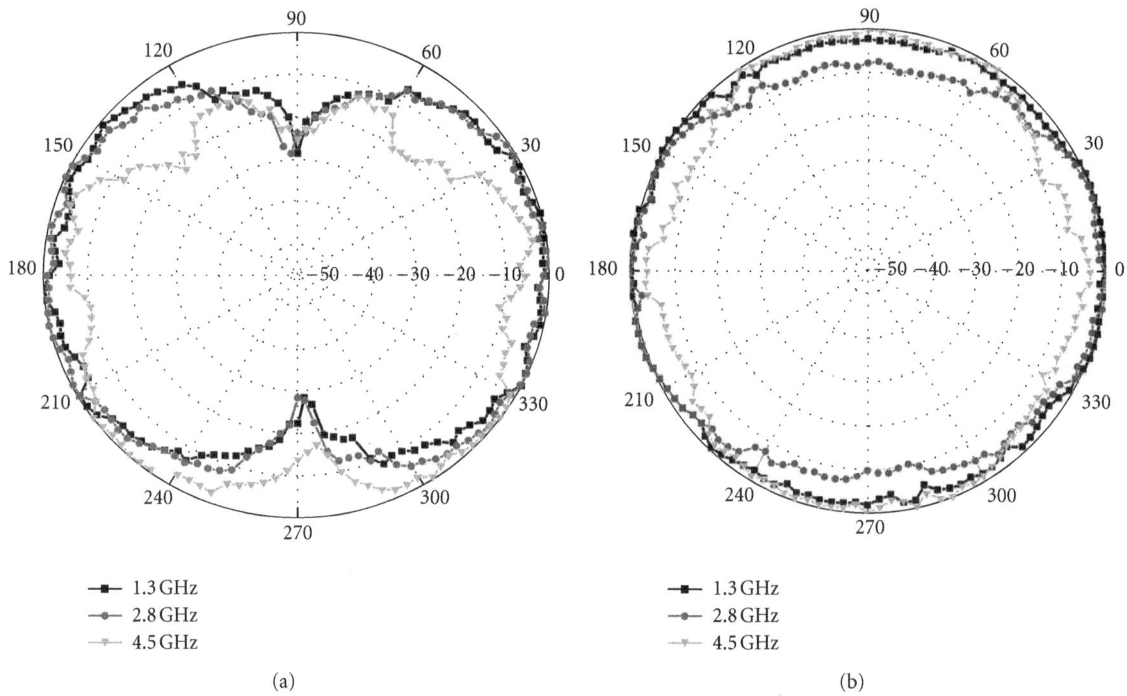

FIGURE 4: Measured copolarized far-field radiation patterns: (a) E pane, (b) H plane.

be A_0), then the area of the next iteration can be computed as

$$A_{k+1} = A_k + \frac{3}{8}\left(\frac{2}{3+2\sqrt{2}}\right)^{k+1} * a^2, \qquad (1)$$

where a is the side of the initial square slot that has an area $A_0 = a^2$.

The overall perimeter for Koch k is given by

$$l_k = \frac{3}{2} \cdot \left(2 - \sqrt{2}\right)^k a + a. \qquad (2)$$

The perimeter increases by $l_k - 4a$, which means the resonance frequency reduction can be achieved. In other words, with the assigned resonance frequency, the volume of the antenna is reduced with the higher iterations.

Figure 3 shows return loss of reference and modified Koch slot antennas for different iterations. As shown in Figure 3, the measured −10 dB impedance bandwidth of the Koch fractal antenna is 3400 MHz (1270–4670 MHz) or about 114%, while the reference one without the fractal iterations is 1290 MHz (1620–2910 MHz) or about 57%. The

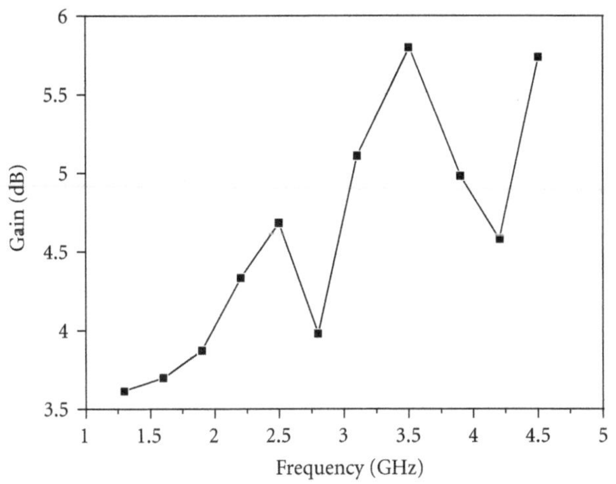

Figure 5: Gain against frequency.

Koch fractal expands the bandwidth to be about 2 times of the reference antenna with a square slot.

Figure 4 is the measured copolarized far-field radiation patterns of novel antenna at E and H planes at different frequencies. The radiation patterns keep relatively symmetrical and stable at both planes in the whole operating frequency band.

Figure 5 is the gain of the fractal slot antenna. The maximum and minimum values are 5.7 and 3.6 dB, respectively, resulting in the maximum gain variation of 2.1 dB over the entire UWB frequency band.

3. Conclusion

A novel CPW-fed Koch fractal slot antenna has been proposed and investigated. The application of the Koch fractal slot enhances the impedance bandwidth obviously. This design is a breakthrough in expanding the bandwidth of the slot antenna and can be directly applied to UWB communications. This fractal technique can also be expected to apply other antennas with square or rectangular slots to enhance the impedance bandwidth.

References

[1] S. S. Zhong and X. L. Liang, "Progress in ultra-wideband planar antennas," *Journal of Shanghai University*, vol. 11, no. 2, pp. 95–101, 2007.

[2] H. T. Nguyen, S. Noghanian, and L. Shafai, "Microstrip patch miniaturization by slots loading," in *Proceedings of the IEEE Antennas and Propagation Society International Symposium and USNC/URSI Meeting*, pp. 215–218, Washington, DC, USA, 2005.

[3] C. Y. Huang and D. Y. Lin, "CPW-fed bow-tie slot antenna for ultra-wideband communications," *Electronics Letters*, vol. 42, no. 19, pp. 1073–1074, 2006.

[4] J. Y. Chiou, J. Y. Sze, and K. L. Wong, "A broad-band CPW-fed strip-loaded square slot antenna," *IEEE Transactions on Antennas and Propagation*, vol. 51, no. 4, pp. 719–721, 2003.

[5] B. Purahong, P. Jearpradikul, T. Archevapanich, N. Anantrasirichai, and O. Sangaroon, "CPW-fed slot antenna with inset U-strip tuning stub for wideband," in *Proceedings of the International Conference on Control, Automation and Systems*, pp. 1781–1784, Seoul, Korea, October 2008.

[6] R. Chair, A. A. Kishk, K. F. Lee, C. E. Smith, and D. Kajfez, "Microstrip line and CPW fed ultra wideband slot antennas with U-shaped tuning stub and reflector," *Progress in Electromagnetics Research*, vol. 56, pp. 163–182, 2006.

[7] B. Mirzapour and H. R. Hassani, "Size reduction and bandwidth enhancement of snowflake fractal antenna," *IET Microwaves, Antennas and Propagation*, vol. 2, no. 2, pp. 180–187, 2008.

[8] H.-Y. Xu, H. Zhang, G.-Y. Li, and Q.-B. Xu, "An ultra-wideband fractal slot antenna with low backscattering cross section," *Microwave and Technology Letters*, vol. 53, no. 5, pp. 1150–1154, 2011.

[9] B. Mirzapour and H. R. Hassani, "Size reduction and bandwidth enhancement of snowflake fractal antenna," *IET Microwaves, Antennas and Propagation*, vol. 2, no. 2, pp. 180–187, 2008.

[10] R. Kumar and P. Malathi, "Design of dual wide band CPW-fed fractal antenna," in *Proceedings of the International Conference on Emerging Trends in Electronic and Photonic Devices and Systems, ELECTRO '09*, pp. 289–292, December 2009.

[11] Y. B. Thakare and Rajkumar, "Design of fractal patch antenna for size and radar cross-section reduction," *IET Microwaves, Antennas and Propagation*, vol. 4, no. 2, pp. 175–181, 2010.

[12] H. Y. Xu, H. Zhang, G. Y. Li, Q. B. Xu, and K. Lu, "An ultra-wideband fractal slot antenna with low backscattering cross section," *Microwave and Optical Technology Letters*, vol. 53, pp. 1150–1154, 2011.

[13] D. D. Krishna, M. Gopikrishna, C. K. Anandan, P. Mohanan, and K. Vasudevan, "CPW-fed Koch fractal slot antenna for WLAN/WiMAX applications," *IEEE Antennas and Wireless Propagation Letters*, vol. 7, Article ID 2000814, pp. 389–392, 2008.

[14] D. D. Krishna, M. Gopikrishna, C. K. Aanandan, P. Mohanan, and K. Vasudevan, "Compact wideband Koch fractal printed slot antenna," *IET Microwaves, Antennas and Propagation*, vol. 3, no. 5, pp. 782–789, 2009.

[15] H. D. Chen, "Broadband CPW-fed square slot antennas with a widened tuning stub," *IEEE Transactions on Antennas and Propagation*, vol. 51, no. 8, pp. 1982–1986, 2003.

16

Analysis of Horn Antennas in Receiving Mode as an EM Field Sensor

Maifuz Ali[1] and Seong-Ook Park[2]

[1] *Information & Electronics Research Institute, Department of Electrical Engineering, Korea Advanced Institute of Science and Technology (KAIST), Daejon 305-701, Republic of Korea*
[2] *Department of Electrical Engineering, Korea Advanced Institute of Science and Technology (KAIST), Daejon 305-701, Republic of Korea*

Correspondence should be addressed to Maifuz Ali, maifuzali@lycos.com

Academic Editor: Krishnasamy Selvan

Electromagnetic (EM) field measurement uses antenna in receiving mode as EM field sensor (EMFS). FDTD technique is applied to analyze the horn antenna in this mode. The performances of horn antenna as EMFS are predicted in terms of antenna factor (AF), and the results are compared with measurements. New empirical formula for quick calculation of antenna factor of pyramidal horn is also presented. Also, the gain of horn antennas are computed in receiving mode using FDTD technique, and the results are compared with published measured result and far-field gain formula.

1. Introduction

A key issue in the era of proliferating sensitive electronic instruments is the problem of electromagnetic interference (EMI). Accurate calibration of electromagnetic field sensor (EMFS) is required for EM field measurement. The frequency domain measurement performance of the sensor depends on its antenna factor (AF) which is the ratio of the incident electric field on the antenna surface to the received voltage across 50 Ω load resistance [1, 2]. The theoretical simulation of the AF of EMFS has attracted many research interests due to implementation cost and time required to calibrate a sensor experimentally [3–5]. Theoretical analysis of EMFS has been carried out till now using analytical technique and numerical techniques like method of moment (MoM). However, the FDTD technique which has found extensive use in electromagnetics [6, 7], to the best of the authors knowledge, has till now not been used explicitly in computing the AF of an EMFS. Also, the FDTD has the advantage that a single FDTD simulation can provide for both time domain and frequency domain analysis [6]. Further, the FDTD has the advantage that the problem of modeling a new structure is reduced to a problem of mesh generation rather than the complexity of reformulation of integral equations.

Commonly the measuring antennas or EMFS used are dipole and loop antenna. However, these are mostly effective up to a frequency of 1 GHz [3]. Beyond this range, a few waveguide-based sensors have been analyzed using MoM technique. In this work, waveguide-based horn antenna is analyzed in receiving mode for EMFS using standard FDTD technique and the obtained results are compared with measurement. New empirical formula for quick calculation of antenna factor of pyramidal horn is also presented.

In this case the knowledge of the effective area of the antenna is of great importance since the electromagnetic field power transmitted to the measuring unit is simply a product of the Pointing vector and the effective area. The effective area or antenna factor of the horn antenna is directly connected with the gain of the antenna. Also calculations of antenna gain is of interest to a wider group of the researchers. In this work, gain of horn antennas is computed in the receiving mode using the same FDTD technique and compared with the available gain formula in [8] and measured results of [9].

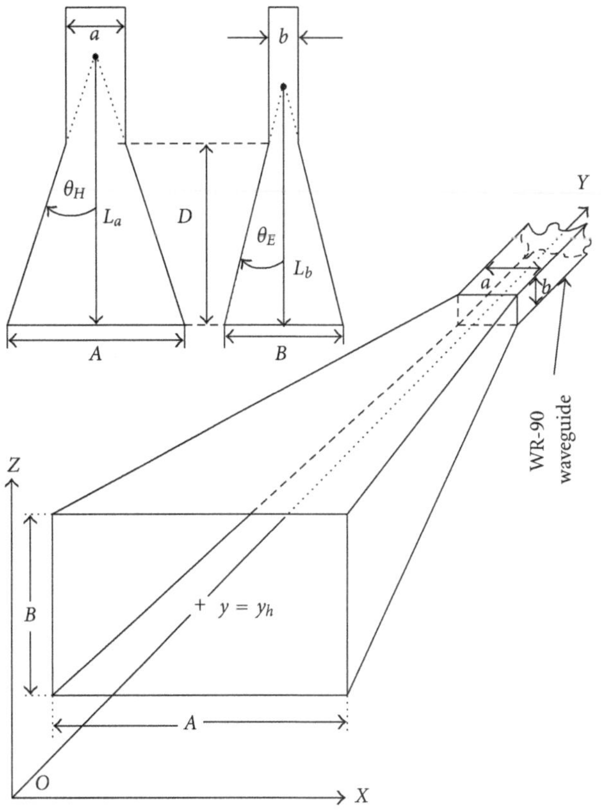

FIGURE 1: Details of geometry of a pyramidal horn antenna.

2. FDTD Formulation of the Problem

The FDTD model uses a uniform space lattice of cubic Yee cells having $\Delta x = \Delta y = \Delta z = \Delta$. A 10Δ-thick unsplit perfectly matched layer (PML) [10, 11] is used as absorbing boundary conditions (ABC) on all six sides of the FDTD lattice. This PML is spaced 5Δ cells from the closest surface of the scatterer. Gaussian pulse [7, 10] is taken as the excitation source

$$E_{z_{i,j,k}}(t) = Ae^{-0.5((t-t_0)/t_\omega)^2}. \quad (1)$$

The Gaussian function maximizes to A at $t = t_0$ and is zero at $t = \pm\infty$. t_ω is the standard deviation and relates the line width at half-height by the relationship

$$t_{1/2} = \sqrt{8\ln(2)}t_\omega = 2.35482t_\omega. \quad (2)$$

The complete geometry of the horn antenna is shown in Figure 1. The tangential electric field components along this structure are set to zero. Linearly polarized (along z-axis) perfectly plane wave of Gaussian pulse propagating along y-axis in free space incidents on the open end of the RW-90 rectangular waveguide-fed horn at $y = y_h$ as shown in Figure 2.

In order to simulate a uniform plane wave within the FDTD lattice, the problem space is divided into the total field and scattered field regions and the antenna is placed within the total field regions. Details of this method given in [10] are used in this work. The plane wave is generated in the XZ-plane, at $y = y_a$. As primary and edge diffracted waves propagate through the horn waveguide structure, the subtractions in the XZ-plane at $y = y_b = y_h + 1$ take place like in [10] but without the area $(A \times B)$ inside the horn at $y = y_b$.

As perfect plane wave and lossless free space are considered, time domain electric field at the aperture of the horn antenna is

$$E_{z_{i,j,k}}(t) = Ae^{-0.5((t-t_0+t')/t_\omega)^2}, \quad (3)$$

where t' is the time shift due to the spacial difference between the antenna aperture and the position where Gaussian pulse is applied into the FDTD lattice. Fourier transform of $E^t_{z_{i,j,k}}$ gives the frequency domain incident electric field, which is

$$E_i(\omega) = F\left\{E_{z_{i,j,k}}(t)\right\}, \quad (4)$$

where $F\{\}$ denotes the Fourier transform.

2.1. Power Flowing through the Waveguide. The time average complex power flowing through the waveguide is given by [3, 12]

$$P^c(\omega) = \int_0^a \int_0^b \frac{1}{2}\left[E(\omega) \times (H(\omega))^*\right] dx\, dz. \quad (5)$$

The integration is carried out over the waveguide cross section.

Although many modes are generated inside horn and enter into the waveguide after traveling few λ distance through the waveguide, all the modes die down except TE_{10} mode. Power calculating at the XZ-plane ($y = y_0$) ensures that the measuring device receives only the dominant mode scattered power. Neglecting all components except E_z and H_x^*, time average power flowing through the waveguide along y-direction is real part of $P^c(\omega)$ which is given by

$$\begin{aligned} P_y(\omega) &= \mathrm{Re}\{P^c(\omega)\} \\ &\approx \mathrm{Re}\left\{\int_0^a \int_0^b \frac{1}{2} E_z(\omega) H_x(\omega)^* \Big|_{y=y_0} dx\, dz\right\}, \end{aligned} \quad (6)$$

where y_0 is the point in which all other modes are died down except TE_{10} and a minimum 3 cell before the PML start within the FDTD space lattice. a and b are the dimensions of the feed waveguide shown in Figure 1.

2.2. Voltage Developed at the Matched Measuring Device. Since most measuring devices have an input impedance of 50 Ω, the voltage measured by these is given by

$$V_L(\omega) = \sqrt{50 \times P_y(\omega)}\,\text{Volts} \quad (7)$$

on the condition that the waveguide transporting this power is well matched with the measuring device.

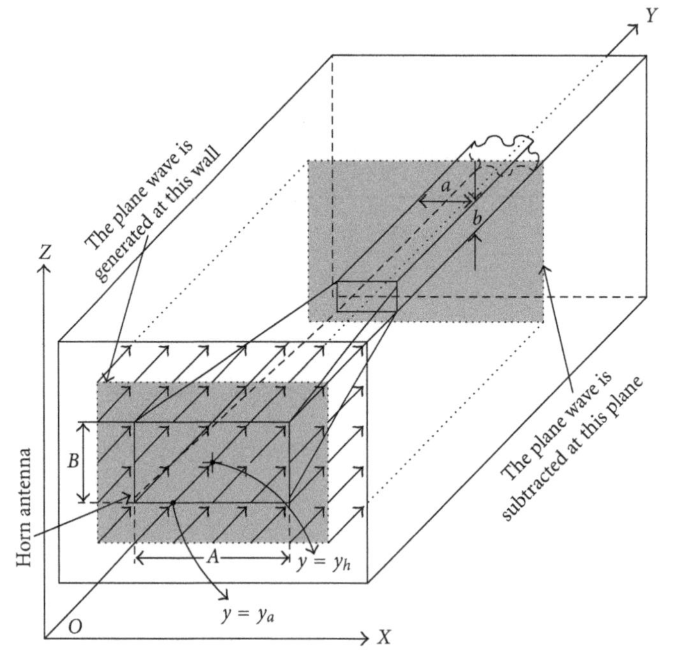

Figure 2: Receiving antenna case. A horn antenna under plane-wave illumination within the FDTD grid.

2.3. Calculation of Far-Field Antenna Factor. To carry out field strength measurements, one typically connects an antenna to a spectrum analyzer. The AF is the parameter that is used to convert the voltage or power reading of the receiver to the field strength incident on the antenna. In terms of an equation, the AF is defined as [13, 14]

$$\mathrm{AF} = \frac{E_i(\omega)}{V_L(\omega)}\,\mathrm{m}^{-1} = 20\log\left|\left(\frac{E_i(\omega)}{V_L(\omega)}\right)\right|\ [\mathrm{dB}(\mathrm{m}^{-1})], \quad (8)$$

where E_i is the incident electric field on the surface of the antenna and V_L is the voltage induced across a 50 Ω load of the receiver.

During the progress of FDTD calculation time domain incident field $E_{z_{i,j,k}}(t)$ at $y = y_h$ and time domain electric field $E_z(t)$ and magnetic field $H_x(t)$ at $y = y_0$ on the surface area $a \times b$ are saved for each time step. The FDTD calculations are continued until all transients are dissipated, so that the Fourier transform yields the steady-state frequency domain response of the antenna. The frequency domain incident electric field is calculated from (4), and the voltage developed across 50 Ω load $V_L(\omega)$ is obtained from (7). Finally, AF of the horn antenna is evaluated using (8). FDTD predicted AF flowing through this procedure takes the account of mutual coupling effects [15].

2.4. Calculation of Gain in Receiving Mode. A few widely used primary incident wave models for pyramidal horn antenna analysis are compared on the basis of gain calculations. The analysis based on edge-wave diffraction theory in [9] shows that these models agree well for high-gain horn with small flare angles but that the differences are more profound for lower gain horn, where the flare angles are larger. In this work, gain of various pyramidal horn antennas is calculated in receiving mode using FDTD technique.

Since the field strength can, alternatively, be measured in terms of antenna gain, the antenna factor can also be expressed as a function of antenna gain. If only the radiation far field is taken into account for antenna factor calculations, the antenna factor of the sensor is related to the gain by the formula [15, 16]

$$\mathrm{AF} = \frac{9.734}{\lambda\sqrt{G_r(\theta,\phi)}}, \quad (9)$$

where $G_r(\theta,\phi)$ is the antenna gain as a function of the angle of arrival of the energy and λ is the free space wavelength in meter unit. From (8) and (9) gain of pyramidal horn is given by

$$G = \frac{94.751}{\lambda^2}\left(\frac{V_L(\omega)}{E_i(\omega)}\right)^2, \quad (10)$$

where $E_i(\omega)$ and $V_L(\omega)$ are given in (4) and (7), respectively.

3. Empirical Formula for Antenna Factor of Horn Antenna

The gain of pyramidal horn antenna is calculated from [8, equation (18.4.2)], given by

$$G = \frac{4\pi}{\lambda^2}(AB)e(\sigma_a,\sigma_b). \quad (11)$$

A and B are the dimensions of the horn shown in Figure 1 and $e(\sigma_a,\sigma_b)$ is the aperture efficiency given in [8, equation (18.4.3)]. Other far-field gain formulas like Schelkunoff's horn-gain formula of [17] can also be used. But due to the availability of readily programs in [8] and accuracy, (11) is used for gain calculation in this work. From (9) and (11), AF of horn antenna is

$$\mathrm{AF} = 20\log_{10}\left|\frac{9.734}{\sqrt{4\pi ABe(\sigma_a,\sigma_b)}}\right| [\mathrm{dB}(\mathrm{m}^{-1})]. \quad (12)$$

4. Numerical Results and Discussions

Simulations are carried out in FDTD spatial grid with uniform cell size of $\Delta x = \Delta y = \Delta z = \Delta(\cong \lambda_{\min}/24) = 1.0\,\mathrm{mm}$, where $\lambda_{\min} = 24.0\,\mathrm{mm}$ is the wavelength at maximum frequency 12.4 GHz.

$\Delta t = 1.67\,ps$ is calculated from [10, equation (1.11)]. A linearly polarized (along z-axis) perfectly plane wave of Gaussian pulse having significant frequency content in a frequency range from 8 GHz to 12.5 GHz of maximum amplitude $A = 1.0\,\mathrm{V/m}$, $t_0 = 80.65\,\mathrm{ps}$, and $t_w = 10.08\,\mathrm{ps}$ of (1) falls upon the open end of the horn at $y = y_h$.

The AF of two WR-90 waveguide-fed horns of different size are computed and compared with the experimental results. The experiment was carried out over the frequency range of 8 GHz to 12.5 GHz in step of 0.1 GHz. The experimental setup and procedure used in this work is given in detail in [3], and for this reason it is not repeated here.

4.1. Antenna Factor of Pyramidal Horn Antennas

4.1.1. Horn-1. The dimensions of the first horn analyzed as an EMFS are $A = 14\,\mathrm{cm}$, $B = 12\,\mathrm{cm}$, $D = 18.5\,\mathrm{cm}$, $\theta_H = 17.57°$ and $\theta_E = 16.53°$. FDTD computed AF is compared with the AF calculated from (12) and experimental result shown in Figure 3.

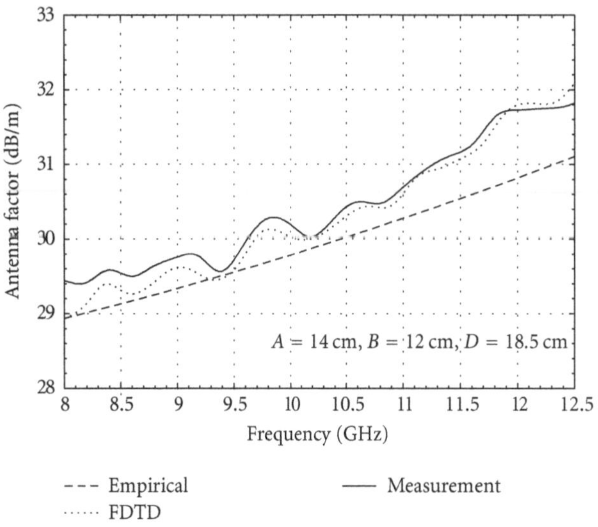

FIGURE 3: AF of pyramidal horn is computed using FDTD and empirical formula compared with the measurement.

4.1.2. Horn-2. The dimensions of the second horn analyzed as an EMFS are $A = 9\,\mathrm{cm}$, $B = 7.9\,\mathrm{cm}$, $D = 7.0\,\mathrm{cm}$, $\theta_H = 25.62°$ and $\theta_E = 26.17°$. FDTD computed AF is compared with the AF calculated from (12) and experimental result shown in Figure 4.

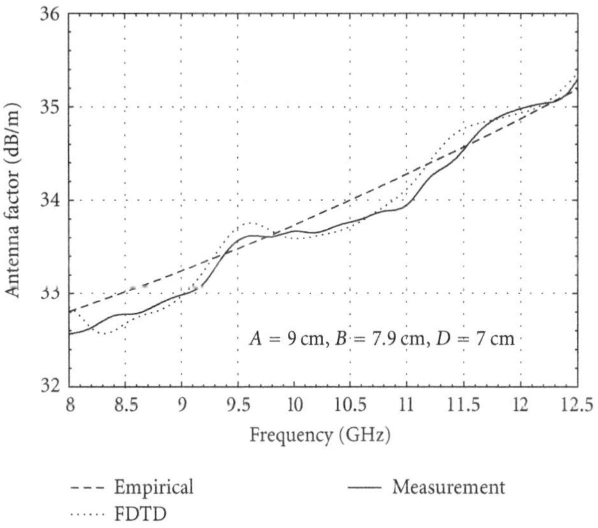

FIGURE 4: AF of pyramidal horn is computed using FDTD and empirical formula compared with the measurement.

From Figures 3 and 4, it is observed that the FDTD computed AF agrees well with the measured results. In Figure 4, AF computed using empirical formula (12) agrees on an average with the measurement and FDTD computed results. But in Figure 3, for comparatively high-gain horn, AF computed using empirical formula (12) has significant deviation from the measurement and FDTD computed results, mainly above 10.5 GHz. As the empirical formula assumes only far-field type fields, the deviation from this empirical formula underlines the effect of non-far-field type fields at the sensing plane of the sensor [3]. AF calculated from the empirical formula agrees with the measurement and FDTD calculation more for low-gain horn than for higher-gain horn.

4.2. Pyramidal Horn Antenna for Cross-Polarized Incident Field Detection. If the horn is rotated $\theta = 90°$ taking y-axis as the axis of rotation, then the horn antenna becomes cross polarized with respect to the incident electric field. Then (6) for calculating time average power flowing through the waveguide along Y-direction becomes

$$P(\omega) \approx \mathrm{Re}\left[\int_0^a\int_0^b -\frac{1}{2}E_x(\omega)(H_z(\omega))^*\Big|_{y=y_0} dx\,dz\right]. \quad (13)$$

AF of the horn-1 and horn-2 of 4.1 are calculated for orthogonal incident electric field in the same way as in Section 4.1 and shown in Figure 5. AF of the horn in this position is greater than 100 dB, though in ideal case AF $= \infty$. In practice, the received power is too low to separate itself from the EM noise present in the environment. For a good receiver the cross-polarization pick up of the antenna

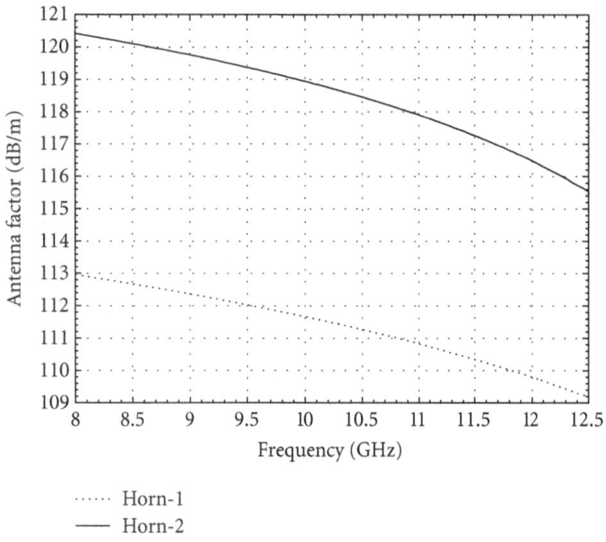

FIGURE 5: AF of pyramidal horn due to the incidence of cross-polarized electric field.

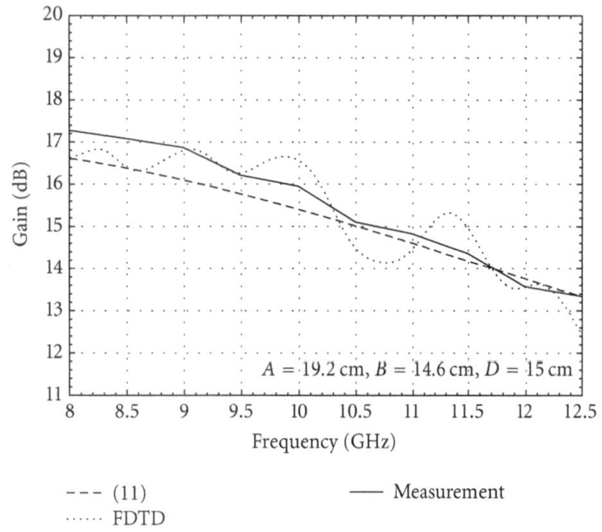

FIGURE 6: Gain of short horn computed using FDTD and compared with (11) and measurement of [9, Figure 2].

is expected to be minimum. Hence, this greater value of cross-polarization isolation, the performance of the antenna, is better as an EM field sensor [18].

4.3. Gain of Pyramidal Horn Antenna in Receiving Mode. The gain of two different pyramidal horn antennas are calculated using FDTD technique of (10) and compared with (11) and measured results of [9].

4.3.1. Horn-1. The dimensions of the first horn are A = 19.2 cm, B = 14.6 cm, D = 15.0 cm, θ_H = 29.423°, and θ_E = 24.355°. The FDTD computed gain is compared with (11) and experimental results of [9, Figure 2] shown in Figure 6.

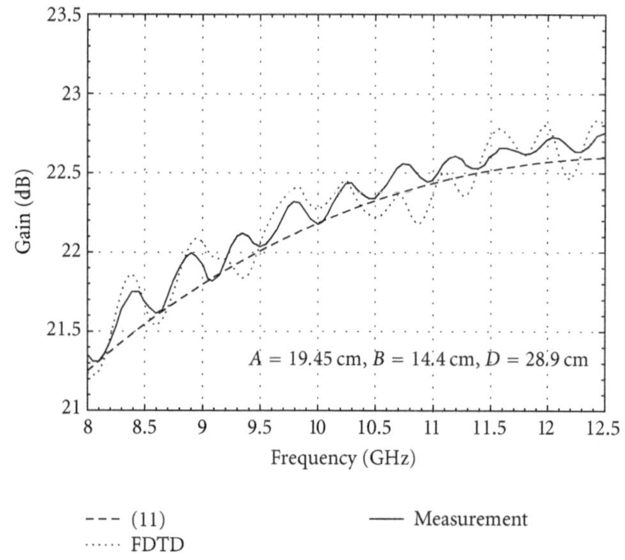

FIGURE 7: Gain of medium horn computed using FDTD and compared with (11) and measurement of [9, Figure 3].

4.3.2. Horn-2. The dimensions of the second horn are A = 19.45 cm, B = 14.4 cm, D = 28.9 cm, θ_H = 16.539°, and θ_E = 13.037°. The FDTD computed gain is compared with (11) and experimental results of [9, Figure 3] shown in Figure 7.

The results in Figure 6 are for horn with larger flare angles (short pyramidal horn), while the results in Figure 7 are for a medium pyramidal horn. From these results, it is seen that flare angles of the horns do not affect FDTD gain computation like in [9] and the results are well agreed with the measured values. Calculated gains using (11) match on an average with the FDTD computed results and measurements of [9], but no oscillation in magnitude is observed.

5. Conclusions

An empirical formula for AF of pyramidal horn antenna has been presented. AF obtained from measurements carried out for horn antennas of different sizes compares favorably with the FDTD predicted AF and AF calculated from the empirical formula (12). AF of the horn antenna is low which is desired for EM field sensor. The voltage developed at the receiver due to the cross-polarized component of incident field on the aperture of the pyramidal horn is found to be low and the resulting AF is high. Further the horn antenna has wide bandwidth and is therefore suitable as efficient EMFS.

The gain of pyramidal horns are computed in receiving mode using FDTD method, and results are compared with (11) and measured results of [9]. It can be concluded that the gain calculated using FDTD technique agrees well with measurement and not depends upon the horn shape like in [9].

Perfect coaxial match over the entire bandwidth is assumed for FDTD simulations. But in practice VSWR $\neq$ 1 over the entire bandwidth. Due to this idealistic assumption, some difference is introduced between the FDTD simulated and measured results.

The amplitude oscillation in FDTD computed AF or gain due to the oscillation of amplitude in power flowing through the fed waveguide, calculated from (6), represents a standing wave component resulting from interference between the primary wave and the edge diffracted waves [19]. However, same variation is also observed in the measurements leading to the conclusion that it is an intrinsic antenna property. But this amplitude oscillation is not observed in the antenna gain using (11). As the empirical formula assumes only far-field type fields, the deviation from the empirical formula underlines the effect of non-far-field type fields.

Acknowledgments

The authors would like to thank Proffessor Subrata Sanyal, Proffessor A. Bhattacharya, and Proffessor Ajoy Chakraborty of the Indian Institute of Technology, Kharagpur, India, for critical discussion from time to time. This research was supported by the Korea Communications Commission (KCC) and National Research foundation (NRF) of Korea under Contract KCA-2011-1191304001-110010100 and 2011-0006570, respectively, and the Brain Korea 21 Project, the School of Information Technology, KAIST in 2011.

References

[1] H. Hosoyama, T. Iwasaki, and S. Ishigami, "Complex antenna factor of a V-dipole antenna with two coaxial feeders for field measurements," *IEEE Transactions on Electromagnetic Compatibility*, vol. 41, no. 2, pp. 154–158, 1999.

[2] S. Ishigami, H. Iida, and T. Iwasaki, "Measurements of complex antenna factor by the near-field 3-antenna method," *IEEE Transactions on Electromagnetic Compatibility*, vol. 38, no. 3, pp. 424–432, 1996.

[3] A. Bhattacharya, S. Gupta, and A. Chakraborty, "Analysis of rectangular waveguide and thick windows as EMI sensors," *Progress in Electromagnetics Research*, vol. 22, pp. 231–258, 1999.

[4] Y. Matsumoto, T. Umeda, A. Nishikata, K. Fujii, Y. Yamanaka, and A. Sugiura, "EMI antenna calibration on an absorber-lined ground plane to determine free-space antenna factor," *IEEE Transactions on Electromagnetic Compatibility*, vol. 45, no. 4, pp. 656–660, 2003.

[5] K. Fujii, S. Harada, A. Sugiura, Y. Matsumoto, and Y. Yamanaka, "An estimation method for the free-space antenna factor of VHF EMI antennas," *IEEE Transactions on Electromagnetic Compatibility*, vol. 47, no. 3, pp. 627–634, 2005.

[6] A. Taflove and S. C. Hagness, *Computational Electromagnetic—The Finite-Difference Time-Domain Method*, Artech House, Boston, UK, 2005.

[7] W. Yu and R. Mittra, *CFDTD: Conformal Finite-Difference Time-Domain Maxwell's Equations Solver: Software and User's Guide*, Artech House, Boston, UK, 2004.

[8] S. J. Orfanidis, *Electromagnetic Waves and Antennas*, Rutgers University, 2008.

[9] G. Mayhew-Ridgers, J. W. Odendaal, and J. Joubert, "On primary incident wave models for pyramidal horn gain calculations," *IEEE Transactions on Antennas and Propagation*, vol. 48, no. 8, pp. 1246–1252, 2000.

[10] D. M. Sullivan, *Electromagnetic Simulation Using The FDTD Method*, IEEE Press, New York, NY, USA, 2000.

[11] D. M. Sullivan, "An unsplit step 3-D PML for use with the FDTD method," *IEEE Microwave and Guided Wave Letters*, vol. 7, no. 7, pp. 184–186, 1997.

[12] R. F. Harrington, *Time-Harmonic Electromagnetic Fields*, McGraw-Hill Book Company, New York, NY, USA, 1961.

[13] T. Iwasaki and K. Tomizawa, "Systematic uncertainties of the complex antenna factor of a dipole antenna as determined by two methods," *IEEE Transactions on Electromagnetic Compatibility*, vol. 46, no. 2, pp. 234–245, 2004.

[14] W. Joseph and L. Martens, "An improved method to determine the antenna factor," *IEEE Transactions on Instrumentation and Measurement*, vol. 54, no. 1, pp. 252–257, 2005.

[15] R. M. Hekert, J. K. Daher, K. P. Ray, and B. Subbarao, "Measurementand modeling of near and far field antenna factor," in *Proceedings of the International Conference on Electromagnetic Compatibility and Interference (INCEMIC '94)*, pp. 237–241, August 1994.

[16] W. S. Cho, M. Kanda, H. J. Hwang, and M. W. Howard, "A disk-loaded thick cylindrical dipole antenna for valuation of an EMC test site from 30 to 300 MHz," *IEEE Transactions on Electromagnetic Compatibility*, vol. 42, no. 2, pp. 172–180, 2000.

[17] K. T. Selvan, "An approximate generalization of schelkunoff s horn-gain formulas," *IEEE Transactions on Antennas and Propagation*, vol. 47, no. 6, pp. 1001–1004, 1999.

[18] S. Ghosh and A. Chakrabarty, "Ultrawideband performance of dielectric loaded T-shaped monopole transmit and receive antenna/EMI sensor," *IEEE Antennas and Wireless Propagation Letters*, vol. 7, Article ID 921335, pp. 358–361, 2008.

[19] J. F. Nye and W. Liang, "Theory and measurement of the field of a pyramidal horn," *IEEE Transactions on Antennas and Propagation*, vol. 44, no. 11, pp. 1488–1498, 1996.

17

Analysis of a Reflectarray by Using an Iterative Domain Decomposition Technique

Carlos Delgado, Josefa Gómez, Abdelhamid Tayebi, Iván González, and Felipe Cátedra

Computer Science Department, University of Alcalá, 28871 Madrid, Spain

Correspondence should be addressed to Carlos Delgado, carlos.delgado@uah.es

Academic Editor: Manuel Arrebola

We present an efficient method for the analysis of different objects that may contain a complex feeding system and a reflector structure. The approach is based on a domain decomposition technique that divides the geometry into several parts to minimize the vast computational resources required when applying a full wave method. This technique is also parallelized by using the Message Passing Interface to minimize the memory and time requirements of the simulation. A reflectarray analysis serves as an example of the proposed approach.

1. Introduction

In the last years we have seen a rising interest in computer simulation tools in several engineering and science fields, due to technological breakthroughs and the development of novel efficient numerical algorithms. Many researchers are involved in the development of new methods focused on the acceleration and improvement of conventional electromagnetic field solvers. The logical consequence of this fact is a great increase in the demand for computer simulation tools.

Addressing Method-of-Moments- (MoM-) based electromagnetic solvers, we have seen a good number of efficient approaches in the last years. Many of these are related to the reduction of the computational requirements of the legacy MoM by only storing the coupling between basis and testing functions that are located in the near field region. There are very well-known methods, like the Multilevel Fast Multipole Algorithm (MLFMA) [1], that compute the matrix-vector products efficiently by aggregating a number of active currents to some specific points, translating them and disaggregating those multipoles in order to take into account the effect of the active basis functions over the testing functions. Other efficient techniques reduce the total number of unknowns by defining extended sets of macrobasis functions [2–6]. Fast matrix compression and low-rank approximation approaches can also be mentioned as efficient methods in terms of memory and CPU time requirements [7, 8]. In addition to the MLFMA there are a number of techniques that also take advantage of the fast evaluation of matrix-vector products in the iterative solution of large problems, among which we can mention the Complex Multipole Beam Approach (CMBA) [9], the Impedance Matrix Localization (IML) technique [10], the Adaptive Integral Method (AIM) [11], or the Multilevel Matrix Decomposition Algorithm (MLMDA) [12]. However, there are some cases for which the use of a full-wave method still poses a bottleneck, even considering these fast approaches, due to the size of the problem, slow convergence, or other issues. In these cases it is sometimes possible to resort to high-frequency methods [13] or hybrid combinations [14] if the nature of the geometrical properties of the problem and the excitation lie within the scope of such techniques. Unfortunately this is not usually the case when analyzing reflectarray antennas. Domain decomposition techniques (DD) allow to limit the maximum size of the system matrix to be solved [15] regardless of the size of the objects involved. It is necessary to include interactions between the domains in order to reach the final solution, which is usually done in an iterative or recursive way [16–18]. There are some works worthwhile to be mentioned that study the convergence problems that may prevent reaching a stop situation [19, 20]. Some typical strategies to avoid this undesirable divergence rely on the definition of a relaxation factor [19, 21], make use of an extension of the original domains [17, 18, 21] and/or use preconditioners [17].

Due to the interest raised by the analysis of reflectarray antennas and similar geometries, a number of works reported in the last years deal specifically with this kind of problems. For instance, [22] presents an efficient technique based on the extension of an adaptive integral method that allows the full-wave analysis of electrically large multilayered printed arrays that have one or more planar metallizations and vertical conductors. The technique proposed in [23] computes the generalized scattering matrix of a dielectric interface with periodic metallizations, which is based on a spectral domain Moment Method, but assuming multiple incident Floquet-harmonics. A full-wave analysis using the finite integration technique is applied in [24], where the mutual coupling between the feeding horn and the elements of the reflectarray is considered. The analysis reported in [25] is based on an application of the Moment Method. In [26] the authors make use of a fast simulation tool based on MoM-MLFMA to analyze a three-layered reflectarray. The work reported in [27] combines MoM-MLFMA with the use of macrobasis functions to address large problems, including reflectarray antennas.

In this work we propose a combination of the MLFMA with a new domain decomposition technique that provides fast analysis of arbitrary structures in complex environments and contains a novel procedure for selecting the interactions between domains that will have a noticeable contribution to the final results. This new domain decomposition method has been validated and compared with the MoM-MLFMA approach. The simulation results of both approaches are compared with measurements of a reflectarray antenna that has been previously designed, fabricated, and measured [28] in the Research Advanced Antenna Technology Communications Research Centre, Canada.

The geometry of the bodies considered in this work is defined using parametric surfaces [29, 30], as well as the mesh required in order to apply the MoM-based approach. We obtain a continuous quadrangular mesh by using the mesher described in [31]. The use of parametric subpatches enables us to conform them to curved surfaces and maintain a high accuracy degree reducing at the same time the sampling rate (which in traditional approaches is considered to be around 10 samples per wavelength). It represents an advantage when compared with other approaches that use a faceted meshing, which loses the real shape of the original bodies. Once the geometry has been discretized, high-order modified rooftop and razor blade functions are used as basis and testing functions, respectively. In addition, the proposed approach also allows the analysis of structures embedded in dielectric slabs. In this case, volumetric rooftops and razor blade functions are defined over curved volumetric elements. For very thin dielectric slabs a surface representation is considered instead of the previously mentioned volumetric approach, and the MoM impedance matrix elements are corrected by a term that depends on the size of the subdomain, the frequency, the electric permittivity, and the thickness of the slab.

Regarding the convergence issues that can be derived from the application of domain decomposition methods, the approach proposed in this work relies on the definition of extensions for each domain. After computing the updated currents over the extended version we only retain those which are included in the original, nonextended domain. We reach a double goal with this strategy: the improvement of the convergence properties in the iterative process, and the correction of the artificial edge behavior of the current. Extension values of about 0.4 to 0.7 wavelengths typically yield good results. The negative side of this strategy is the extra CPU time required to solve the extended problem. However we have found that it only represents a small fraction of the total simulation time for the typical domain sizes considered.

The MPI is used for the parallelization of the approach. This paradigm provides communication functionalities between a set of processes. Therefore, it is possible to address this kind of problems using a multiprocessor computer or cluster, increasing the computational efficiency [32, 33].

The paper is organized as follows. The main features of the proposed approach are addressed in Section 2. Some remarks about the selection of active and passive domains are given in Section 3. Section 4 presents the results obtained when analyzing a specific reflectarray antenna, and Section 5 contains the conclusions derived from this work.

2. Background of the Technique

The domain decomposition approach combines the MoM and MLFMA together with an iterative process used to calculate the interactions between different domains, which will be described in the next section. The first step will be, therefore, a partitioning process of the geometry into several domains. Thus, the total current over the original geometry can be expressed as

$$\vec{J} = \sum_{d=1}^{N_d} \vec{J}_d, \tag{1}$$

where N_d denotes the number of domains. Subsequently, a full wave analysis is individually applied over each domain using the MoM and MLFMA, obtaining the *primary currents*, which are those obtained by considering each block isolated from the rest and the excitation vector given by the external field. Several small problems are solved instead of a very large one, which is the main advantage when compared to the analysis of the whole structure.

The artificial edges derived from the partitioning of the objects in terms of domains may affect noticeably the final results. The solution to this potential burden is to extend the area of each domain and compute the induced currents while considering the extended domain. Figure 1 illustrates this decomposition procedure and one example of an extended domain.

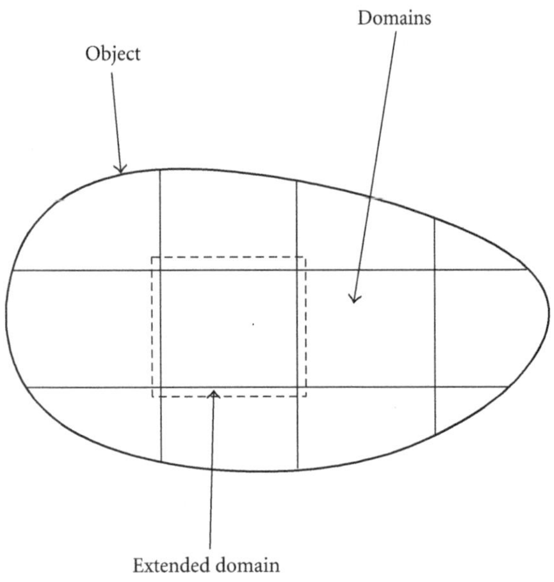

Figure 1: Scheme of the decomposition approach.

Once the currents have been calculated, those that are located inside the extension are discarded. By doing this we avoid the negative effect of the singular behavior of the currents near the artificially introduced edges. According to our experience an extension with a side length between 0.7 to 1.0 wavelengths is enough to obtain good accuracy.

After computing the primary currents an iterative process is performed to calculate the influence of the currents obtained inside one domain over the others. This stage of the process is required in order to update the primary currents calculated with the MoM. In order to compute the interactions between different parts of the geometry, a selective process is necessary to avoid unnecessary calculations. Let us denote as *active domains* the ones that contain previously computed currents which will radiate over other domains of the geometry (these ones called *passive domains*). It can be inferred that a given domain D can serve as active and passive domain at the same time, since the current distribution calculated over D in the kth iteration may produce interactions with other parts of the geometry in the $(k + 1)$th iteration, while other active domains may contribute to update the current distribution on D as well. A ray-tracing approach determines which active domains contribute to the updating current process of a certain passive domain, and therefore only a few computations are required to obtain the final currents. Multiscale problems can take advantage of this methodology because only the resulting fields radiated by an active domain are required to feed the passive domain.

3. The Iterative Selection of Active and Passive Domains

In order to increase the efficiency of the approach, it is very important to discard interactions between active and passive domains when these interactions are negligible. This strategy reduces significantly the CPU time required. Moreover, realistic structures with complex small details can be efficiently analyzed because no geometrical approximation is considered. The immediate benefit of this method is a significant decrease in the required CPU time and memory storage resources.

The total current over a given domain (let us consider domain-i) can be seen as the superposition of the currents computed by a number of interactions between that domain and the rest of the domains contained in the geometry:

$$\vec{J}_i = \vec{J}_i^0 + \sum_{j=1}^{M_i} \vec{J}_i^j, \tag{2}$$

where the term $\vec{J}_i^0$ represents the primary current computed over domain-i. The other terms of the expression indicate the currents due to the interaction between domain-i and the rest of the domains, where M_i stands for the highest order of the interactions with domain-i considered as the passive domain.

In order to compute the interactions between different domains, it is necessary to calculate the fields radiated by the previously computed currents on the rest of the domains over each passive domain, which can be iteratively performed. Thus, the impress field over domain-i due to the kth order effect can be written as

$$\vec{V}_i^k = \beth_i\left\{\sum_{n=1,n\neq i}^{N_d} \vec{J}_n^{k-1}\right\}, \tag{3}$$

where the operator $\beth_i$ computes the field over domain-i due to the previously calculated currents over the rest of the geometry, which serve as active domains. The MLFMA approach is very well suited to obtain $\vec{V}_i^k$. Now it is possible to obtain the current over domain-i due to the kth order effect, $\vec{J}_i^k$, by solving the conventional MoM equation for the isolated domain. This current can be efficiently obtained by utilizing MLFMA with an iterative solver such as the Biconjugate Stabilized Gradient method (BiCGStab).

Expression (3) needs to be modified in order to select those pairs of active and passive domains that will entail a noticeable contribution to the final current, and discard the negligible interactions. In practical applications this procedure allows us to reduce the amount of primary memory and CPU time maintaining the same degree of accuracy. In order to take into account this consideration, the modified expression can be seen as follows:

$$\vec{V}_i^k = \beth_i\left\{\sum_{n=1,n\neq i}^{N_d} \aleph_i\left(\vec{J}_n^{k-1}\right)\right\}, \tag{4}$$

where we have introduced the function $\aleph_i(\vec{J}_n^{k-1})$ that takes a zero value if the interaction between domain-i and domain-n is negligible and takes the value of $\vec{J}_n^{k-1}$ if such interaction entails a noticeable contribution to the current distribution over the passive domain. For the selection of the passive domains which interact with an active one it is important to consider the possibility of total or partial eclipse situations, as well as the shape of the pattern radiated by the active currents. We utilize the following mechanisms to accomplish this task.

First, using the multipole aggregation of the current over the active domain we determine the main directions in which such domain radiates, discarding the passive domain if it is not located in the path of the main set of output directions. This set of directions is determined by setting a threshold over the strongest field value.

In addition to the previous procedure we also consider a ray-tracing algorithm, very successfully used in high-frequency methods, known as the Angular Z-Buffer (AZB) [34, 35]. With the AZB the space is divided into angular regions (anxels), and the surfaces of the geometry are classified depending on which anxels contain them. This information is stored in the AZB matrices in a preprocessing step. As an example, let us consider a domain of the problem called D_1. The AZB matrix associated to D_1 contains a number of cells which are assigned to different angular ranges, since one of the dimensions of such matrix makes reference to the theta position of the angular range and the other dimension indicates its phi position. For the generation of the AZB matrix, different ray tubes are shot over D_1 (depending on the nature of the external excitation) and the output directions are obtained. The output ray tubes are then checked for intersections with other domains. If some rays belonging to the same tube intersect with different domains, these are processed in order to analyze the possibility of eclipse effects. Only the nonhidden domains are included in the AZB matrix of D_1. Further information can be found in great detail in [36]. This technique is very well suited to be used in combination with the previous approach based on the multipole expansion of the currents over the active domain, because the former gives the initial tube rays to be considered for the search of passive domain candidates to interact with the active domain.

Thus, for a given direction (from the main output radiation directions of the active domain) the AZB yields the passive domains where there can be a strong coupling, discarding the negligible interactions. This can be extended to consider multiple effects between domains, as shown in Figure 2 where D_1, D_2, and D_3 are domains contained in the geometry.

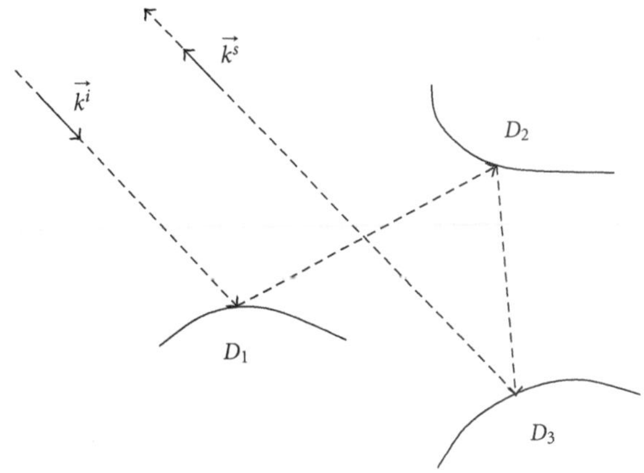

Figure 2: Example of ray tracing for multiple effects.

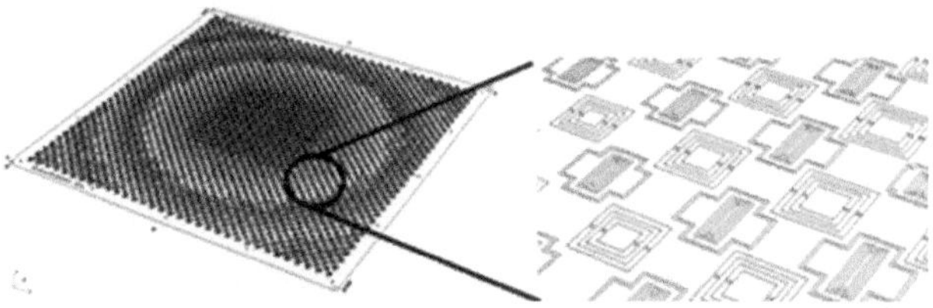

Figure 3: Schematic view of the original reflectarray.

4. Analysis of a Reflectarray

In this work we have modified and analyzed a reflectarray antenna for applications such as radar or long-distance communication. Some preliminary results of the numerical approach presented here can be seen in [37]. The design of the broadband reflectarray is presented in [28] and provides high radiation efficiency, approximately 60%. Regarding its construction, flat reflectors are often defined by several layers of periodical metallic patches and dielectric slabs over a ground plane [38]. Although microstrip reflectarray antennas are narrow in bandwidth, they are well suited for dual-frequency operation [39]. In addition to the original reflectarray considered here, there are other works in the literature presenting dual-band reflectarrays [40–43]. Currently, reflectarray antennas are widely used as an alternative to conventional reflector antennas due to several advantages such as reduced weight and volume, mechanical robustness, compatibility with active devices, low cost, and short manufacturing time.

Figure 3 shows the original reflectarray and the shape of the unit cells. The individual elements are designed to scatter the incident field while impressing the appropriate phase shifts in such a way as to form a plane wave front that propagates in a prescribed direction or, more generally, to create the required radiation pattern. The dimensions of the whole square structure are 40 cm × 40 cm, and the reflective surface is located over the XY plane.

The modifications we have introduced consist of a new compact feed system comprising an offset cassegrain structure where the horn is located under the plane of the reflectarray, for which it is necessary to open a hole on the reflectarray to feed the subreflector. This configuration can be seen in Figure 4. The feeding antenna is a pyramidal horn polarized along the X-axis. Figure 4 shows the location of the feed antenna over the reflectarray, and Figure 5 displays its far-field radiation pattern.

The aim of the modification provided in this work is the generation of a more compact layout for this antenna, while maintaining its original features up to a certain degree. The Cassegrain structure allows this goal, with good radiation properties as the final result.

The reflectarray is composed of a dielectric layer over a metallic ground plane, and the metallic strips are located over the dielectric. The period of the reflective elements is defined as 12 × 12 mm. The substrate has a thickness of 3.175 mm and a permittivity of 2.17. The simulation

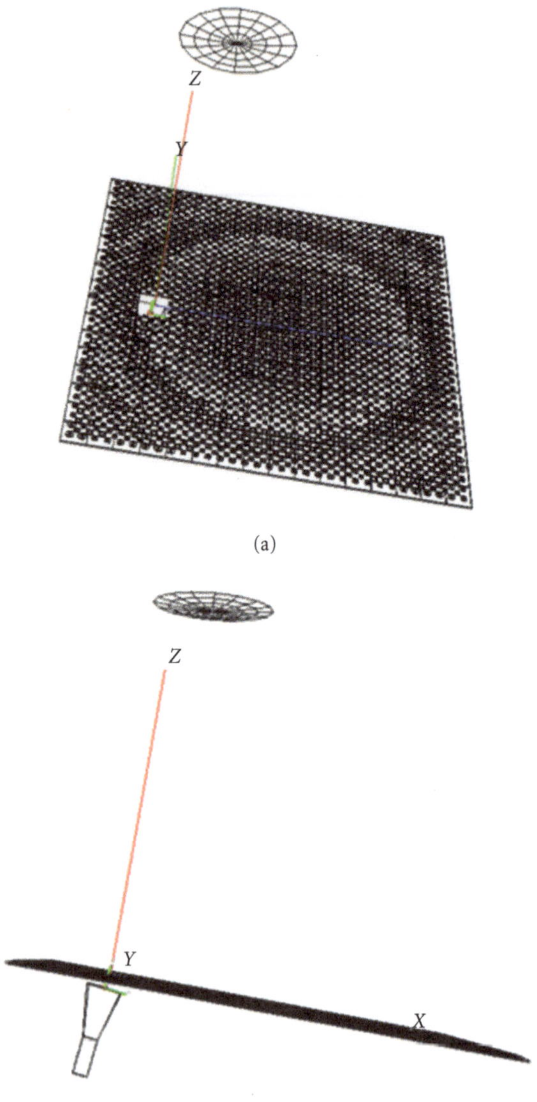

FIGURE 4: Geometrical views of the modified reflectarray.

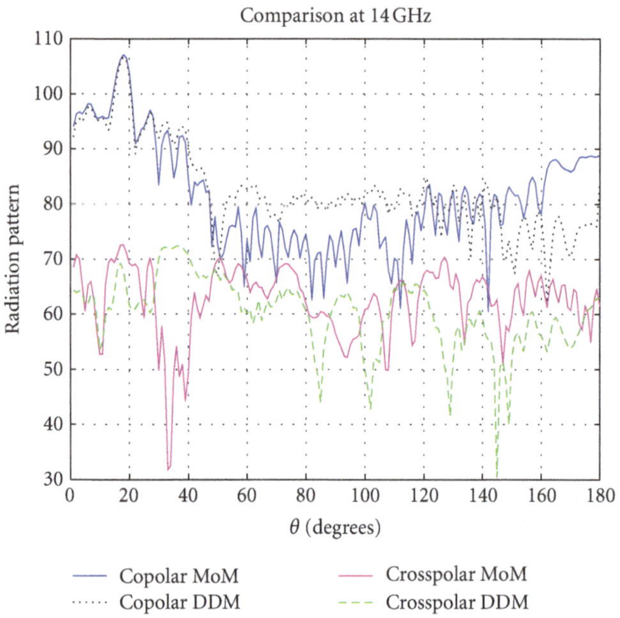

FIGURE 5: Comparison of radiation patterns applying the Moment Method and the domain decomposition approach at 14 GHz.

FIGURE 6: 3D pattern view.

has been performed considering a superficial thin layer approximation for the dielectric region.

The reflectarray was originally designed to radiate in the direction defined by $\theta = 16^\circ$ and $\varphi = 0^\circ$ at 14 GHz. The horn is located at $x = y = z = 0$ and feeds a subreflector whose center is $x = y = 0$, $z = 0.21$ and radius is 0.106 m. The focal length of the subreflector is 0.215 m, and the offset is -0.02 m. We have computed the radiation patter of the geometry using two different approaches, namely, the full wave technique given by the MoM-MLFMA and the domain decomposition approach described in this work. With the latter method we have considered 6 domains in the geometry: the first one comprises the horn, the second domain contains the subreflector, and the reflectarray is partitioned into 4 domains. The CPU-time required by the MoM is 4806 seconds, while the Domain decomposition approach took 987 seconds to obtain the results. The total number of unknowns when analyzing the reflectarray antenna at 14 GHz is 271636 unknowns. Around 27% of that amount (73720 unknowns) belongs to the dielectric part. However, the efficiency of the developed technique compared to the conventional full-wave analysis increases when the electrical size of the structure is even higher.

Figure 5 depicts the comparison between the radiation patterns obtained by these two approaches at 14 GHz. The direction of the main beam is now given by $\theta = 18^\circ$ and $\varphi = 0^\circ$ at both frequencies. A good agreement can be appreciated, with some discrepancies for the lower levels, due to the fact that, even considering the offset, there is still some blockage owing to the location of the horn and subreflector. A 3D view of the pattern obtained by the proposed approach can be seen in Figure 6, and the current distribution obtained

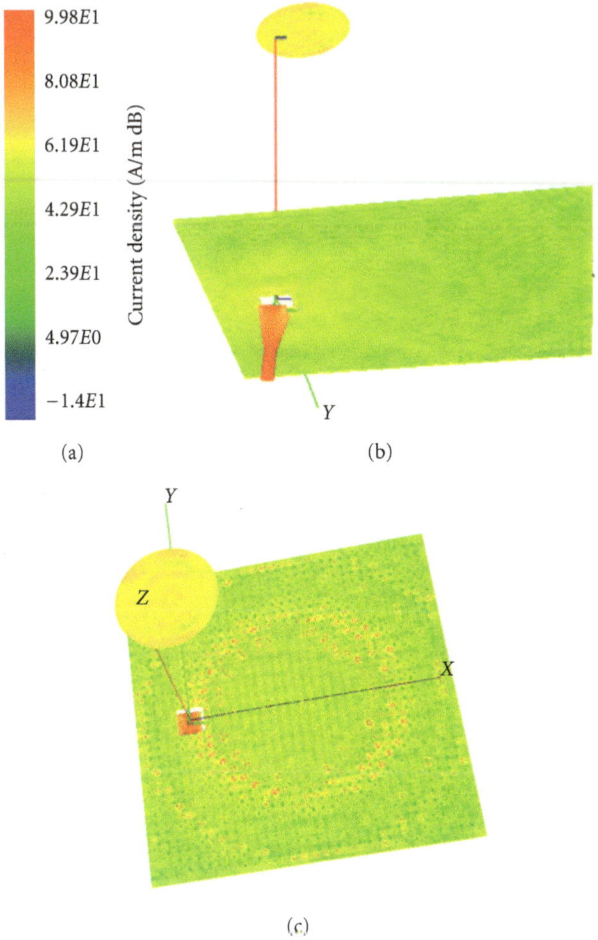

Figure 7: Current distribution.

is shown in Figure 7. In Figures 8 and 9 we compare the radiation pattern obtained with the modified configuration with the original configuration. It is possible to appreciate the differences due to the effect of the subreflector blockage and the hole required for the new horn placement, but the overall behaviour of this more compact configuration can be of interest for a number of practical applications.

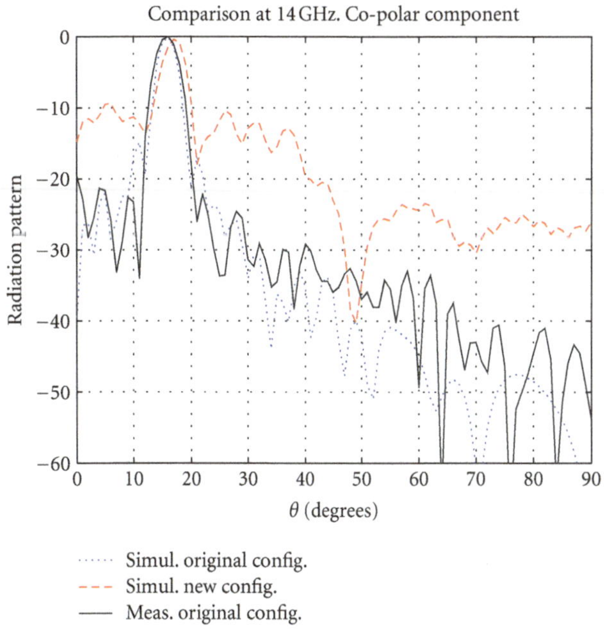

Figure 8: Radiation pattern of the original and modified configurations. Copolar component.

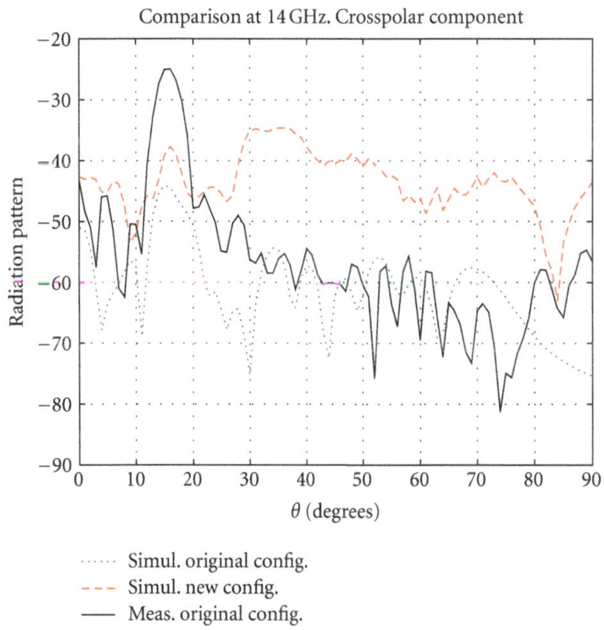

Figure 9: Radiation pattern of the original and modified configurations. Crosspolar component.

5. Conclusion

An efficient approach based on the MoM-MLFMA combined with a domain decomposition scheme has been presented and validated comparing to the full-wave solution. This approach considerably reduces the memory and time requirements by solving several small problems instead of a single large one. The predictions given by both methods show a good agreement. The MPI paradigm was incorporated to increase the performance in the simulations.

Acknowledgments

This work has been supported in part by the Comunidad de Madrid Project S-2009/TIC1485, by the Castilla-La Mancha Project PPII10-0192-0083, by the Spanish Department of Science, Technology Projects TEC 2010-15706 and CONSOLIDER-INGENIO no. CSD-2008-0068, and by University of Alcalá Project UAH2011/EXP-014. The authors thank Dr. Reza Chaharmir for the original model and measurements of the reflectarray antenna reported in this paper.

References

[1] W. C. Chew, J. Jin, E. Michielssen, and J. Song, *Fast and Efficient Algorithms in Computational Electromagnetics*, Arctech House, 2001.

[2] E. García, C. Delgado, I. G. Diego, and M. F. Cátedra, "An iterative solution for electrically large problems combining the characteristic basis function method and the multilevel fast multipole algorithm," *IEEE Transactions on Antennas and Propagation*, vol. 56, no. 8, pp. 2363–2371, 2008.

[3] C. Delgado, M. F. Cátedra, and R. Mittra, "Efficient multilevel approach for the generation of characteristic basis functions for large scatters," *IEEE Transactions on Antennas and Propagation*, vol. 56, no. 7, pp. 2134–2137, 2008.

[4] C. Delgado, M. F. Cátedra, and R. Mittra, "Application of the characteristic basis function method utilizing a class of basis and testing functions defined on NURBS patches," *IEEE Transactions on Antennas and Propagation*, vol. 56, no. 3, pp. 784–791, 2008.

[5] C. Craeye, "A fast impedance and pattern computation scheme for finite antenna arrays," *IEEE Transactions on Antennas and Propagation*, vol. 54, no. 10, pp. 3030–3034, 2006.

[6] L. Matekovits, V. A. Laza, and G. Vecchi, "Analysis of large complex structures with the synthetic-functions approach," *IEEE Transactions on Antennas and Propagation*, vol. 55, no. 9, pp. 2509–2521, 2007.

[7] K. Zhao, M. Vouvakis, and J.-F. Lee, "The adaptive cross approximation algorithm for accelerated method of moments computations of EMC problems," *IEEE Transactions on Electromagnetic Compatibility*, vol. 47, no. 4, pp. 763–773, 2005.

[8] A. Heldring, J. M. Rius, J. M. Tamayo, J. Parrón, and E. Ubeda, "Multiscale compressed block decomposition for fast direct solution of method of moments linear system," *IEEE Transactions on Antennas and Propagation*, vol. 59, no. 2, pp. 526–536, 2011.

[9] A. Boag and R. Mittra, "Complex multipole beam approach to electromagnetic scattering problems," *IEEE Transactions on Antennas and Propagation*, vol. 42, no. 3, pp. 366–372, 1994.

[10] F. X. Canning, "The impedance matrix localization (IML) method for moment-method calculations," *IEEE Antennas and Propagation Magazine*, vol. 32, no. 5, pp. 18–30, 1990.

[11] E. Bleszynski, M. Bleszynski, and T. Jaroszewicz, "AIM: adaptive integral method for solving large-scale electromagnetic scattering and radiation problems," *Radio Science*, vol. 31, no. 5, pp. 1225–1251, 1996.

[12] E. Michielssen and A. Boag, "Multilevel evaluation of electromagnetic fields for the rapid solution of scattering problems," *Microwave and Optical Technology Letters*, vol. 7, no. 17, pp. 790–795, 1994.

[13] E. F. Knott, "A progression of high-frequency RCS prediction techniques," *Proceedings of the IEEE*, vol. 73, no. 2, pp. 252–264, 1985.

[14] U. Jakobus and F. M. Landstorfer, "Improved PO-MM hybrid formulation for scattering from three-dimensional perfectly conducting bodies of arbitrary shape," *IEEE Transactions on Antennas and Propagation*, vol. 43, no. 2, pp. 162–169, 1995.

[15] K. R. Umashankar, S. Nimmagadda, and A. Taflove, "Numerical analysis of electromagnetic scattering by electrically large objects using spatial decomposition technique," *IEEE Transactions on Antennas and Propagation*, vol. 40, no. 8, pp. 867–877, 1992.

[16] D. Torrungrueng and E. H. Newman, "The multiple sweep method of moments (MSMM) analysis of electrically large bodies," *IEEE Transactions on Antennas and Propagation*, vol. 45, no. 8, pp. 1252–1258, 1997.

[17] W. D. Li, W. Hong, and H. X. Zhou, "An IE-ODDM-MLFMA scheme with DILU preconditioner for analysis of electromagnetic scattering from large complex objects," *IEEE Transactions on Antennas and Propagation*, vol. 56, no. 5, pp. 1368–1380, 2008.

[18] W. D. Li, W. Hong, and H. X. Zhou, "Integral equation-based overlapped domain decomposition method for the analysis of electromagnetic scattering of 3D conducting objects," *Microwave and Optical Technology Letters*, vol. 49, no. 2, pp. 265–274, 2007.

[19] D. Çolak, R. J. Burkholder, and E. H. Newman, "On the convergence properties of the multiple sweep method of moments," *Applied Computational Electromagnetics Society Journal*, vol. 22, no. 2, pp. 207–218, 2007.

[20] J. Laviada, M. R. Pino, and F. Las-Heras, "Convergence of certain class of enhanced spatial domain decompositions," *Microwave and Optical Technology Letters*, vol. 53, no. 10, pp. 2388–2393, 2011.

[21] J. Laviada, M. R. Pino, and F. Las-Heras, "On the influence of domain extensions and the relaxation parameter on spatial domain decomposition methods," in *Proceedings of the 6th European Conference on Antennas and Propagation (EUCAP '12)*, Prague, Czech Republic, March 2012.

[22] P. De Vita, F. De Vita, A. Di Maria, and A. Freni, "An efficient technique for the analysis of large multilayered printed arrays," *IEEE Antennas and Wireless Propagation Letters*, vol. 8, pp. 104–107, 2009.

[23] C. Wan and J. A. Encinar, "Efficient computation of generalized scattering matrix for analyzing multilayered periodic structures," *IEEE Transactions on Antennas and Propagation*, vol. 43, no. 11, pp. 1233–1242, 1995.

[24] S. H. Zainud-Deen, A. M. Abd-Elhady, A. A. Mitkees, and A. A. Kishk, "Design of dielectric resonator reflectarray using full-wave analysis," in *Proceedings of the 26th National Radio Science Conference (NRSC '09)*, pp. 1–9, New Cairo, Egypt, March 2009.

[25] C. Tienda, M. Arrebola, J. A. Encinar, and G. Toso, "Analysis of parabolic reflectarray in dual-reflector configuration," in *Proceedings of the 4th European Conference on Antennas and Propagation (EuCAP '10)*, Barcelona, Spain, April 2010.

[26] I. González, E. García, F. S. De Adana, and M. F. Cátedra, "Monurbs: a parallelized fast multipole multilevel code for analyzing complex bodies modeled by NURBS surfaces," *Applied Computational Electromagnetics Society Journal*, vol. 23, no. 2, pp. 134–142, 2008.

[27] E. García, C. Delgado, L. Lozano, I. González, and F. Cátedra, "An efficient hybrid scheme combining the characteristic basis function method and the multilevel fast multipole algorithm for solving bistatic RCS and radiation problems," *Progress in Electromagnetic Research B*, vol. 34, pp. 327–343, 2011.

[28] M. R. Chaharmir, J. Shaker, and N. Gagnon, "Broadband dual-band linear orthogonal polarisation reflectarray," *Electronics Letters*, vol. 45, no. 1, pp. 13–14, 2009.

[29] G. Farin, *Curves and Surfaces for Computer Aided Geometric Design*, Academic Press, 1988.

[30] L. Valle, F. Rivas, and M. F. Cátedra, "Combining the moment method with geometrical modelling by NURBS surfaces and Bezier patches," *IEEE Transactions on Antennas and Propagation*, vol. 42, no. 3, pp. 373–381, 1994.

[31] J. Moreno, L. Lozano, M. J. Algar, I. González, and F. Cátedra, "Pre-processing and meshing optimization for electromagnetic tools," in *Proceedings of IEEE International Symposium

on Antennas and Propagation (APSURSI '11), pp. 317–320, Spokane, Wash, USA, July 2011.

[32] W. Gropp, E. Lusk, and A. Skjellum, *Using MPI Portable Parallel programming with the Message-Passing Interface*, The MIT Press, 1999.

[33] D. J. Ludick and D. B. Davidson, "Investigating efficient parallelization techniques for the Characteristic Basis Function Method (CBFM)," in *Proceedings of the International Conference on Electromagnetics in Advanced Applications (ICEAA '09)*, pp. 400–403, Torino, Italy, September 2009.

[34] M. F. Cátedra and J. Pérez, *Cell Planning for Wireless Communications*, Artech House, Boston, Mass, USA, 1999.

[35] F. S. De Adana, O. Gutiérrez, I. González, L. Lozano, and M. F. Cátedra, *Practical Applications of Asymptotic Techniques in Electromagnetics*, Artech House, 2010.

[36] F. Cátedra, L. Lozano, I. González, E. García, and M. J. Algar, "Efficient techniques for accelerating the ray-tracing for computing the multiple bounce scattering of complex bodies modeled by flat facets," *Applied Computational Electromagnetics Society Journal*, vol. 25, no. 5, pp. 395–409, 2010.

[37] C. Delgado, E. García, and F. Cátedra, "Application of a hybrid domain decomposition approach for the analysis of large reflector antennas," in *Proceedings of the 5th European Conference on Antennas and Propagation (EUCAP '11)*, pp. 846–848, Rome, Italy, April 2011.

[38] M. H. Jamaluddin, R. Gillard, R. Sauleau, P. Dumon, and L. Le Coq, "Reflectarray element based on strip-loaded dielectric resonator antenna," *Electronics Letters*, vol. 44, no. 11, pp. 664–665, 2008.

[39] M. Mussetta, P. Pirinoli, P. T. Cong, M. Orefice, and R. E. Zich, "Optimization of a dual-band reflectarray antenna," in *Proceedings of IEEE International Symposium on Antennas and Propagation (APSURSI '10)*, pp. 1–4, Toronto, Canada, July 2010.

[40] S. H. Hsu, C. Han, J. Huang, and K. Chang, "An offset linear-array-fed Ku/Ka dual-band reflectarray for planet cloud/precipitation radar," *IEEE Transactions on Antennas and Propagation*, vol. 55, no. 11, pp. 3114–3122, 2007.

[41] J. Huang, C. Han, and K. Chang, "A cassegrain offset-fed dual-band reflectarray," in *Proceedings of IEEE International Symposium on Antennas and Propagation (APSURSI '06)*, pp. 2439–2442, Albuquerque, NM, USA, July 2006.

[42] P. Pirinoli, P. T. Cong, M. Mussetta, and M. Orefice, "Dual-band reflectarray antenna: design and experimental validation," in *Proceedings of the International Conference on Electromagnetics in Advanced Applications (ICEAA '09)*, pp. 1024–1027, Torino, Italy, September 2009.

[43] M. R. Chaharmir, J. Shaker, and N. Gagnon, "Broadband dual-band linear orthogonal polarisation reflectarray," *Electronics Letters*, vol. 45, no. 1, pp. 13–14, 2009.

SAR Experiments Using a Conformal Antenna Array Radar Demonstrator

Peter Knott, Thomas Bertuch, Helmut Wilden, Olaf Peters, Andreas R. Brenner, and Ingo Walterscheid

Fraunhofer FHR, Fraunhofer Street 20, 53343 Wachtberg, Germany

Correspondence should be addressed to Peter Knott, peter.knott@fhr.fraunhofer.de

Academic Editor: P. Mohanan

Conformal antenna arrays have been studied for several years but only few examples of applications in modern radar or communication systems may be found up to date due to technological difficulties. The objective of the "Electronic Radar with Conformal Array Antenna" (ERAKO) demonstrator system which has been developed at the Fraunhofer Institute for High Frequency Physics and Radar Techniques (FHR) is to demonstrate the feasibility of an active electronically scanned antenna for conformal integration into small and medium sized airborne platforms. For practical trials the antenna has been adapted for operation with the Phased Array Multifunctional Imaging Radar (PAMIR) system developed at the institute. The antenna in combination with the PAMIR front-end needed to undergo a special calibration procedure for beam forming and imaging post-processing. The present paper describes the design and development of the conformal antenna array of the demonstrator system, its connection to the PAMIR system and results of recently conducted synthetic aperture radar (SAR) experiments.

1. Introduction

Conformal antenna arrays integrated into the surface of a nonplanar part of modern aircrafts, vehicles, or ships are of great interest for certain applications where planar arrays or reflector antennas have definite drawbacks. Some of the potential advantages are improved aerodynamics, increased payload, large field of view (LFOV), and low observability [1]. Despite these advantages, the usage of conformal array technology in commercial applications is still comparably rare. Electromagnetic modelling of non-planar antenna arrays is a much more complex task than for their planar counterparts because analytical refinement and array periodicity cannot generally be used to simplify the physical relations. Also, the technology needed for fabrication, assembly, and feeding of flush mounted antenna elements on single or double curved surfaces is still comparably complex and costly.

The objective of the "Electronic Radar with Conformal Array Antenna" (ERAKO) demonstrator system which has been developed at Fraunhofer FHR is the development of a single curved antenna array using existing multilayer technology [2, 3]. The system is based on a wing or fin type elliptical profile with a strong degree of curvature and designed to demonstrate multifunction radar. Possible modes of operation include detection, acquisition, and tracking of multiple targets as well as synthetic aperture radar (SAR). For an overview over important system parameters of the demonstrator's antenna array see Table 1.

2. Conformal Antenna Array

The aperture of the ERAKO Antenna array is single curved in the horizontal plane (xy-plane) and planar along the z-axis. For the geometry of the conformal antenna array, an elliptical cross-section with a large axial ratio ($a/b = 4$) and a strong degree of curvature at the tip has been chosen. This configuration may be seen as an approximation for an antenna mounted circumferentially to the wing or fin of an aircraft.

The antenna aperture is subdivided into 30 linear sub-arrays oriented along the z-axis. Each sub-array consists of several planar dielectric and conducting layers. Four

Figure 1: Subarray consisting of four patch antenna elements with common feed.

Figure 2: ERAKO antenna front-end showing the 30 sub-array columns. The polarization of the antenna elements is indicated.

Table 1: ERAKO system parameter overview.

Design/center frequency	9.4 GHz
Rel. 10 dB-bandwidth	>15%
Mutual coupling	< −20 dB
Polarization	Horizontal
Sub-array gain	12 dBi
Scanning range (1st segment)	±90°

horizontally polarized microstrip patch antenna elements designed at a centre frequency of 9.4 GHz are coupled to an underlying feed network via slots in the ground plane and each sub-array is connected to a common coaxial connector via its fixed parallel feed network. A single sub-array column is shown in Figure 1 and has a size of 107 × 16 × 10 mm including a conducting copper foil on a foam layer on the backside used to shield the feed network and prevent backward radiation.

Figure 2 shows the antenna aperture with the sub-arrays arranged in a facetted way and attached to a metallic case. Out of these sub-arrays, 24 are connected to a conformal combined feed and calibration network and may be actively operated. The outer three columns at each side of the antenna are terminated and serve as dummies for reducing array edge effects. The aperture dimensions are 330 × 200 × 165 mm, and the minimum radius of curvature is 83 mm.

The scattering parameters of the embedded antenna elements have been obtained using an Agilent E8364B network analyzer. As an example, the comparison of the element matching and mutual coupling between neighbouring antenna elements for one of the centre sub-arrays (columns) and one of the side elements including the calibration network is shown in Figure 3. It can be seen that, with and without a protecting radome, the 10 dB-bandwidth of the antenna elements well exceeds 1 GHz (>10%). Compared to the isolated antenna design, the resonance frequency is slightly shifted and reflection coefficient levels are slightly increased due to multiple imperfect transitions between different lines inside the calibration network. Mutual coupling between neighbouring columns is low, the coupling between two neighbouring columns (S_{12}) does not exceed −20 dB, and −25 dB between elements further apart (S_{13}) over the full frequency range.

For array beamforming, the embedded element patterns of all the columns in the antenna aperture have been measured in an anechoic chamber at FHR over the frequency range of interest. The mutually normalized embedded element patterns of one of the centre sub-array columns (number 15) and two of the side columns (#4 and #27) in the azimuth plane (E-plane) measured during Tx operation at the centre frequency (9.45 GHz) are shown in Figure 4 for the co-(horizontal) polarization. The embedded element far field patterns show the broad diagrams typical for patch antenna elements with a maximum gain of approximately 12 dBi for the central column and exhibit some ripples due to the facetted aperture design, mutual coupling effects, and the radome influence. The level of cross-polarization isolation is below −20 dB over most part of the azimuth angle range (not shown in the figure). The element pattern of the central element exhibits a wide and almost symmetric shape, which is due to its prominent position and orientation. At the same time, the element patterns of the side elements show their maxima at different angles and are rather distorted. The different maxima positions are due to their rotated orientation dictated by the curved aperture. The distorted shapes can be explained by considering the mounting of the conformal array on the radar system's front-end, as can be observed in Figure 6. The strong decay of the element patterns on one side is due to the shadowing of the front-end, whereas the gradual decay on the opposite side is due to the shadowing of the conformal array itself. In the latter case, a strong ripple is observed which is because of the interference of several contributions of wave propagating around the aperture and radiating at many different positions. To obtain optimum radiation performance from the antenna, different methods suited for pattern synthesis of conformal antenna arrays have been tested and applied [4].

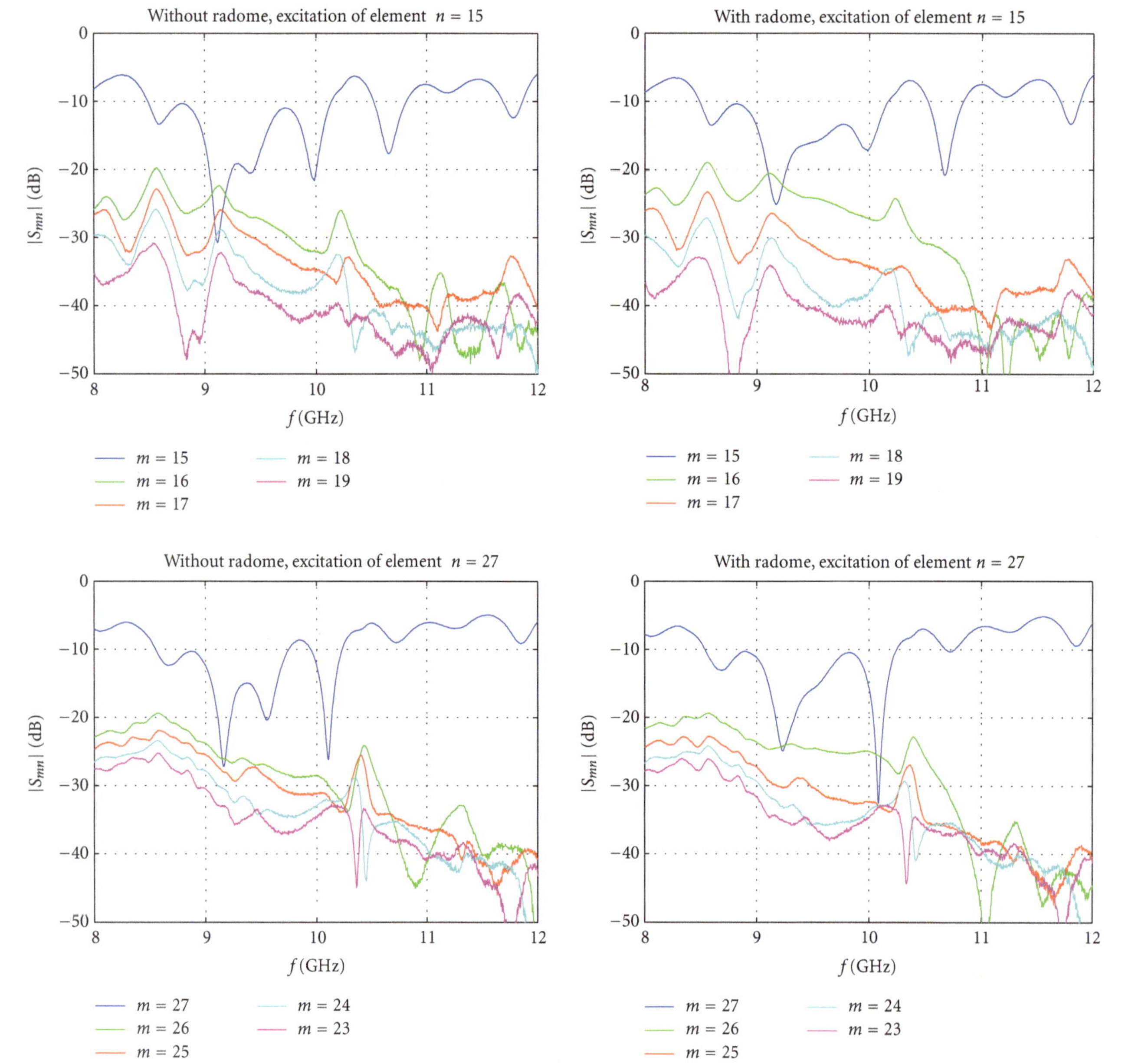

FIGURE 3: S-parameter measurement results for one of the centre elements (#15, top) and one of the side elements (#27, bottom) with (left) and without (right) radome (matching and mutual coupling to neighbouring elements).

3. The ERAKO System

For radar applications, it is foreseen to connect each of the 24 subarrays to a T/R module channel. Because the thermal properties of the electronics may change significantly during operation, a calibration network has been included at the backside of the aperture. Eight sub-arrays are connected to one calibration network. The planar substrates of the sub-arrays and the calibration networks form 90° angles.

For each sub-array port, a small fraction of the transmit power of −30 dB including the phase information is redirected to a calibration port via this network and specially designed vertical directional couplers [5]. Eight sub-array columns are connected to one dedicated calibration port. Using the back-coupled signal deviations between the amplitude and phase relations between the T/R modules may be compensated. The combined feed and calibration network assembled in the ERAKO antenna front-end as seen from below is shown in Figure 5. The backward couplers in three different groups and their connectors and the ports of the dummy elements at the sides of the array can be seen.

These calibration ports are to be used for an on-line calibration procedure for the Transmit-Receive Modules during long-term radar operation. Each of the three ports can be connected to an additional channel of the digital multichannel receiver. The procedure is typically conducted

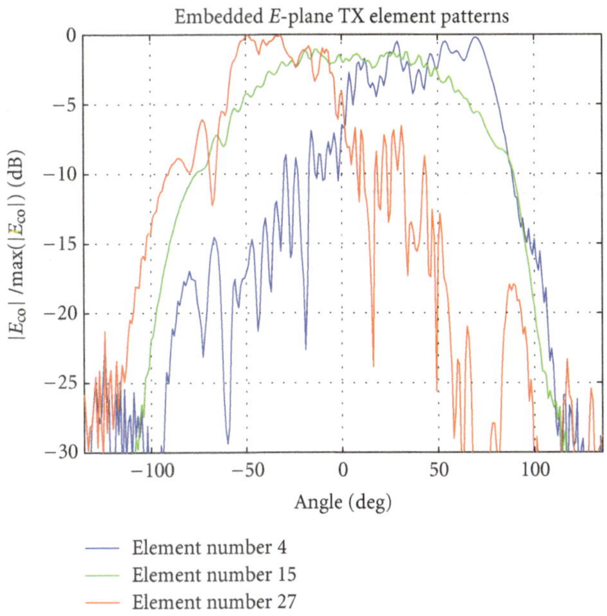

Figure 4: Mutually normalized embedded element patterns measured in TX operation of one of the centre elements (#15) and two of the side elements (#4 and #27) of the array aperture at the centre frequency ($f = 9.45$ GHz) for the copolarisation.

Figure 5: Feed and calibration network of the ERAKO antenna array seen from below.

Figure 6: PAMIR front-end including ERAKO antenna installed on experimental truck.

such that a signal is transmitted from a single selected channel while all other transmitters are switched off. Inside the feed network, this signal is coupled back into one of the three designated calibration ports, as illustrated in Figure 10. By comparing the actual amplitude and phase shift "received"/detected at the calibration channel to the desired values which have been pre-determined in a similar fashion during an off-line procedure, changes in the behaviour of the active electronic components of the selected channel's transmitter may be detected, which could be caused by temperature drift or ageing. The desired values for steering the array radiation pattern into specific directions are determined using the measurements described in Section 3.2. For bigger antenna arrays, larger numbers of sub-array channels may be combined into one calibration path.

Using this calibration information, corrected phase and amplitude settings may be determined and applied to compensate for the error. For the characterisation of the full array, this step must be repeated sequentially for all 24 active antenna channels. The procedure may be applied in both directions for the transmitting and receiving parts of the electronics. For the operation of the radar system, it is essential to find an appropriate strategy for the time distribution of the calibration signals. Without delaying the target acquisition or tracking this may be done, for example, by including single calibration pulses (or short sequences) in between radar pulse trains every few seconds only.

3.1. Experimental Setup. The Phased Array Multifunctional Imaging Radar (PAMIR) experimental system was originally developed for high resolution long range surveillance with highest flexibility to demonstrate multimode operation [6]. The system serves as airborne sensor for Synthetic Aperture Radar (SAR) imaging with a resolution of up to 10 cm and for detecting moving targets in clutter with a minimum speed of 1 m/s. This multifunctionality can only be achieved with an Active Electronically Scanned Array (AESA) antenna. For this purpose, an experimental radar platform was designed that may be flexibly adopted to changing system requirements.

One of two mockups that were developed for the PAMIR system is a cylindrical aircraft pod containing a linear antenna array with a length of 4.25 m which optimizes the Ground Moving Target Indicator (GMTI) capability of the system for detecting very small target velocities. In addition to the large case used for airborne experiments, there is also a shorter version with a length of approximately 1 m, which can accommodate up to three RF subgroups (in total 48 antenna elements). This case can be installed at the top of an experimental truck vehicle in order to conduct SAR trials. For the experiments in combination with the conformal ERAKO antenna array, these two systems were combined.

The ERAKO antenna was mechanically and electronically connected to two RF subgroups (with 16 channels each) requiring the construction of a special frame structure as support. Since the radar data is measured at a velocity of

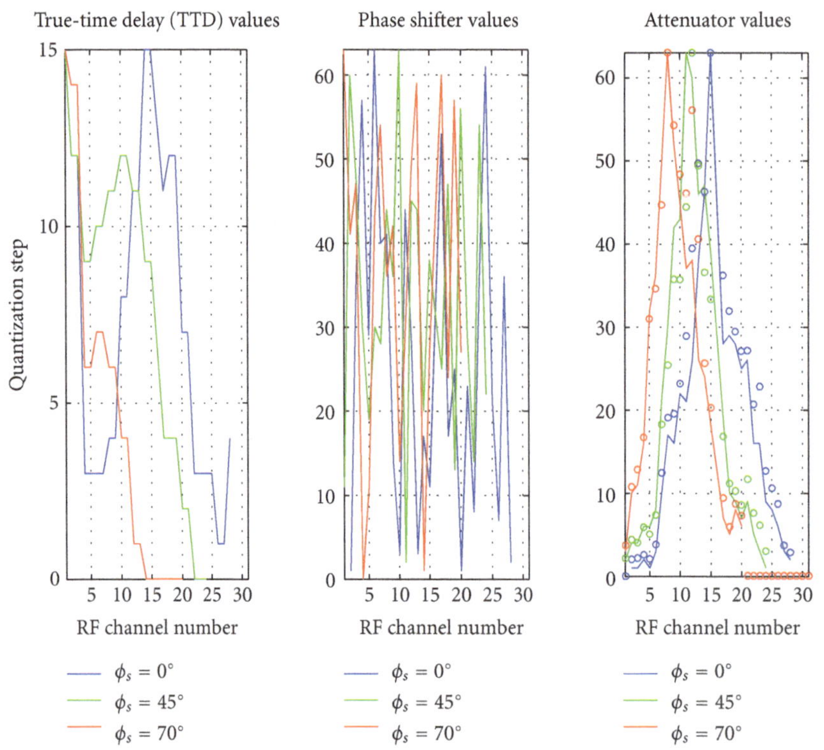

FIGURE 7: Experimentally optimised settings for the true-time delay (TTD), phase shifter, and attenuator settings in RX-operation for three different electronic scan angles. Ideal attenuator values are denoted by circles.

approx. 80 km/h, the ERAKO antenna had to be protected against environmental influences such as wind pressure, rain, or flying insects. The aperture of the antenna has been enclosed in a conformal radome made of glass epoxy. Measurements of reflection and mutual coupling coefficients between the individual antenna elements have shown only a very small influence due to the radome. The PAMIR front-end including the conformal antenna mounted at the top of the truck is shown in Figure 6.

3.2. Array Calibration Process. The ERAKO SAR experiments should be conducted at different steering angles in azimuth of the electronically scanned array antenna with respect to the nose-on direction of the antenna. Especially large scanning angles (e.g., 70°) are of great importance because they are not normally available for typical planar phased array antennas and demonstrate some of the advantages of conformal array antennas on curved apertures. The aim of the calibration procedure was to find optimum settings for the electronic components in each channel of the ERAKO system and to produce the desired array far field for each scanning angle. In transmit (TX) mode, the individual channels should operate at maximum output power while in receive (RX) mode the outer elements should amplify the received signals at lower levels than the ones at the centre for low side lobes in the RX radiation pattern. Therefore, the T/R module settings had to be determined separately for the RX and TX case, respectively.

Each channel of the PAMIR electronics includes a phase shifter with a resolution of four bits and amplitude attenuators with a resolution of six bits; all quantisation steps are in linear scale. In addition, there are electronically switchable true time delays (TTD) for two adjacent channels which can be used to stabilise the phase front of the radiation pattern over frequency for wideband signal operation.

The settings of the individual channels of PAMIR in the transmit (TX) and in the receive mode (RX) were determined for each scan angle in an iterative off-line calibration process, based on the measured embedded element patterns and the frequency response of complex transmission parameters. As an example, Figure 7 shows the experimentally optimised settings for the true-time delay (TTD), phase shifter, and attenuator settings in RX operation for three different electronic scan angles (0°, 45°, and 70°). Because of the finite number of quantisation steps and unavoidable non-linearities, small deviations from the ideal attenuator values determined previously in the beamforming process [7] had to be taken into account. In these cases, ideal values are denoted by circles.

Subsequently, the parameters resulting from the calibration procedure were verified in the anechoic chamber where the performance of the antenna array was measured. The resulting normalized array radiation patterns in RX mode measured over frequency at different scanning angles of 0°, 45°, and 70°, respectively, are shown in Figure 8 as an example. The upper images show the variation of the normalised array pattern over frequency and the lower

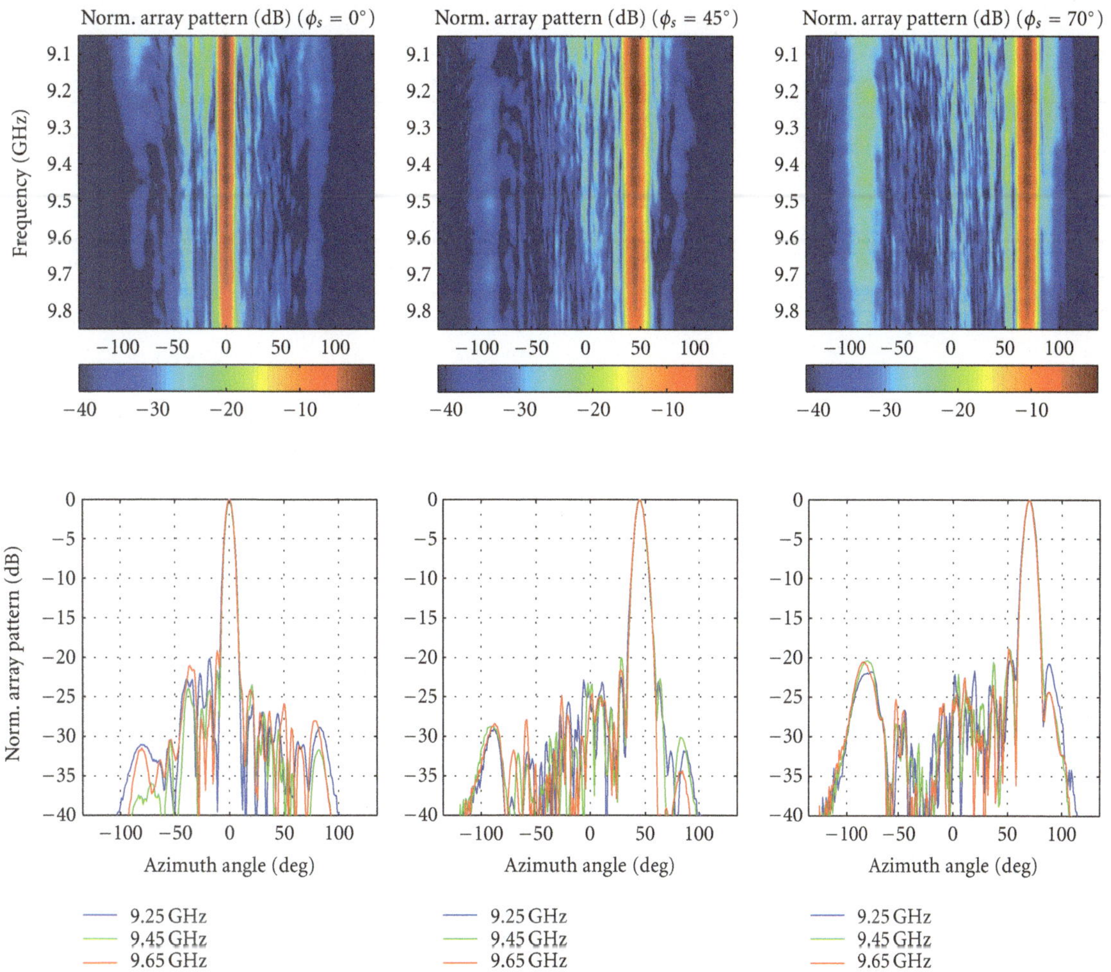

FIGURE 8: Measured array radiation pattern at scanning angles (0°, 45°, and 70°) in RX operation and its variation over frequency.

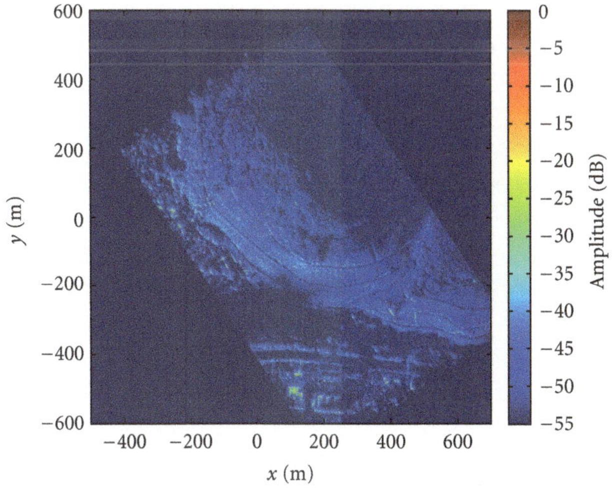

FIGURE 9: Example of a SAR image with a resolution of 50 cm × 50 cm.

diagrams are selected cross-sections for frequencies 9.25, 9.45, and 9.65 GHz for comparison. It is worthwhile to note that the main beam direction is very stable and the side-lobe level remains below −20 dB for almost all frequencies, even for large scan angles.

An important parameter for scanning array antennas is the so-called scan loss. Measurements have shown that this parameter is 1.28 dB with respect to a scanning angle of 0° in case of the RX operation and 0.86 dB for the scanning angles of 45° and 70°, respectively. In the TX case, the values are 0.57 dB and 1.35 dB. These are significantly lower than for an ideal planar array where values of 1.5 dB and 4.7 dB can be expected for the considered scanning angles. This shows clearly the potential superiority of the conformal array over the planar array concerning the scan loss. Despite of the fact that the frequency range of the PAMIR system had to be reduced to match the available bandwidth of the antenna elements, the calibration measurements were performed in the frequency range from 9.05 GHz to 9.85 GHz. However, only a total bandwidth of 380 MHz has been used for the radar experiments.

4. SAR Experiments

The experimental mockup calibrated using the procedure described above was used to conduct several SAR

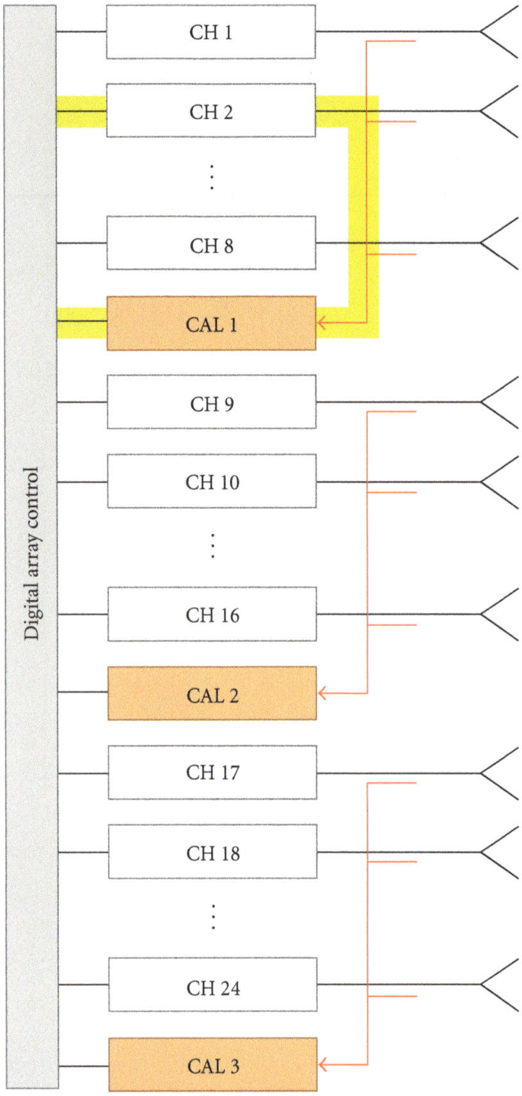

FIGURE 10: In the on-line calibration procedure, a signal is transmitted to a single antenna channel (e.g., CH 2) and fed back to a calibration channel by means of the calibration network to detect changes in the properties of active components.

experiments [8]. The PAMIR system including the ERAKO antenna array was installed on the roof of the experimental truck. Side-looking SAR measurements have successfully been made for different scan directions while driving over a road bridge. An example of a resulting SAR image for a resolution of 50 cm × 50 cm taken at a scan angle of 70° is shown in Figure 9.

5. Conclusions

A conformal antenna array has been connected to the electronics of a wideband multifunctional imaging radar system to perform active radar experiments. For the process of calibration, an iterative off-line method has been developed for finding the settings of the individual channels taking into account the special properties of the curved aperture of the conformal antenna. Subsequently, the optimum settings have been used to form the desired array radiation patterns and several SAR experiments have been conducted with very good results.

With the experiments described above, the feasibility and potential of a conformal antenna array for active radar operation has been shown. They are an interesting alternative to planar arrays on platforms where the antenna shape is dictated by the curvature of the vehicle, for example, small unmanned aerial vehicles (UAVs). While the scope of the present paper only covers the topic of imaging radar, other modes of operation with electronic scanning capability are also possible. Hopefully, such demonstrations will help to increase the number of applications of conformal antenna arrays for radar or communication systems in the future.

References

[1] L. Josefsson and P. Persson, *Conformal Array Antenna Theory and Design*, vol. 29 of *IEEE Press Series on Electromagnetic Wave Theory*, John Wiley & Sons, New York, NY, USA, 2006.

[2] T. Bertuch, "A facetted cylindrically curved array for a conformal radar demonstrator," in *Proceedings of the European Workshop on Conformal Antennas (EWCA'03)*, pp. 95–998, Bonn, Germany, October 2003.

[3] P. Knott and T. Bertuch, "Design and development of a radar demonstrator with conformal antenna array," in *Proceedings of the European Workshop on Conformal Antennas (EWCA'05)*, Stockholm, Sweden, May 2005.

[4] R. J. Mailloux, *Phased Array Antenna Handbook*, SciTech Publishing, Mendham Borough, NJ, USA, 1994.

[5] T. Bertuch, "Backward directional coupler for weak coupling of double-sided microstrip lines using longitudinal double-slots," *Electronics Letters*, vol. 40, no. 5, pp. 309–310, 2004.

[6] A. R. Brenner and J. H. G. Ender, "Demonstration of advanced reconnaissance techniques with the airborne SAR/GMTI sensor PAMIR," *IEE Proceedings: Radar, Sonar and Navigation*, vol. 153, no. 2, pp. 152–162, 2006.

[7] P. Knott, "Antenna design and beamforming for a conformal antenna array demonstrator," in *Proceedings of the IEEE Aerospace Conference*, Big Sky, Mont, USA, March 2006.

[8] T. Bertuch, P. Knott, H. Wilden, and O. Peters, "SAR experiments using a conformal antenna array radar demonstrator," in *Proceedings of the European Conference on Synthetic Aperture Radar (EuSAR'10)*, Aachen, Germany, June 2010.

Sparse Planar Array Synthesis Using Matrix Enhancement and Matrix Pencil

Mei-yan Zheng,[1] Ke-song Chen,[1] Hong-gang Wu,[2] and Xian-pan Liu[3]

[1] *School of Electronic Engineering, University of Electronic Science and Technology of China, Chengdu 611731, China*
[2] *The Second Research Institute of Civil Aviation of China, Chengdu 610041, China*
[3] *School of Astronautics and Aeronautics, University of Electronic Science and Technology of China, Chengdu 611731, China*

Correspondence should be addressed to Mei-yan Zheng; zhengmeiyan123456@126.com

Academic Editor: Tat Yeo

The matrix enhancement and matrix pencil (MEMP) plays important roles in modern signal processing applications. In this paper, MEMP is applied to attack the problem of two-dimensional sparse array synthesis. Firstly, the desired array radiation pattern, as the original pattern for approximating, is sampled to form an enhanced matrix. After performing the singular value decomposition (SVD) and discarding the insignificant singular values according to the prior approximate error, the minimum number of elements can be obtained. Secondly, in order to obtain the eigenvalues, the generalized eigen-decomposition is employed on the approximate matrix, which is the optimal low-rank approximation of the enhanced matrix corresponding to sparse planar array, and then the ESPRIT algorithm is utilized to pair the eigenvalues related to each dimension of the planar array. Finally, element positions and excitations of the sparse planar array are calculated according to the correct pairing of eigenvalues. Simulation results are presented to illustrate the effectiveness of the proposed approach.

1. Introduction

The question of nonuniform planar array synthesis with fewer number of elements for a desired beam pattern has been a popular concern since early in the 1970s. Reducing the number of elements in planar arrays has long been a problem both practical and theoretical significance being the subject of concern; for example, satellite communication system has limitation on the weight of antenna systems and subsystem accessories which wishes to use the minimum number of elements. Furthermore, fewer elements can lower the cost obviously and simplify the antenna system.

The synthesis of nonuniform array elements' positions, excitations, and phases is a complicated nonlinear problem which contains a number of decision variables. So far, there have been many technologies used to synthesize the nonuniform arrays, including optimization algorithm (such as genetic algorithm (GA) [1], differential evolution algorithm (DEA) [2], analytical methods [3, 4], matrix pencil method (MPM) [5], dynamic programming, and particle swarm optimization (PSO)) and other comprehensive technologies. However, some optimization algorithms such as GA, DEA, and PSO are suitable for solving nonlinear optimization of high dimension complex function, but they are very time consuming for their intrinsic iterative mechanism. Many other comprehensive approaches based on iteration do not guarantee that all variables are global optimal solutions. The new non iterative methods, called MPM [6, 7] and MEMP [8], have been successfully used by Yingbo Hua, Tapan K. S. and Odilon Pererira in the parameters estimation of a sum of complex exponentials, such as two-dimensional frequencies [8], azimuth/elevation angel [9], and two-dimensional DOA [10]. These years Liu and others find a new application in the synthesis of nonuniform array for MPM [5, 11].

MPM is qualified to the synthesis of sparse planar array which is designed to a separable planar array by a product of two orthogonal linear sparse arrays [5]. It can guarantee the best features in the two principal plane patterns in this particular case, but it cannot at the same time guarantee that others have the same characteristics.

Further expansion of MPM to the synthesis of inseparable distribution sparse planar array in this paper, just as in signal processing applications, the new method similarly called matrix enhancement and matrix pencil (MEMP) in synthesis of sparse arrays. The organization of this paper is listed as follows. In Section 2, the optimal array model is presented. In Section 3, the idea of matrix enhancement and matrix pencil is introduced, and then the method will be applied to estimate the minimum number of elements and to calculate the locations and excitations of the new array. Section 4 introduces the essential steps of MEMP. Some numerical results are presented in Section 5 to show the efficiency of the proposed method. Finally, some conclusions are drawn in Section 6.

2. The Optimal Array Model

Assume an array of N discrete similarly oriented identical elements arranged arbitrarily in a 3-D space, as shown in Figure 1. Denote the excitation in the ith element located at (r_i, θ_i, ϕ_i) as R_i.

According to the superposition theorem of electromagnetic waves, we can write the following array factor [12]:

$$F(\theta, \varphi) = \sum_{i=1}^{N} R_i \exp\left[j(\phi_i + kr_i \cos\alpha_i)\right], \quad (1)$$

where $k = 2\pi/\lambda$, λ is the wavelength. Consider,

$$\cos\alpha_i = \sin\theta \sin\theta_i \cos(\phi - \phi_i) + \cos\theta\cos\theta_i, \quad (2)$$

where $0 \le \theta \le \pi$, and $0 \le \phi \le 2\pi$ are the azimuth and elevation, respectively.

For the planar array, (1) can be simplified as

$$F(u, v) = \sum_{i=1}^{N} R_i \exp\left[jk\left(d_x u + d_y v\right)\right], \quad (3)$$

where d_x and d_y are the abscissa and ordinate of the ith element, and $u = \sin\theta\cos\phi$, $v = \sin\theta\sin\phi$.

3. MEMP to Synthesize Planar Array

3.1. Estimation of the Minimum Number of Elements. Matrix enhancement and matrix pencil is to use as few elements as possible to form a new planar array to approximate the desired pattern. Thus, the optimal mathematical is described as

$$\min \quad \{Q\}$$
$$\text{st.} \quad \left\{\text{Min}\left\|F(u, v) - \sum_{i=1}^{Q} R'_i \exp\left[jk\left(d'_x u + d'_y v\right)\right]\right\|\right\} \le \varepsilon, \quad (4)$$

where R'_i, d'_x, and d'_y are the excitation, abscissa, and ordinate of the new planar array. $L = 2$ because this paper uses the total least squares criteria.

Firstly, to sample the desired pattern from $u = -1, v = -1$ to $u = 1, v = 1$ uniformly, then $u_m = m\Delta = m/N$, $m = -N, \ldots, 0, \ldots, N$; $v_n = n\Delta = n/N$, $n = -N, \ldots, 0, \ldots, N$, which has $(2N+1)(2N+1)$ sampling points.

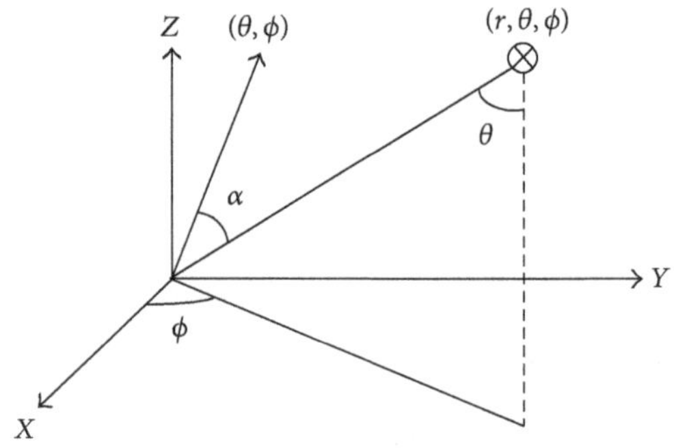

FIGURE 1: Reference coordinates of an arbitrary array.

The value of any sampling point is

$$f(m, n) = \sum_{i=1}^{N} R_i y_i^m z_i^n, \quad (5)$$

where $y_i = e^{jkd_x\Delta}$ and $z_i = e^{jkd_y\Delta}$.

Secondly, to construct an enhanced matrix $\mathbf{X}_e$ using the sampling point data,

$$\mathbf{X}_e = \begin{bmatrix} \mathbf{X}_0 & \mathbf{X}_1 & \cdots & \mathbf{X}_{2N-K+1} \\ \mathbf{X}_1 & \mathbf{X}_2 & \cdots & \mathbf{X}_{2N-K+2} \\ \vdots & \vdots & & \vdots \\ \mathbf{X}_{K-1} & \mathbf{X}_K & \cdots & \mathbf{X}_{2N} \end{bmatrix}, \quad (6)$$

where

$$\mathbf{X}_m = \begin{bmatrix} x(m;0) & x(m;1) & \cdots & x(m;2N-L+1) \\ x(m;1) & x(m;2) & \cdots & x(m;2N-L+2) \\ \vdots & \vdots & & \vdots \\ x(m;L-1) & x(m;L) & \cdots & x(m;2N) \end{bmatrix}, \quad (7)$$

where $x(m;n) = f(m-N; n-N)$. $\mathbf{X}_e$ is a Hankel block matrix and $\mathbf{X}_m$ is a Hankel matrix. The choice of parameters $\{K, L\}$ must satisfy the following conditions:

$$(K-1)L \ge N,$$
$$K(L-1) \ge N, \quad (8)$$
$$(2N+2-K)(2N+2-L) \ge N.$$

Thirdly, to perform the singular value decomposition (SVD) of the enhanced matrix $\mathbf{X}_e$,

$$\mathbf{X}_e = \sum_{i=1}^{\min} \sigma_i \mathbf{u}_i v_i^H$$
$$= \mathbf{U}_s \Sigma_s \mathbf{V}_s^H + \mathbf{U}_n \Sigma_n \mathbf{V}_n^H \quad (9)$$

$\min = \min\{KL, (2N+2-K)(2N+2-L)\}$ which is the smaller dimension of $\mathbf{X}_e$, and $\mathbf{U}_s$, Σ_s, and $\mathbf{V}_S$ contain the N principal eigenvalues and eigenvectors. $\mathbf{U}_n$, Σ_n, and $\mathbf{V}_n$ contain the remaining eigenvalues and eigenvectors. Specifically,

$$\mathbf{U}_s = [\mathbf{u}_1, \mathbf{u}_2, \ldots, \mathbf{u}_N], \tag{10a}$$

$$\Sigma_s = \operatorname{diag}\{\sigma_1, \sigma_1, \ldots, \sigma_N\}, \tag{10b}$$

$$\mathbf{V}_S = [v_1, v_2, \ldots, v_N], \tag{10c}$$

$$\mathbf{U}_n = [\mathbf{u}_{N+1}, \mathbf{u}_{N+2}, \ldots, \mathbf{u}_{\min}], \tag{10d}$$

$$\Sigma_n = \operatorname{diag}\{\sigma_{N+1}, \sigma_{N+2}, \ldots, \sigma_{\min}\}, \tag{10e}$$

$$\mathbf{V}_n = [v_{N+1}, v_{N+2}, \ldots, v_{\min}], \tag{10f}$$

where $\sigma_1 \geq \sigma_2 \geq \cdots \geq \sigma_{\min}$. The rank of enhanced matrix $\mathbf{X}_e$ is equal to the number of nonzero singular values. Generally, an array of N elements has N non-zero singular values; that is, $\sigma_i > 0$ $(i = 1, \ldots, N)$, $\sigma_i = 0$ $(i > N + 1)$. So, Σ_n is equal to zero. Equation (9) can be simplified to

$$\mathbf{X}_e = \mathbf{U}_s \Sigma_s \mathbf{V}_s^H. \tag{11}$$

However, some literature results [13] show that the number of important singular values is much smaller than the total number of array elements, that is the unimportant elements' contributions can be replaced by a combination of other elements. Therefore, we can discard the unimportant singular values to get a low-rank approximation of the enhanced matrix, which corresponds to the new array with fewer elements. The usual approach is to set these small singular values to zero. That is,

$$\mathbf{X}_Q = \mathbf{U}_s \Sigma_Q \mathbf{V}_s^H, \tag{12}$$

where $\Sigma_Q = \operatorname{diag}\{\sigma_1, \sigma_1, \ldots, \sigma_Q, 0, \ldots, 0\}$, $Q \leq N$.

In the actual synthesis of array, the minimum value of Q can be determined by the following formula [11]:

$$Q = \min \left\{ q; \mid \frac{\sqrt{\sum_{i=q+1}^{N} \sigma_i^2}}{\sqrt{\sum_{i=1}^{q} \sigma_i^2}} < \varepsilon \right\}, \tag{13}$$

where ε is a small positive number, and the choice of ε depends on the approximation between the reconstructed pattern and the original pattern.

3.2. The Solutions of Eigenvalues y_i and z_i. The eigenvalues are obtained by constructing two matrixs and solving the generalized eigenvalue of the matrixes [7, 8], then utilizing the ESPRIT algorithm in the literature [14] and pairing algorithm to estimate the correct pairing of y_i and z_i.

3.2.1. Extracting y_i. Once the low rank matrix $\mathbf{X}_Q$ is obtained, to solve the eigenvalue y_i is equal to solve the generalized eigenvalue of the following equation:

$$(\mathbf{X}_{Q,f} - y\mathbf{X}_{Q,l})\, v = 0, \tag{14}$$

where $\mathbf{X}_{Q,f}$ and $\mathbf{X}_{Q,l}$ are obtained by deleting the first L rows and the last L rows of $\mathbf{X}_Q$. To solve (14) is equal to solve the generalized eigenvalue of the following equation:

$$(\mathbf{U}_2 - y\mathbf{U}_1)\, v = 0, \tag{15}$$

where $\mathbf{U}_2$ and $\mathbf{U}_1$ are obtained by deleting the first L rows and the last L rows of $\mathbf{U}_Q$ which contains only Q dominant left-singular vectors of $\mathbf{U}_s$ in (11).

Therefore, it can be obviously observed that the matrix pencil $\mathbf{U}_2 - y\mathbf{U}_1$ can be expressed as

$$(\mathbf{U}_2 - y\mathbf{U}_1) = \mathbf{E}_1 (\mathbf{Y}_d - y\mathbf{I})\, \mathbf{T}, \tag{16}$$

where $\mathbf{Y}_d$ is a diagonal matrix composed of the eigenvalues of $\{y_i,\ i = 1, \ldots, Q\}$.

3.2.2. Extracting z_i. In order to extract another set of $\{z_i,\ i = 1, \ldots, Q\}$, firstly introduce a permutation matrix $\mathbf{P}$ as

$$\begin{aligned} \mathbf{P} = [&\mathbf{p}^T(1), \mathbf{p}^T(1+L), \ldots, \mathbf{p}^T(1+(K-1)L), \\ &\mathbf{p}^T(2), \mathbf{p}^T(2+L), \ldots, \\ &\mathbf{p}^T(1+(K-1)L), \ldots, \ldots, \mathbf{p}^T(L), \\ &\mathbf{p}^T(L+L), \ldots, \mathbf{p}^T(1+(K-1)L)\,]^T. \end{aligned} \tag{17}$$

The element $\mathbf{p}(i)$ is a $KL \times 1$ vector with one at the ith position and zero everywhere else. Left multiplying $\mathbf{U}_s$ by $\mathbf{P}$, we can get

$$\mathbf{U}_{sP} = \mathbf{P}\mathbf{U}_s. \tag{18}$$

According to (15), we can conclude that to solve the eigenvalues z_i is equal to solve the generalized eigenvalues of the following equation:

$$(\mathbf{U}_{2P} - z\mathbf{U}_{1P})\, v = 0, \tag{19}$$

where $\mathbf{U}_{2P}$ and $\mathbf{U}_{1P}$ are obtained by deleting the first K rows and the last K rows of $\mathbf{U}_{QP}$ which contains only Q dominant left-singular vectors of $\mathbf{U}_{sP}$ in (18).

Therefore, it can be obviously observed that the matrix pencil $\mathbf{U}_{2P} - z\mathbf{U}_{1P}$ can be expressed as

$$(\mathbf{U}_{2P} - z\mathbf{U}_{1P}) = \mathbf{E}_{1P} (\mathbf{Z}_d - z\mathbf{I})\, \mathbf{T}, \tag{20}$$

where $\mathbf{Z}_d$ is a diagonal matrix composed of the eigenvalues of $\{z_i,\ i = 1, \ldots, Q\}$.

3.2.3. Pairing. According to (16) and (20), we can deduce that

$$\begin{aligned} \mathbf{F}_1 &= \mathbf{U}_1^{+}\mathbf{U}_2 = \mathbf{T}^{-1}\mathbf{Y}_d\mathbf{T}, \\ \mathbf{F}_2 &= \mathbf{U}_{1P}^{+}\mathbf{U}_{2P} = \mathbf{T}^{-1}\mathbf{Z}_d\mathbf{T}. \end{aligned} \tag{21}$$

It can be seen from (21) that the pairing information of the eigenvalues is included in $\mathbf{T}$.

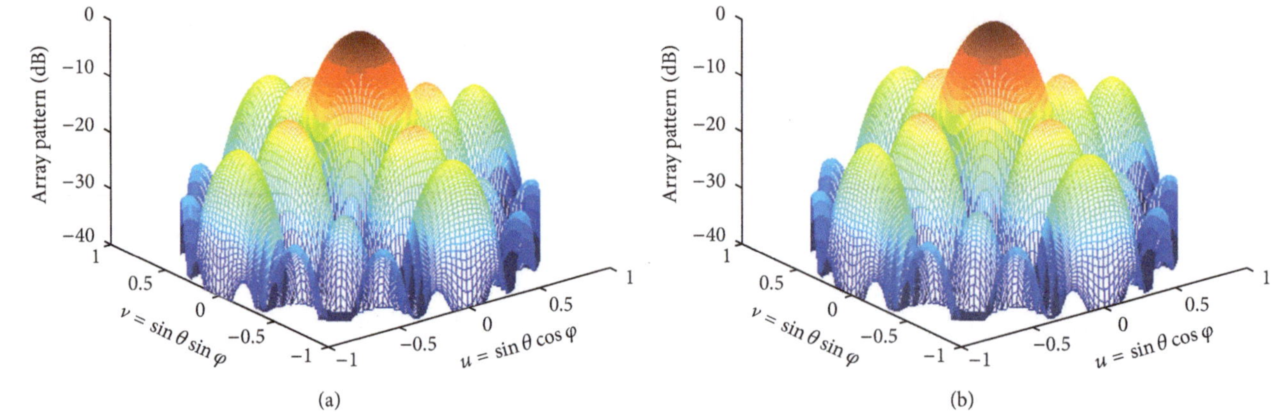

FIGURE 2: (a) The desired radiation pattern. (b) The reconstructed radiation pattern.

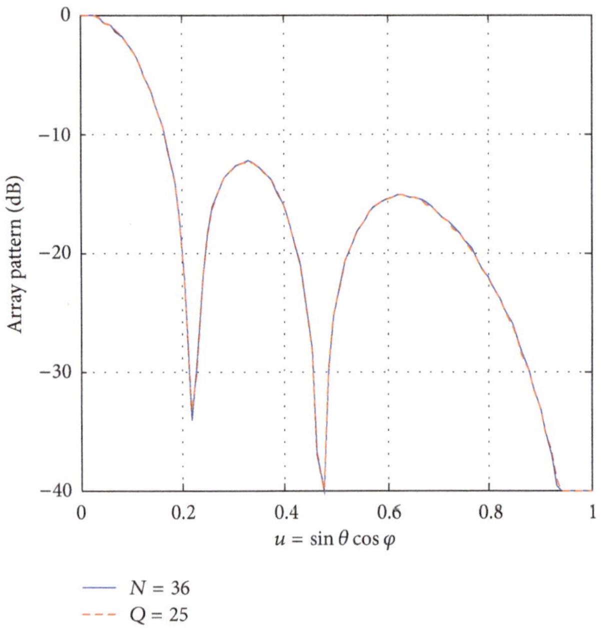

FIGURE 3: The cross-section of the pattern.

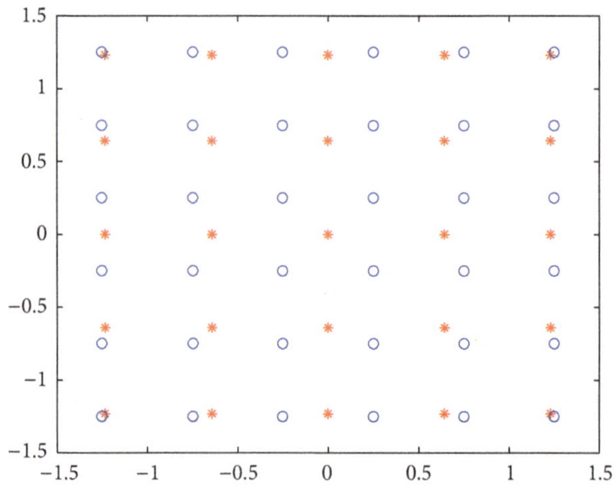

FIGURE 4: The elements' location.

The eigenvalue decomposition of a linear combination of $\mathbf{F}_1$ and $\mathbf{F}_2$ is computed to obtain a transformation diagonalizing of these two matrixs as follows:

$$\beta \mathbf{F}_1 + (1-\beta)\mathbf{F}_1 = \mathbf{T}^{-1}\mathbf{D}\mathbf{T}. \tag{22}$$

The transformation $\mathbf{T}$ is applied to $\mathbf{F}_1$ and $\mathbf{F}_2$ to obtain the matrixs:

$$\mathbf{Y}_d = \mathbf{T}\mathbf{F}_1\mathbf{T}^{-1}, \qquad \mathbf{Z}_d = \mathbf{T}\mathbf{F}_2\mathbf{T}^{-1}. \tag{23}$$

The values in the diagonal position of $\mathbf{Y}_d$ and $\mathbf{Z}_d$ determine the pairing.

3.3. To Solve Elements' Excitations and Locations. Once y_i and z_i are obtained, the locations can be solved according to (14) in the literature [13]; that is,

$$d'_x = \frac{1}{jk\Delta}\ln(y_i), \qquad d'_y = \frac{1}{jk\Delta}\ln(z_i). \tag{24}$$

As pointed in [9], a variety of synthetical examples show that, for the symmetrical pencil beam pattern, the imaginary parts of d'_i are much smaller than the real parts. For example, the imaginary parts of d'_i are ten orders smaller than the real parts for the synthesis of traditional Chebyshev planar pattern. Under this case, we can use the real parts of d'_i directly. The elements' locations are calculated as follows:

$$d'_x = \frac{1}{jk\Delta}\ln(\hat{y}_i), \qquad d'_y = \frac{1}{jk\Delta}\ln(\hat{z}_i), \tag{25}$$

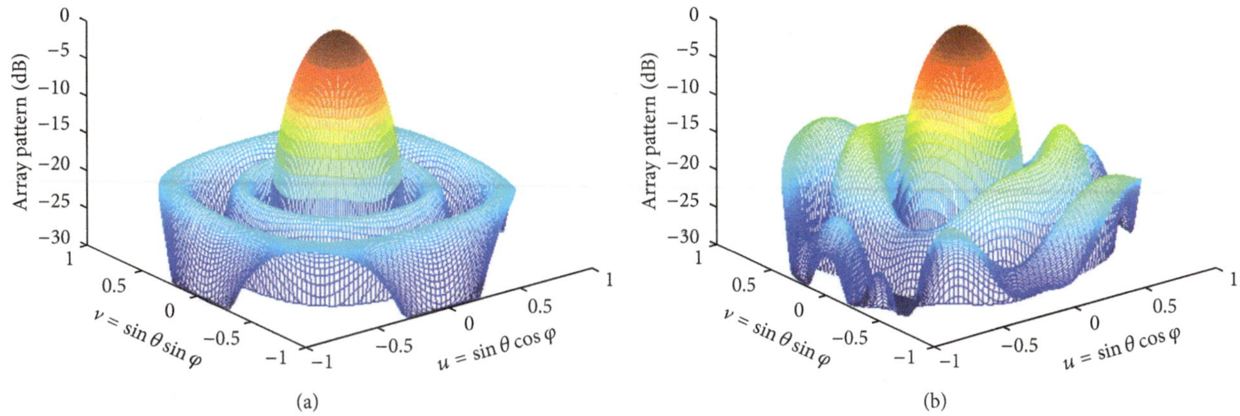

FIGURE 5: (a) The Chebyshev radiation pattern. (b) The reconstructed radiation pattern.

where

$$\widehat{y}_i = \frac{y_i}{|y_i|}, \qquad \widehat{z}_i = \frac{z_i}{|z_i|}. \tag{26}$$

The elements' excitations can be obtained as follows:

$$\mathbf{R} = \mathbf{E}_L^{+}\mathbf{X}_e\mathbf{E}_R^{+}. \tag{27}$$

The diagonal elements of matrix $\mathbf{R}$ are R_i $(i = 1, \ldots, Q)$ where

$$\mathbf{E}_L^{+} = \left(\mathbf{E}_L^H\mathbf{E}_L\right)^{-1}\mathbf{E}_L^H,$$
$$\mathbf{E}_R^{+} = \mathbf{E}_R^H\left(\mathbf{E}_R\mathbf{E}_R^H\right)^{-1}, \tag{28}$$

where

$$\mathbf{E}_L = \begin{bmatrix} \mathbf{Z}_L\mathbf{Y}_d^{-N} \\ \mathbf{Z}_L\mathbf{Y}_d^{-N+1} \\ \vdots \\ \mathbf{Z}_L\mathbf{Y}_d^{K-N-1} \end{bmatrix}, \tag{29}$$

$$\mathbf{E}_R = \left[\mathbf{Z}_R, \mathbf{Y}_d\mathbf{Z}_R, \ldots, \mathbf{Y}_d^{2N-K+1}\mathbf{Z}_R\right],$$

where

$$\mathbf{Z}_L = \begin{bmatrix} (\widehat{z}_1)^{-N} & (\widehat{z}_2)^{-N} & \cdots & (\widehat{z}_Q)^{-N} \\ (\widehat{z}_1)^{-N+1} & (\widehat{z}_2)^{-N+1} & \cdots & (\widehat{z}_Q)^{-N+1} \\ \vdots & \vdots & & \vdots \\ (\widehat{z}_1)^{L-1-N} & (\widehat{z}_2)^{L-1-N} & \cdots & (\widehat{z}_Q)^{L-1-N} \end{bmatrix}, \tag{30}$$

$$\mathbf{y}_d = \operatorname{diag}\left\{\widehat{y}_1, \widehat{y}_2, \ldots, \widehat{y}_Q\right\}, \tag{31}$$

$$\mathbf{Z}_R = \begin{bmatrix} 1 & \widehat{z}_1 & \cdots & (\widehat{z}_1)^{2N-L+1} \\ 1 & \widehat{z}_2 & \cdots & (\widehat{z}_2)^{2N-L+1} \\ \vdots & \vdots & & \vdots \\ 1 & \widehat{z}_Q & \cdots & (\widehat{z}_Q)^{2N-L+1} \end{bmatrix}. \tag{32}$$

4. The Steps of MEMP

The essential steps of MEMP are summarized as follows.

Step 1. To sample the desired pattern uniformly and to construct an enhanced matrix $\mathbf{X}_e$ by the sampling data; $\mathbf{X}_e$ is shown as (6).

Step 2. To perform the singular value decomposition (SVD) of enhanced matrix $\mathbf{X}_e$ and to calculate the singular values and left singular vector $\mathbf{U}_s$.

Step 3. According to (13), to determine the minimum number of elements Q.

Step 4. According to (16) and (20), to extract the eigenvalues y_i and z_i, then to pair them as (22).

Step 5. According to (25) and (27), to solve the elements' excitations and locations.

5. Simulation Results

In order to illustrate the effectiveness of MEMP, this paper gives two examples to reduce the number of elements and let the reconstructed array keep the characteristics of the original array.

5.1. Synthesis of Planar Array with Excitation of One. Suppose, there is a rectangular planar array with 6 × 6 elements. The elements are arranged in a rectangular grid, $d_x = d_y = \lambda/2$, and the elements' excitations are one. In this example, there are (2×36+1)(2×36+1) sampling points (by further increasing the sampling points will not improve the approximate accuracy of the pattern). The parameters $K = L = 37$. According to (13), the required minimum number of elements is $Q = 25$ when $\varepsilon = 10^{-5}$. The desired pattern and the reconstructed pattern are shown in Figure 2. The obtained sidelobe levels are −13.1525 dB and −13.1522 dB, respectively. Figure 3 compares the cross-section of the reconstructed pattern and the desired pattern. The figure shows that the array with $Q = 25$ unequally spaced distribution of the elements can basically

TABLE 1: The locations (excitations) of uniformly and nonuniformly spaced arrays.

The locations of uniformly spaced array			The locations (excitations) of nonuniformly spaced array		
0.25, 1.25	0.75, 1.25	1.25, 1.25	0, 1.2308 (1.4579)	0.6423, 1.2308 (1.3936)	1.2308, 1.2308 (1.2226)
0.25, 0.75	0.75, 0.75	1.25, 0.75	0, 0.6423 (1.6619)	0.6423, 0.6423 (1.5886)	1.2308, 0.6423 (1.3936)
0.25, 0.25	0.75, 0.25	1.25, 0.25	0, 0 (1.7386)	0.6423, 0 (1.6619)	1.2308, 0 (1.4579)

TABLE 2: The locations (excitations) of Chebyshev and nonuniformly spaced array.

The locations (excitations) of Chebyshev planar array				
0.25, 1.25 (0.3654)		0.75, 1.25 (0.1822)		1.25, 1.25 (0.0364)
0.25, 0.75 (0.2640)		0.75, 0.75 (0.3917)		1.25, 0.75 (0.1822)
0.25, 0.25 (0.4503)		0.75, 0.25 (0.2640)		1.25, 0.25 (0.3645)
The locations (excitations) of nonuniform planar array				
−1.2319, −1.2463 (0.0538)	−0.6505, −1.2237 (0.3215)	−0.0914, −1.1991 (0.7001)	0.3336, −1.1513 (0.8382)	0.9598, −0.9605 (0.5520)
−1.2237, −0.7295 (0.2870)	−0.6472, −0.6505 (0.6144)	0, −0.6238 (0.6207)	0.6472, −0.5680 (0.9560)	1.2237, −0.3341 (0.5179)
−1.1513, −0.2376 (0.6953)	−0.5680, 0 (0.9462)	0, 0 (0.7766)	0.5680, 0 (0.8475)	1.1513, 0.2376 (0.3763)
−1.1981, 0.3341 (0.6218)	−0.5971, 0.6505 (0.6486)	0, 0.5680 (0.8748)	0.5971, 0.7295 (0.8196)	1.1981, 0.6238 (0.6182)
−0.9598, 0.9605 (0.5523)	−0.3336, 1.2237 (0.6272)	0.0914, 1.1991 (0.7050)	0.6505, 1.1513 (0.4298)	1.2319, 1.2463 (0.0526)

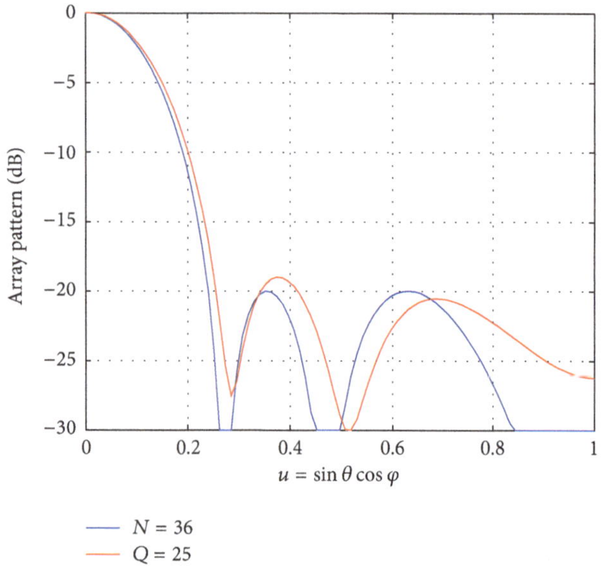

FIGURE 6: The cross section of the pattern.

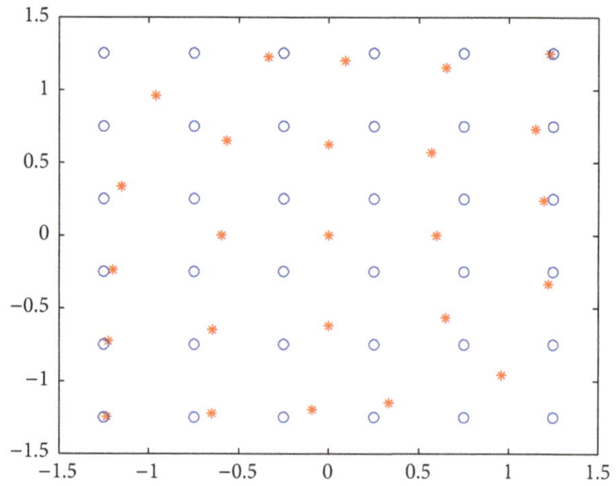

FIGURE 7: The elements' locations.

reconstruct the equally spaced distribution of the elements $N = 36$. The example is to save 30% of the array elements. The elements' locations of the equally and unequally spaced are shown in Figure 4. The equally spaced elements' locations and the unequally spaced elements' excitations and locations are shown in Table 1. Because the locations and excitations are symmetrically distributed, only the locations and excitations in the first quadrant are given.

5.2. Synthesis of Chebyshev Planar. Suppose that there is a chebyshev planar array with 6×6 elements. The elements are arranged in a rectangular grid, $d_x = d_y = \lambda/2$, and the required sidelobe level is −20 dB. In this example, there are $(2 \times 36 + 1)(2 \times 36 + 1)$ sampling points, and the parameters $K = L = 37$. Chebyshev planar array has the limitation of the number of elements in every row which is equal to the number of elements in every column which so, we use 5×5 elements to approximate the desired Chebyshev planar array. The desired pattern and the reconstructed pattern are shown in Figure 5. The obtained sidelobe level is −19.3058 dB. Figure 6 compares the cross section of the reconstructed pattern with the desired pattern. Although there is a slight elevation (0.694 dB) of the sidelobe level, we can save 30% elements. The elements' locations of the equally and unequally spaced are shown in Figure 7. The elements' excitations and locations of equally and unequally spaced arrays are shown in Table 2.

6. Conclusion

A synthesis method to sparse planar array which is based on MEMP is presented. Compared to stochastic optimization algorithm, MEMP is a non-iterative algorithm, which is suitable for designing of sparse array with the requirements of narrow beam, low sidelobe level. In addition, compared to MPM, MEMP can guarantee the best features of either plane. The synthesis of planar array in the paper does not consider the mutual coupling [15, 16] between elements. But there exists mutual coupling between actual elements. Therefore, the synthesis results gotten under the condition of unconsidering mutual coupling can be solved by adjusting the elements' excitations or locations. The discussion of this paper enriches the application of MEMP to the synthesis of planar array and provides a valuable reference in engineering applications dealing with complexity of problems.

Acknowledgment

This work was supported by the National Natural Science Foundation of China (nos. 60736045 and 60702070).

References

[1] K. Chen, X. Yun, Z. He, and C. Han, "Synthesis of sparse planar arrays using modified real genetic algorithm," *IEEE Transactions on Antennas and Propagation*, vol. 55, no. 4, pp. 1067–1073, 2007.

[2] D. G. Kurup, M. Himdi, and A. Rydberg, "Synthesis of uniform amplitude unequally spaced antenna arrays using the differential evolution algorithm," *IEEE Transactions on Antennas and Propagation*, vol. 51, no. 9, pp. 2210–2217, 2003.

[3] B. P. Kumar and G. R. Branner, "Design of unequally spaced arrays for performance improvement," *IEEE Transactions on Antennas and Propagation*, vol. 47, no. 3, pp. 511–523, 1999.

[4] B. P. Kumar and G. R. Branner, "Generalized analytical technique for the synthesis of unequally spaced arrays with linear, planar, cylindrical or spherical geometry," *IEEE Transactions on Antennas and Propagation*, vol. 53, no. 2, pp. 621–634, 2005.

[5] K. Yang, Z. Zhao, and Y. Liu, "Synthesis of sparse planar arrays with matrix pencil method," in *Proceedings of the International Conference on Computational Problem-Solving (ICCP '11)*, pp. 82–85, 2011.

[6] T. K. Sarkar and O. Pereira, "Using the matrix pencil method to estimate the parameters of a sum of complex exponentials," *IEEE Antennas and Propagation Magazine*, vol. 37, no. 1, pp. 48–55, 1995.

[7] Y. Hua and T. K. Sarkar, "Matrix pencil method for estimating parameters of exponentially damped/undamped sinusoids in noise," *IEEE Transactions on Acoustics, Speech, and Signal Processing*, vol. 38, no. 5, pp. 814–824, 1990.

[8] Y. Hua, "Estimating two-dimensional frequencies by matrix enhancement and matrix pencil," *IEEE Transactions on Signal Processing*, vol. 40, no. 9, pp. 2267–2280, 1992.

[9] J. Wang and J. J. Zhou, "Modified MEMP method for 2D scattering center measurement based on GTD model," in *Proceedings of the International Conference on Microwave and Millimeter Wave Technology (ICMMT '08)*, pp. 987–990, April 2008.

[10] C. Jian, S. Wang, and L. Lin, "2-D DOA estimation by MEMP based on L-shape array," in *Proceedings of the 8th International Conference on Signal Processing (ICSP '06)*, vol. 1, pp. 16–20.

[11] Y. Liu, Z. Nie, and Q. H. Liu, "Reducing the number of elements in a linear antenna array by the matrix pencil method," *IEEE Transactions on Antennas and Propagation*, vol. 56, no. 9, pp. 2955–2962, 2008.

[12] D. K. Cheng, "Optimization techniques for antenna arrays," *Proceedings of the IEEE*, vol. 59, no. 12, pp. 1664–1674, 1971.

[13] E. K. Miller and D. M. Goodman, "A pole-zero modeling approach to linear array synthesis I. The unconstrained solution," *Radio Science*, vol. 18, no. 1, pp. 57–69, 1983.

[14] S. Rouquette and M. Najim, "Estimation of frequencies and damping factors by two-dimensional ESPRIT type methods," *IEEE Transactions on Signal Processing*, vol. 49, no. 1, pp. 237–245, 2001.

[15] Y. Yu and H. T. Hui, "Design of a mutual coupling compensation network for a small receiving monopole array," *IEEE Transactions on Microwave theory and Techniques*, vol. 59, no. 9, pp. 2241–2245, 2011.

[16] B. H. Wang, H. T. Hui, and M. S. Leong, "Decoupled 2D direction of arrival estimation using compact uniform circular arrays in the presence of elevation-dependent mutual coupling," *IEEE Transactions on Antennas and Propagation*, vol. 58, no. 3, pp. 747–755, 2010.

20

Printed Modified Bow-Tie Dipole Antenna for DVB/WLAN Applications

Ching-Chih Hung, Chia-Mei Peng, and I-Fong Chen

Department of Electronic Engineering, Institute of Computer and Communication Engineering, Jinwen University of Science and Technology, No. 99 An-Chung Road, Hsintien City 23154, Taipei, Taiwan

Correspondence should be addressed to I-Fong Chen; ex0206@just.edu.tw

Academic Editor: Bal Singh Virdee

A printed modified bow-tie dipole antenna which consists of asymmetric-feed and inserted slots is presented to apply to the DVB and WLAN systems. This antenna combines omnidirectional radiation pattern, broad bandwidth, and band rejection in an easy way to fabricate. Experimental results of the constructed prototype indicate that the VSWR 2.5 : 1 bandwidths achieve 166.7%, 28.57%, and 23.63% at 660 MHz, 2450 MHz, and 5500 MHz, respectively.

1. Introduction

An increasingly popular paradigm for wireless communications considers clusters of cooperating networks, which offer the user seamless access through a range of technologies such as digital video broadcasting (DVB, UHF band: 470–862 MHz) and wireless local area network (WLAN, ISM bands: 2.4–2.4835 GHz and 5.15–5.825 GHz) [1]. The interworking wireless technologies, such as WLAN with DVB networks, provide the user with a return link for interactive TV and high bandwidth asymmetric downlink for data transfer to terminals operating in the heterogeneous roaming environment. Therefore, the desired antenna design must be able to cover both DVB and WLAN systems at the same time and to include dual-band/broad bandwidth, simple impedance matching to the feed line, and low profile. The bow-tie dipole antenna has become the main candidate for the above application since its bandwidth performances are reasonably better than other alternatives [2]. However, it is difficult for the traditional printed bow-tie dipole antenna to meet multiband operation for respective communication systems. The solution is to design a dual- or multiband integrated antenna for wireless product applications. For this purpose, in this paper, we introduce a printed modified bow-tie dipole antenna which combines band-rejection slots and asymmetric-feed structure to yield three operating bands [3]. The printed modified bow-tie dipole arms adjustable in position will yield an asymmetrical structure. The impedance matching is obtained through tuning the shifted distance of the arms [4]. The band rejection is obtained by inserting some slots in the dipole arms [5–9]. Details of the specification of the proposed designs and the experimental results of constructed prototype are presented and discussed in the following sections.

2. Antenna Structure and Design

Figure 1 shows the proposed asymmetric modified bow-tie dipole antenna for DVB and WLAN applications. The presented antenna structure is composed of a modified bow-tie dipole radiating element section of length L, width W, flare angle θ, and shifted distance D, which is printed on a 1.6 mm thick FR4 glass epoxy substrate (the relative permittivity is 4.3). The fundamental resonant mode of modified bow-tie dipole is designed to occur at about 660 MHz (the center frequency of the lower operating band: 470–862 MHz). The dipole length and width can be determined from the half- and quarter-wave lengths of the resonant frequencies. Note that the flare angle of modified bow-tie dipole is a function of operating bandwidth. In general, the bandwidth of the lower operating band increases with an increase in the flare angle. By determining appropriate dimensions (L, W, D, and θ)

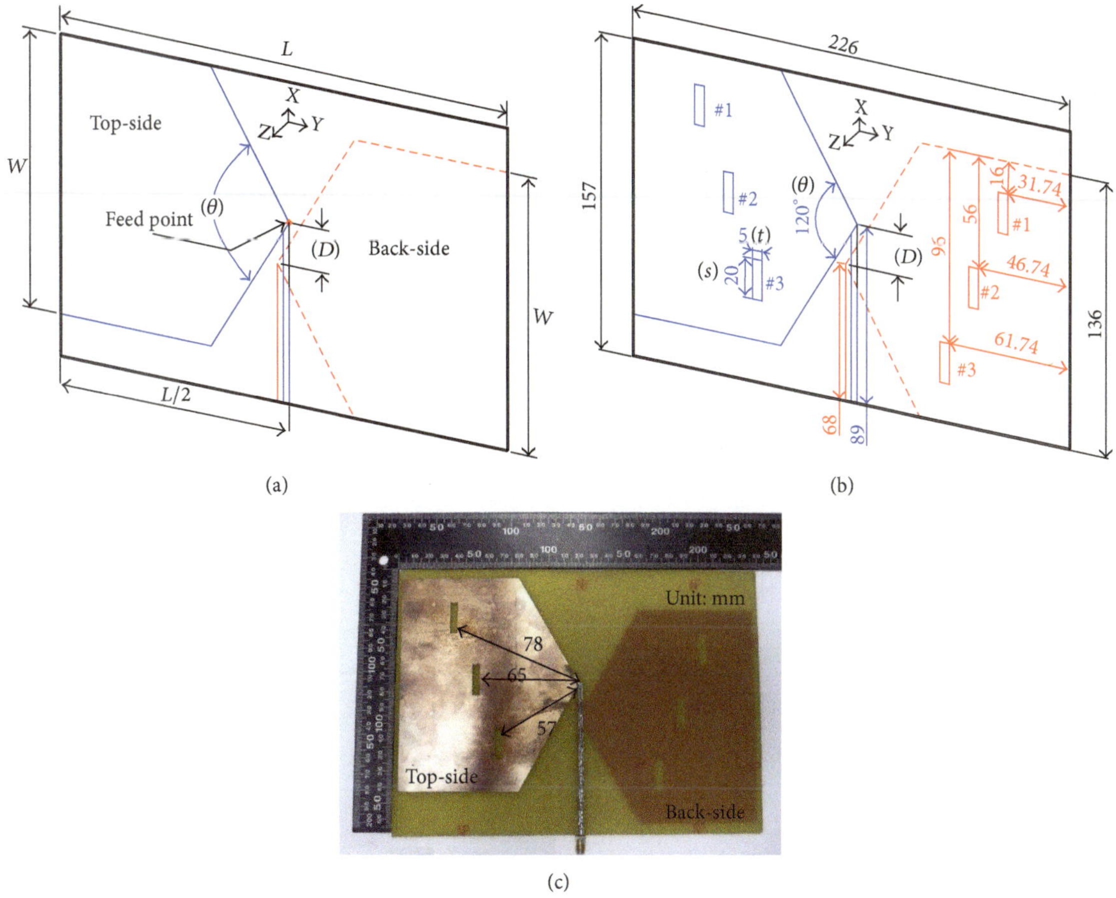

FIGURE 1: (a) Geometry of the original antenna. (b) Geometry of the proposed antenna. (c) Implemented prototype of the proposed antenna.

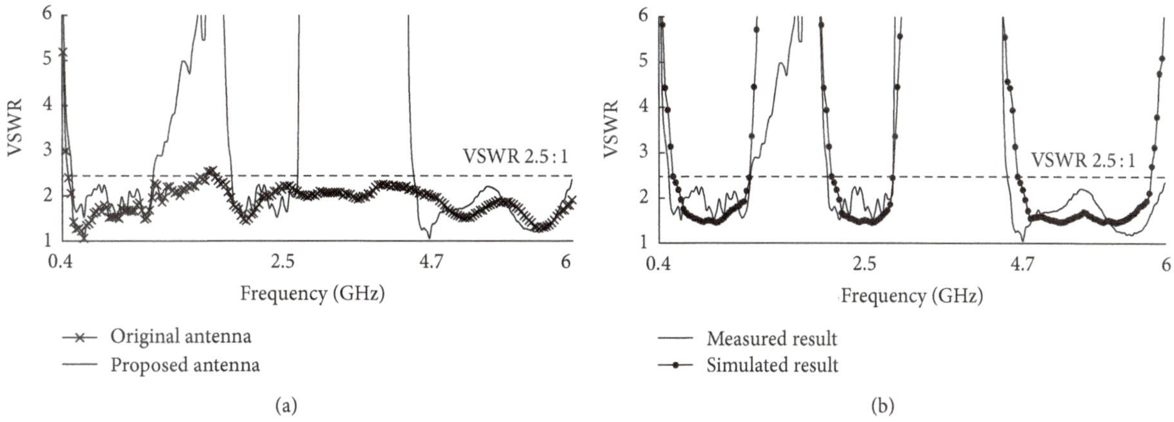

FIGURE 2: (a) Measured VSWR against frequency of the original and proposed antenna. (b) Measured and simulated VSWR against frequency of the proposed antenna.

of the antenna structure, good impedance matching of the printed modified bow-tie dipole can be obtained across an extended bandwidth. A parallel-strip feed excites the modified bow-tie dipole arms with shifted distance D to be of asymmetric structure as shown in Figure 1(a), denoted as the original antenna. The shape of the original antenna is designed for wideband operation. For fitting in with a 50 Ω feed system and to obtain an extended bandwidth, the shifted distance D should be $\lambda/20$ [4]. The band rejection of the dipole structure is obtained by inserting some slots in the dipole arms. Note that the size and number of the slots are not identical. The impedance matching at UHF band (470–862 MHz) and dual ISM band (2.4 GHz and 5.8 GHz) can be tuned by adjusting the size of the slots on the dipole arms.

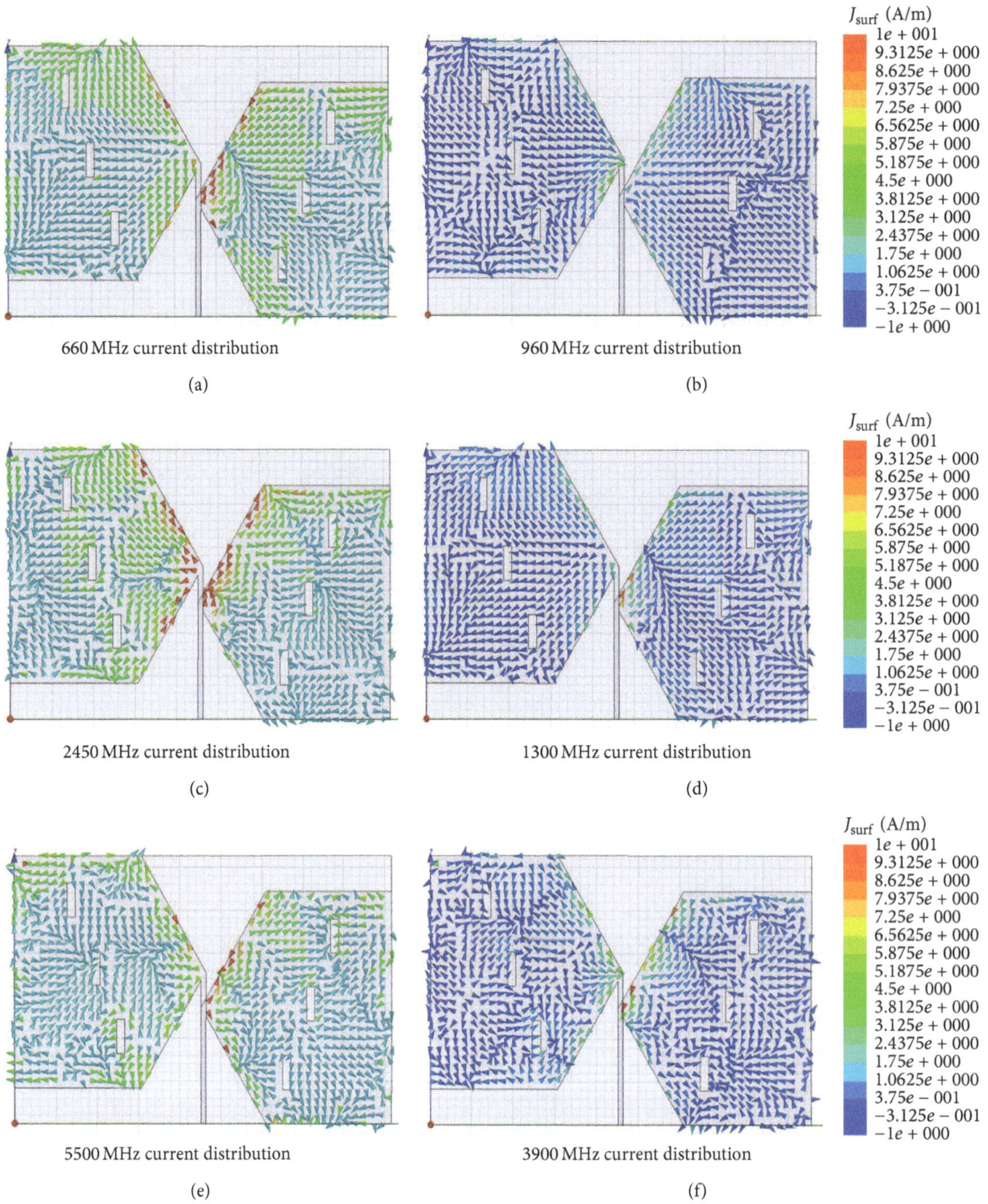

FIGURE 3: Simulated current density versus frequency on the modified bow-tie dipole arms.

3. Experimental Results and Discussion

In the experiment, the parallel-strip feed of the proposed antenna is connected to a 50 Ω SMA connector. By using the described design procedure, the original antenna is constructed to operate the VSWR ≤2.5 in the range of 450 MHz to 6 GHz, as shown in Figure 2(a), with modified bow-tie dipole-arm's size: length L = 22.6 cm, width W = 13.6 cm, flare angle θ = 120°, and shifted distance D = 2.1 cm [4]. Based on the experimental studies, the effects of key parameters (e.g., slots location, slots sizes, and slots numbers) are tested and optimized. In this paper, there are three same size insertion slots on the dipole arms. The slots are located at quarter-wave length of the undesired frequencies from the feed point (i.e., about 78 mm, 65 mm, and 57 mm, resp., at 960 MHz, 1150 MHz, and 1300 MHz). The optimal length (s) and width (t) of the slots are 20 mm and 5 mm, respectively. Figure 2(a) also shows the VSWR plot of the multiband antenna as a result of this design. Figure 2(b) shows the measured and simulated VSWR plot of the proposed antenna and the VSWR ≤2.5 bandwidths are 1100 MHz (166.71%), 700 MHz (28.57%), and 1300 MHz

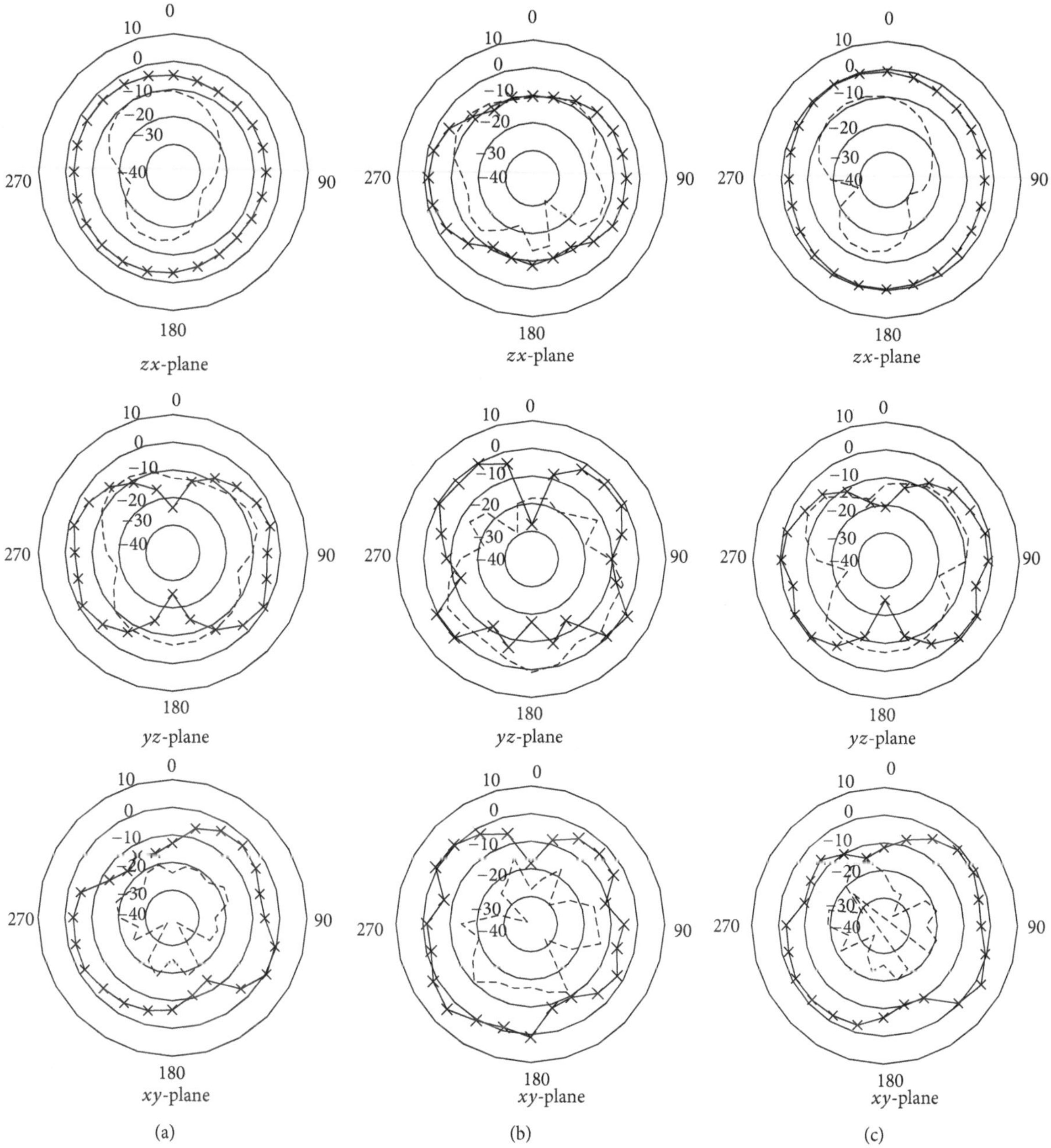

FIGURE 4: Measured radiation patterns for the proposed antenna (-×- copolar.—cross-polar.). (a) 660 MHz, (b) 2450 MHz, and (c) 5500 MHz.

(23.63%) at 660 MHz, 2450 MHz, and 5500 MHz, which meet the required bandwidths of the DVB and WLAN systems. The simulated results are obtained by using the Ansoft HFSS. We can also find that a good agreement between the simulation and the measurement is obtained. The simulation results of the current density on the proposed antenna are shown in Figure 3. Because the impedance mismatching at 960 MHz, 1300 MHz, and 3900 MHz results in the surface current density being lower than 660 MHz, 2450 MHz, and 5500 MHz, the slots are located at quarter-wave length of the undesired frequencies from the feed point that has also been verified.

Figure 4 presents the measured radiation patterns for free space in the yz-plane (E-plane) and the zx-plane (H-plane). The maximum gains in the E-plane and H-plane are 0.13 dBi, 1.27 dBi, and 1.04 dBi and 0.09 dBi, 0.04 dBi, and 0.11 dBi, respectively, at 660 MHz, 2450 MHz, and 5500 MHz. The gain response with frequency is shown in Figure 5, indicating that the effect of band rejection is remarkable. Table 1 shows the measured antenna gains and 3D pattern efficiency within the operating bands of the proposed antenna. Stable radiation patterns are observed. The total 3D pattern efficiency is defined as (gain/directivity) × 100%, which was done by using pattern integration employing the ETS-Lindgren anechoic chamber. Acceptable radiation characteristic for the practical applications is obtained for the proposed antenna. The omni-directional feature of the proposed antenna can also be observed from the horizontal plane. The operating

TABLE 1: The measured antenna gains within the operating bandwidth of the proposed antenna.

Frequency (MHz)	H-plane (zx-plane)		E-plane (yz-plane)	3D pattern
	Max. gain	Avg. gain	Max. gain	Efficiency (%)
470	−2.37	−4.43	0.19	51.22
660	0.09	−3.77	0.13	56.56
860	0.91	−3.03	0.94	58.71
2400	0.06	−3.76	0.43	66.45
2450	0.04	−3.61	1.27	65.52
2500	0.06	−3.15	1.43	68.72
5150	1.01	−2.76	0.93	69.71
5500	0.11	−3.31	1.04	71.23
5850	1.09	−3.05	1.76	70.52

Max.: maximum and Avg: average.

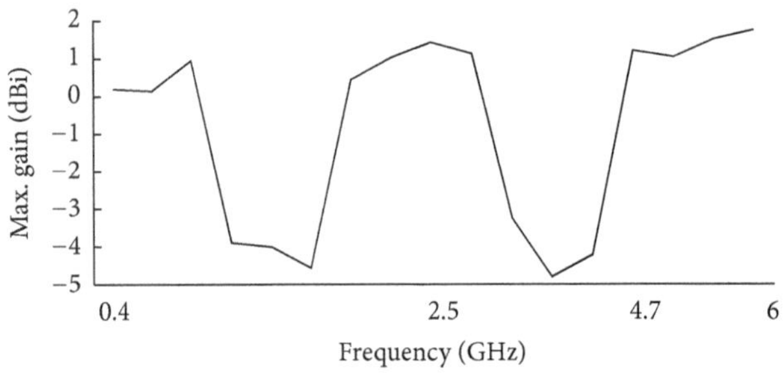

FIGURE 5: Measured antenna gain versus frequency.

bandwidth of the proposed antenna with usable broadside radiation patterns is consistent with the specification of DVB and WLAN systems.

The effect on the antenna performance by varying the insertion-slot numbers and dimensions has been studied, and the results are described below. The structure of insertion slot is a rectangular slot, as shown in Figure 1(b). The corresponding characteristics of resonant frequency, input impedance, and bandwidth are a function of the geometrical parameters of the insertion-slot structure. The experimental results of varying insertion-slot numbers and sizes were obtained by measurement, as shown in Figure 6. According to the above analyses, it is obvious that the inserted slots are used for the band rejection. Figure 6(a) shows impedance characteristics with various numbers of inserted slots, which control the operating bands. Measurement results show that by inserting three slots, the desired operating bands can be obtained. In the UHF band, an increase in the insertion-slot length (s) leads to an increase in impedance bandwidth. Contrarily, the insertion-slot length (s) has a negative effect on impedance bandwidth in the dual-ISM band; the measurement results are shown in the Figure 6(b). On the other hand, an increase in the insertion-slot width (t) leads to a decrease in impedance bandwidth in the UHF band, while in the dual-ISM band, there is a positive relation between insertion-slot width (t) and impedance bandwidth, as shown in Figure 6(c). The optimal slot sizes of length (s) and width (t) are 20 mm and 5 mm, which will provide the appropriate operating band for DVB and WLAN system applications.

4. Conclusion

It has been demonstrated that a printed modified bow-tie dipole provides multiband operation. By correctly choosing the width of the slots, three bandwidths, defined for a VSWR lower than 2.5, achieving 166.71%, 28.57%, and 23.63%, respectively, at 660 MHz, 2450 MHz, and 5500 MHz, can be obtained. The contribution of this paper is to implement a simple and easy fabrication antenna for practical DVB and WLAN applications. The locations of the inserted slots on the dipole arms have been confirmed. The slots are located at quarter-wave length of the undesired frequencies from the feed point. Measurement results show that the structure indeed offers broad bandwidth and band rejection characteristics. Although this antenna was designed for DVB and dual ISM-band applications, its design concept can be extended to other frequency bands of interest.

Acknowledgments

This work was sponsored by the National Science Council, Taiwan, under the Contract no. 100-2221-E-228-001 and Wieson Corp. Tech. (http://www.wieson.com/), Taiwan,

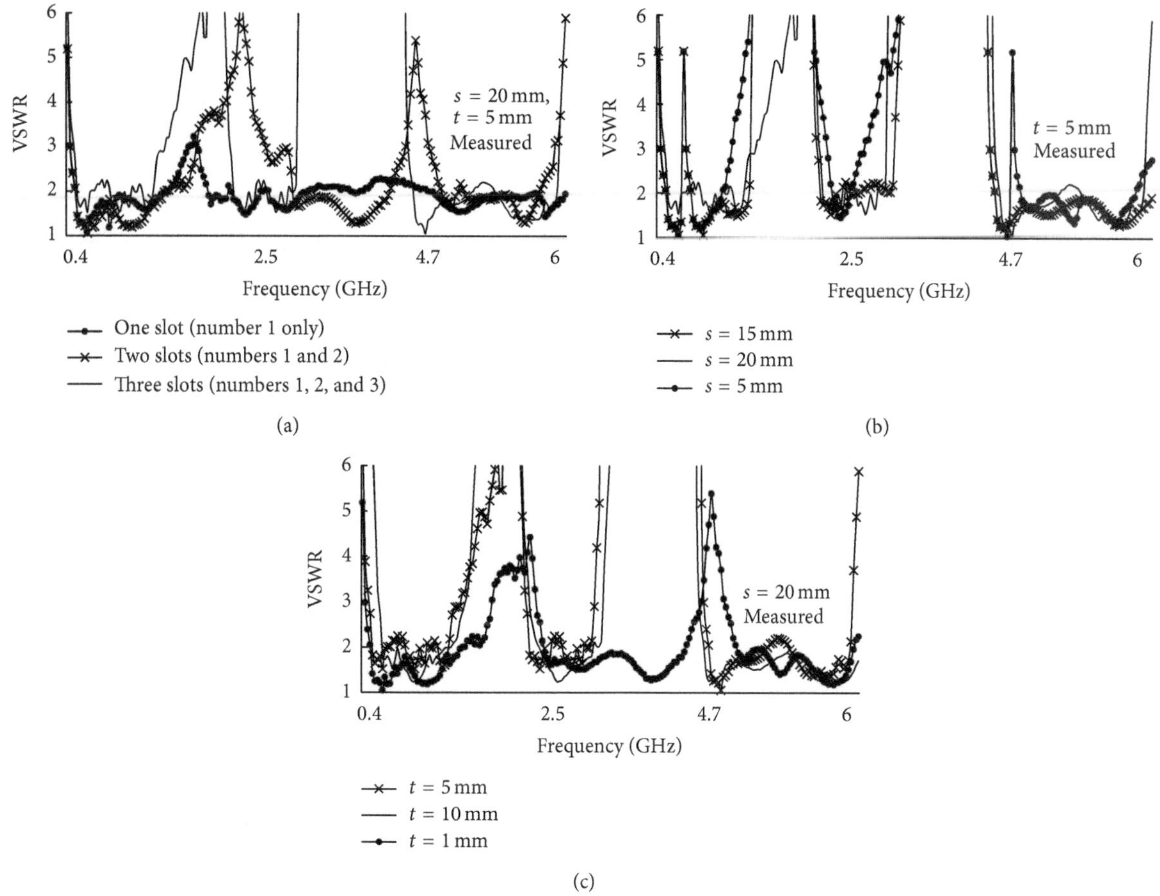

FIGURE 6: Measured VSWR as a function of (a) the inserting-slot numbers, (b) the inserting-slot length (s), and (c) the inserting-slot width (t). Other dimensions are the same as given in Figure 1.

under Contract no. JW100-F-411-148. The authors also appreciate the reviewer's comments that improved the quality of this paper.

References

[1] D. Morris, P. Pangalos, and A. H. Aghvami, "Mobile terminal location discovery schemes for a "Beyond 3G" inter-worked network," in *Proceedings of the sixth IEEE International Symposium on WoWMom*, pp. 401–406, June2005.

[2] J. D. Kraus and R. J. Marhefka, *Antennas*, McGraw-Hill, New York, NY, USA, 2002.

[3] R. Jordan and C. T. Abdallah, "Wireless communications and networking: an overview," *IEEE Antennas and Propagation*, vol. 44, no. 1, pp. 185–193, 2002.

[4] C.-M. Peng, I.-F. Chen, C.-C. Hung, and H.-C. Chen, "Printed modified bow-tie dipole antenna for multi-band applications," *IEICE Transactions on Communications*, vol. 92, no. 4, pp. 1404–1405, 2009.

[5] Q.-X. Chu and Y.-Y. Yang, "3.5/5.5 GHz dual band-notch ultra-wideband antenna," *Electronics Letters*, vol. 44, no. 3, pp. 172–174, 2008.

[6] K. Chung, S. Hong, and J. Choi, "Ultrawide-band printed monopole antenna with band-notch filter," *IET Microwaves, Antennas and Propagation*, vol. 1, no. 2, pp. 518–522, 2007.

[7] D.-O. Kim, N.-I. Jo, H.-A. Jang, and C.-Y. Kim, "Design of the ultrawideband antenna with a quadruple-band rejection characteristics using a combination of the complementary split ring resonators," *Progress in Electromagnetics Research*, vol. 112, pp. 93–107, 2011.

[8] H. H. Yong, M. S. Deokong, and C. Kyoung:, "Wideband bow-tie monopole antenna having an improved band rejection characteristic," Korea Patent: KR 20080858969 B1, 2008.

[9] A. A. Gheethan and D. E. Anagnostou, "Dual band-reject UWB antenna with sharp rejection of narrow and closely-spaced bands," *IEEE Transactions on Antennas and Propagation*, vol. 60, no. 4, pp. 2071–2076, 2012.

Statistical Modeling of Ultrawideband Body-Centric Wireless Channels Considering Room Volume

Miyuki Hirose, Hironobu Yamamoto, and Takehiko Kobayashi

Wireless Systems Laboratory, Tokyo Denki University, 5 Senju-Asahi-Cho, Adachi-Ku, Tokyo 120-8551, Japan

Correspondence should be addressed to Miyuki Hirose, miyuki@grace.c.dendai.ac.jp

Academic Editor: César Briso Rodríguez

This paper presents the results of a statistical modeling of onbody ultrawideband (UWB) radio channels for wireless body area network (WBAN) applications. Measurements were conducted in five different rooms. A measured delay profile can be divided into two domains; in the first domain ($0 < t \leq 4$ ns) there is either a direct (for line of sight) or diffracted (for nonline of sight) wave which is dependent on the propagation distance along the perimeter of the body, but essentially unrelated to room volume, and the second domain ($t > 4$ ns) has multipath components that are dominant and dependent on room volume. The first domain was modeled with a conventional power decay law model, and the second domain with a modified Saleh-Valenzuela model considering the room volume. Realizations of the impulse responses are presented based on the composite model and compared with the measured average power delay profiles.

1. Introduction

Wireless onbody area communication technologies are significant for both medical and nonmedical applications. Ultrawideband (UWB) technologies have been considered for use in wireless body area networks (WBANs) because of their possible low power consumption and antimultipath capabilities. Numerous studies have been carried out on UWB propagation characterization and the modeling of indoor UWB communication channels. A number of measurements relating to WBAN have been carried out to characterize and model on- and offbody UWB propagation in either a radio anechoic chamber or a specific room type [1–4]. The conventional UWB propagation loss model in these studies, however, did not consider the impact of surrounding environments. Since multipaths (in particular, the reflected waves from floors, walls, and ceilings) depend strongly on room volume, it is necessary to evaluate the variation of propagation characteristics in various environments. To address this problem, we measured UWB (3.1–10.6 GHz) radio propagation around the human body in a radio anechoic chamber and four different rooms and proposed a new UWB propagation loss model depending on the room volume [5]. In this study, time-domain statistical channel model will be the presented based on the same measurement campaign as [5].

As for statistical modeling of the channel impulse response, Fort et al. [3] separated the WBAN propagation channels into two parts: (1) diffraction around the body and (2) reflections off of nearby scatterers then back at the body, and modeled the second part using a modified Saleh-Valenzuela (SV) model [6]. The applicable area of the modified SV model in [6], however, was limited to wireless personal area networks not including human bodies. Roblin [7] scrutinized the separability of channels for various scenarios in three different rooms, concluded that UWB channels can be separated in the case of a relatively larger room, but it has not established a channel model. We also divided the channel responses into two parts which were then modeled by power decay law and a modified SV model depending on the room volume.

2. Measurement Setup

The measurement campaigns were conducted in five parallelepiped rooms as shown in Figure 1. The dimensions of Room A (a radio anechoic chamber) were measured

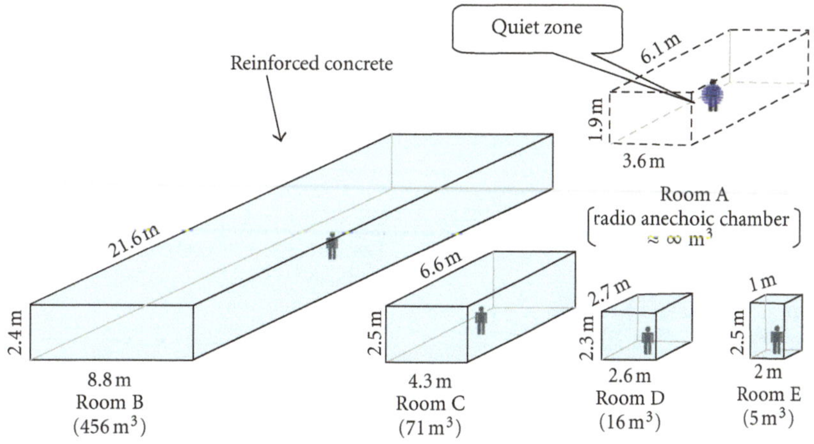

FIGURE 1: The outline of five rooms used for experiments.

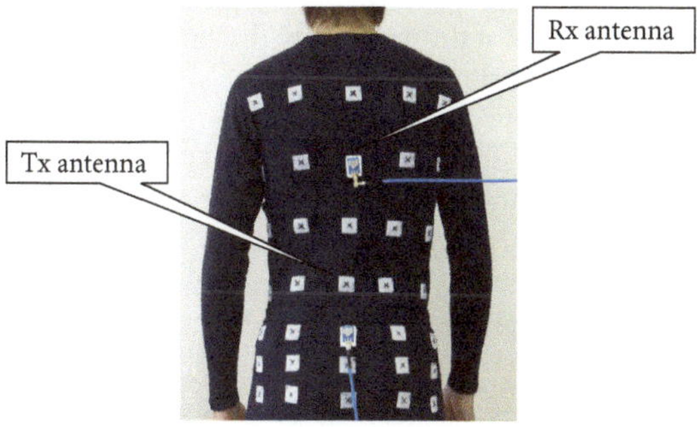

FIGURE 2: Placement of transmitting and receiving antennas on the body. Rectangular patches on the clothes are fabric hook-and-loop fasteners to fix the antennas.

between the apexes of the radio absorbers paneled on all surfaces. The radio anechoic chamber can be considered as a room extending to an infinite volume (i.e., free space) in terms of radio propagation. Rooms B to E were made of reinforced concrete, and their floors, walls, and ceilings were mostly covered with, respectively, linoleum, wallpaper, and plasterboard, all of which were lossy dielectrics. The measurements were carried out using a human subject (adult male, 1.72 m tall and 56 kg). The subject stood upright with the feet shoulder width apart in either a quiet zone of the radio anechoic chamber or the center of Rooms B to E. The UWB (3.1–10.6 GHz) propagation losses were measured with a vector network analyzer (VNA) between onbody meander line antennas [8]. The voltage standing wave ratio of the antennas was less than 2.5 between 3.1 and 10.6 GHz, and the omnidirectionality in the horizontal plane was within 3 dB in a free space. The transmitting antenna was fixed on the center back waist of the subject and placed at a height of 1.0 m from the floor, as shown in Figure 2. The receiving antenna was placed at approximately 100 mm intervals on the torso. Both antennas were vertically polarized and separated 10 mm from the subject body. When the receiving antenna was placed on the back of the subject's body, the path was roughly line of sight (LOS), and when on the front, it was non-LOS (NLOS). In total 69 receiving points around the torso were employed. The transmitting and receiving antennas were fed via coaxial cables, perpendicular to each other in configuration without crossing to reduce undesired cable coupling [9]. The calibration was conducted between the feeding points with a coaxial through adaptor. The frequency-domain transfer function (size = 1,024 = 751 measured within the 7.5-GHz bandwidth + 273 zero padding) was inversely Fourier transformed into a delay profile with the use of a rectangular window. Major specifications of the measurements are listed in Table 1.

TABLE 1: Specifications of the propagation measurements [5].

Bandwidth	3.1–10.6 GHz
Frequency sweeping points by VNA	751 points, 10-MHz interval
Calibration	Internal function of the VNA
Antennas	Meanderline UWB antennas [8]

3. Measurement Results and Modeling

Examples of the delay profiles when the receiving antenna was placed on the center chest (NLOS) and the back side (LOS) of the subject are presented in Figure 3. An increase in total received power was observed when the room volume was decreased (see Appendix A). This was attributed to the more affluent multipaths from the nearby floor, walls, and ceiling in Rooms B to E. The dominant propagation path in Room A (the radio anechoic chamber) was either a direct or a diffracted (around the body) wave, and thus the total reception power is lower than that in the other rooms. With decreasing room volume, mean free path lengths decreased, the power component contained in the multipaths increased, and consequently the total received power increased.

3.1. Division of Propagation Channels. A delay profile can be treated by dividing it into two domains, in the same way as [3, 7]: the first (approximately arriving time $0 < t \leq 4$ ns)

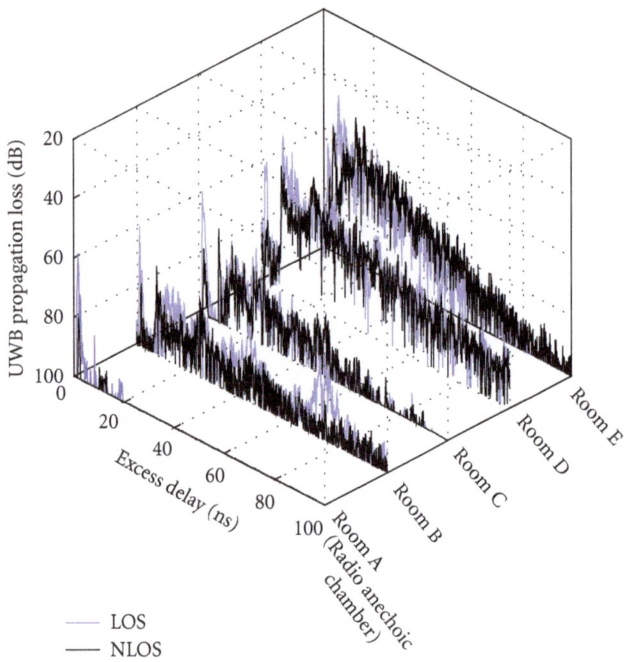

Figure 3: Example of the delay profiles measured in Rooms A to E [5].

and second ($t > 4$ ns) domains, as schematically shown in Figure 4. The first domain represents the contribution of the human alone, consisting of either direct (for LOS) or diffracted (for NLOS) wave measured in free space or radio anechoic chambers. And the second domain represents the contribution of the surrounding environments, consisting of remaining multipath components, which depend on room volume. Justification for dividing the profiles at $t = 4$ ns is given in Appendix B.

3.2. Statistical Analysis of the First Domain. The channel response in the first domain ($0 < t \leq 4$ ns) can be represented by

$$h_1(t) = h_{10} \cdot \left(\frac{d}{d_0}\right)^n \cdot \delta(t - t_0), \quad (1)$$

where h_{10} is the propagation gain at the reference distance d_0 (= 0.1 m), d is the propagation distance along the perimeter of the body, n is the propagation loss exponent, t_0 is the arrival time of the first wave, and $\delta(\cdot)$ is the Dirac delta function. The arrival time t_0 is proportional to d. Equation (1) represents a special case (when the room volume $V = \infty$) of the previously proposed UWB propagation loss model depending on room volume [5] (see Appendix C). The values of h_{10} and n in (1) were found to be 4.3×10^{-4} and 3.8 for LOS and 3.2×10^{-5} and 5.1 for NLOS, respectively, from the data of PL_{dB} shown in Figure 5. The statistics of the $h_1(t)$ followed lognormal distribution with a standard deviation of 4.4 dB (±0.5 dB) and 3.4 dB (±0.5 dB), for LOS and NLOS, respectively, where the values in the parentheses indicate 95% confidence intervals.

Table 2: Parameters of the first arriving multipath component.

	NLOS	LOS
ν	−2.8	−3.0
β_0	59.1	52.4

3.3. Statistical Analysis of the Second Domain. The second domain ($t > 4$ ns) can be represented by a modified SV model [6] based on a cluster concept of rays:

$$h_2(t) = \sum_{l=0}^{\infty}\sum_{k=0}^{\infty} \beta_{k,l}\delta(t - T_l - \tau_{k,l}), \quad (2)$$

where $\{\beta_{k,l}\}$ are the multipath gain coefficients, $\{T_l\}$ is the delay of the lth cluster, and $\{\tau_{k,l}\}$ is the delay of the kth multipath component relative to the lth cluster arrival time (T_l). Delay profiles measured in Rooms B, C, D, and E indicated that rays arrived in clusters, as shown in Figure 6, where the abscissas of the graphs are drawn in antilogarithm. While Fort et al. stated cluster interval times fit to the Weibull distribution [3], in all our cases, the arrival time intervals of the clusters were found to follow an exponential distribution by using Kolmogorov-Smirnov (K-S) test with a 95% confidence interval. This means that cluster arrivals are modeled as a Poisson arrival process with a fixed rate of Λ [1/ns]. Within each cluster, subsequent rays also arrived according to a Poisson process with another fixed rate of λ [1/ns]. The distribution of the cluster and ray arrival times are given by

$$\begin{aligned} p(T_l \mid T_{l-1}) &= \Lambda \exp[-\Lambda(T_l - T_{l-1})], \quad l > 0, \\ p(\tau_{k,l} \mid \tau_{(k-1),l}) &= \lambda \exp[-\lambda(\tau_{k,l} - \tau_{(k-1),l})], \quad k > 0, \end{aligned} \quad (3)$$

where Λ and λ are cluster arrival rate and ray arrival rate within each cluster, respectively. The IEEE 802.15.4a channel model [6] used a lognormal distribution rather than a Rayleigh distribution adopted in the original S-V model [10] for the multipath gain coefficients $\beta_{k,l}$. We also adopted a lognormal distribution for $\beta_{k,l}$ because of a better fitting to the measured data. The average power of both the clusters and the rays within the clusters are assumed to decay exponentially, such that the average power of the multipath component at a given delay $T_l + \tau_{k,l}$ is given by

$$\left\langle \beta_{k,l}^2 \right\rangle = \left\langle \beta_{0,0}^2 \right\rangle \exp\left(-\frac{T_l}{\Gamma}\right) \exp\left(-\frac{\tau_{k,l}}{\gamma}\right), \quad (4)$$

where $\langle \beta_{0,0}^2 \rangle$ is the expected value of the power of the first arriving multipath component, Γ is the delay exponent of the clusters, and γ is the decay exponent of the rays within a cluster. The first arriving multipath detected in measured delay profiles is lower with decreased room volume, as shown in Figure 7. The first multipath component, $\langle \beta_{0,0}^2 \rangle$, can be represented by

$$\left\langle \beta_{0,0}^2 \right\rangle = \beta_0 \cdot \left(\sqrt[3]{V}\right)^{\nu}. \quad (5)$$

The values of β_0 and ν are listed in Table 2.

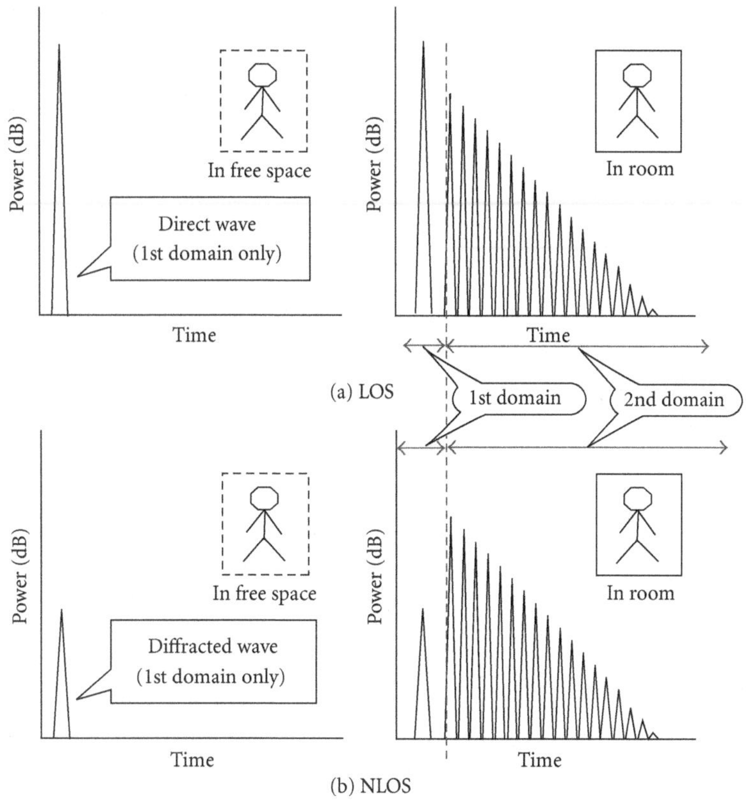

FIGURE 4: Conceptual diagram of the division of the delay profiles: (a) LOS and (b) NLOS. The first domain represents the delay profiles measured in free space or radio anechoic chambers and contains a direct or diffracted wave along the body. The second domain consists of the remaining multipath components.

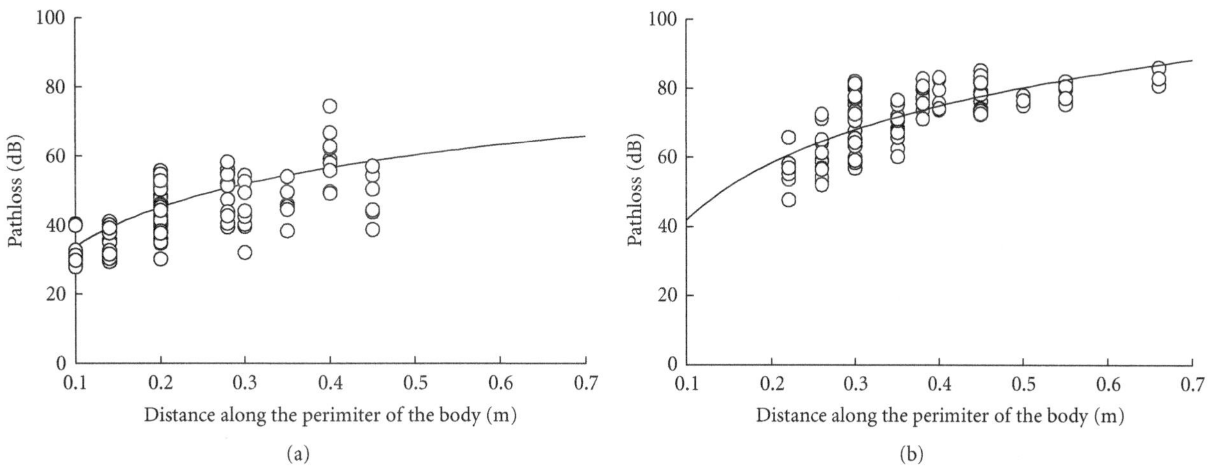

FIGURE 5: Ultrawideband pathloss onbody antennas measured in a radio anechoic chamber: (a) LOS and (b) NLOS.

The values of Λ, λ, Γ, and γ were derived from the delay profile data measured in Rooms B, C, D, and E. Figures 8 and 9 present those values against $V^{1/3}$ along with regression lines. Note that $V^{1/3}$ represents the mean free path length of the rays traveling within a room having a finite (or infinite) volume V. The cluster arrival time rate Λ[1/ns] is approximately 0.08, while the ray arrival time rate λ [1/ns] is 0.4 for both LOS and NLOS. While the arrival rates Λ and λ exhibited no apparent dependence on $V^{1/3}$ or LOS/NLOS scenarios as shown in Figure 8, the power decay factors Γ and γ slightly increased with $V^{1/3}$, as shown in Figure 9. The propagation distances (and therefore propagation losses) of rays increase with the room volume, and therefore the decay factors increase. The slope was steeper for the NLOS than for the LOS cases. The dependence of the cluster power-decay factor and

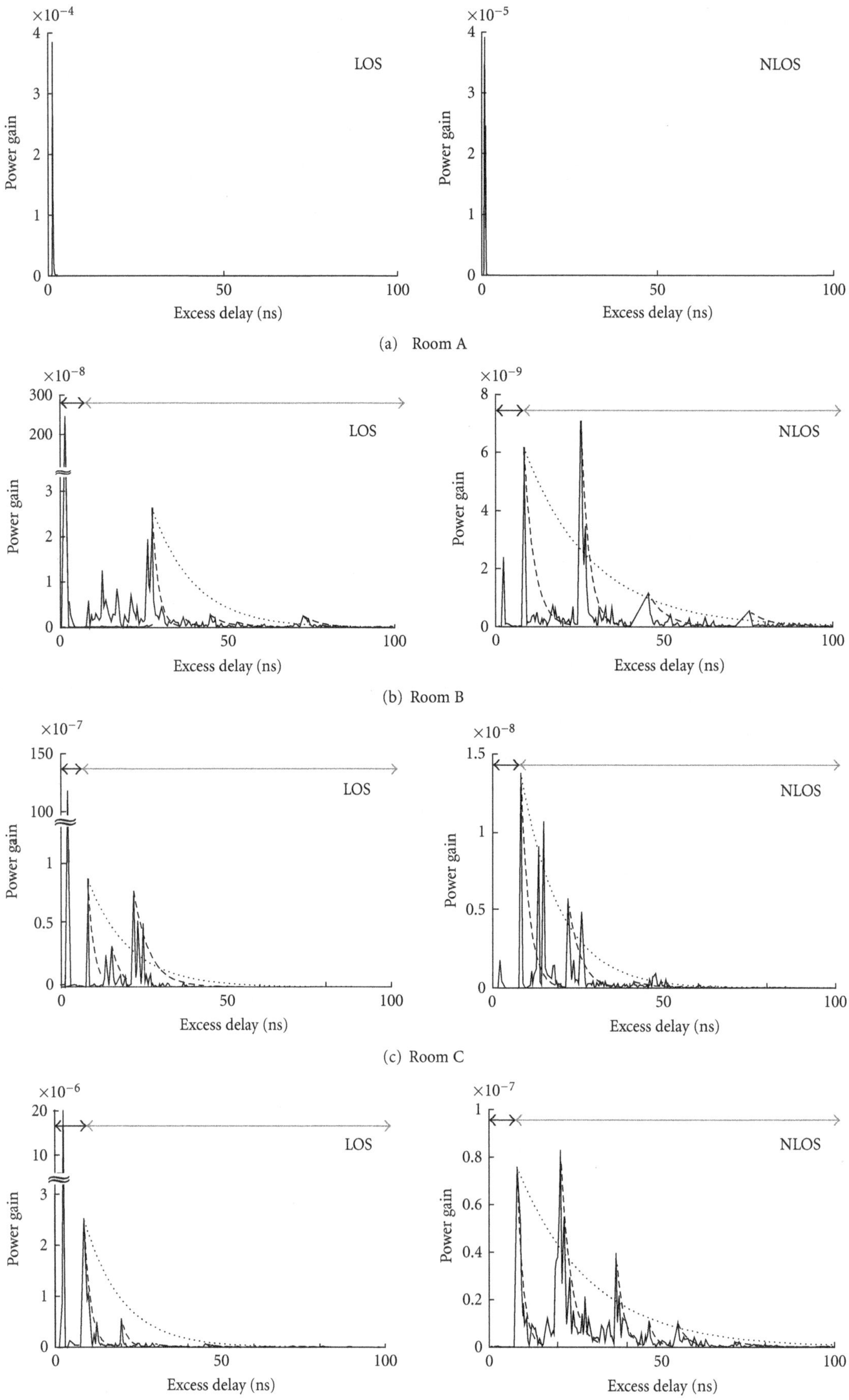

(a) Room A

(b) Room B

(c) Room C

(d) Room D

FIGURE 6: Continued.

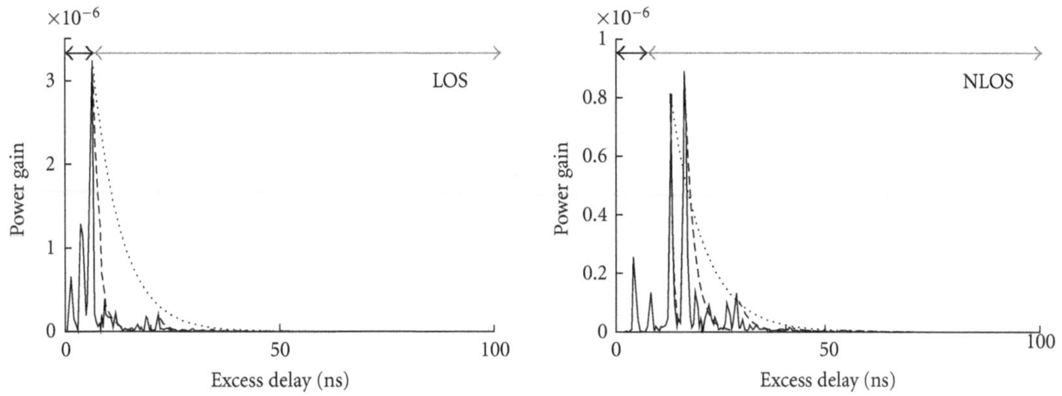

(e) Room E

Figure 6: Examples of the delay profiles measured in Rooms A to E. Profiles for LOS were measured at the center of the back and those for NLOS at the center of the chest. The dashed lines represent exponential power decay of the rays and the clusters. The black and grey arrows indicate the first and the second domains.

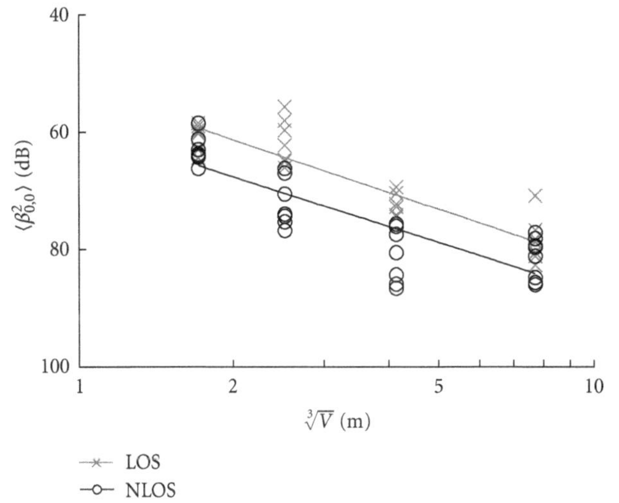

Figure 7: UWB propagation loss of the first ray within the first cluster against $V^{1/3}$.

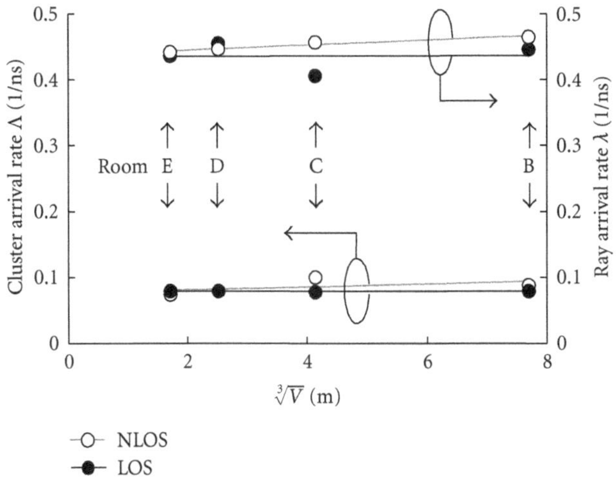

Figure 8: Arrival rates of clusters and rays within clusters.

Table 3: The parameter of decay factor functions for the cluster and the ray within the cluster.

		NLOS	LOS
Cluster	Γ' [ns/m]	1.25	0.70
	Γ_0 [ns]	10.0	7.1
Ray	γ' [ns/m]	0.28	0.11
	γ_0 [ns]	2.4	3.0

the ray power-decay factor on $V^{1/3}$, depicted in Figure 9, is formulated by

$$\Gamma = \Gamma_0 + \Gamma' \cdot \sqrt[3]{V}, \qquad \gamma = \gamma_0 + \gamma' \cdot \sqrt[3]{V}, \tag{6}$$

where Γ_0 and γ_0 are values of Γ and γ when imaginarily $V = 0$, Γ' and γ' are the slope of the cluster and the ray within the cluster against $V^{1/3}$, respectively. The values of Γ_0, γ_0, Γ', and γ' are listed in Table 3. Although the effect of shadowing has not been considered in this paper, it can be included in (2) after the same method as adopted in [6].

4. Realization of Onbody UWB Channels Based on the Composite Model

A composite statistical UWB channel model between onbody antennas is formulated by summing the models described in Section 3. A realization is calculated upon providing input data—whether the path is either LOS or NLOS—d (the distance between the antennas along the perimeter of the body), and the room volume, as shown in Figure 10. Once a number of realizations of the channel responses are calculated randomly, they are then served to estimate transmission performances (e.g., average bit error rates) and/or system capacity of communication systems, detection and false alarm rates of radar systems, and so forth, by simulation.

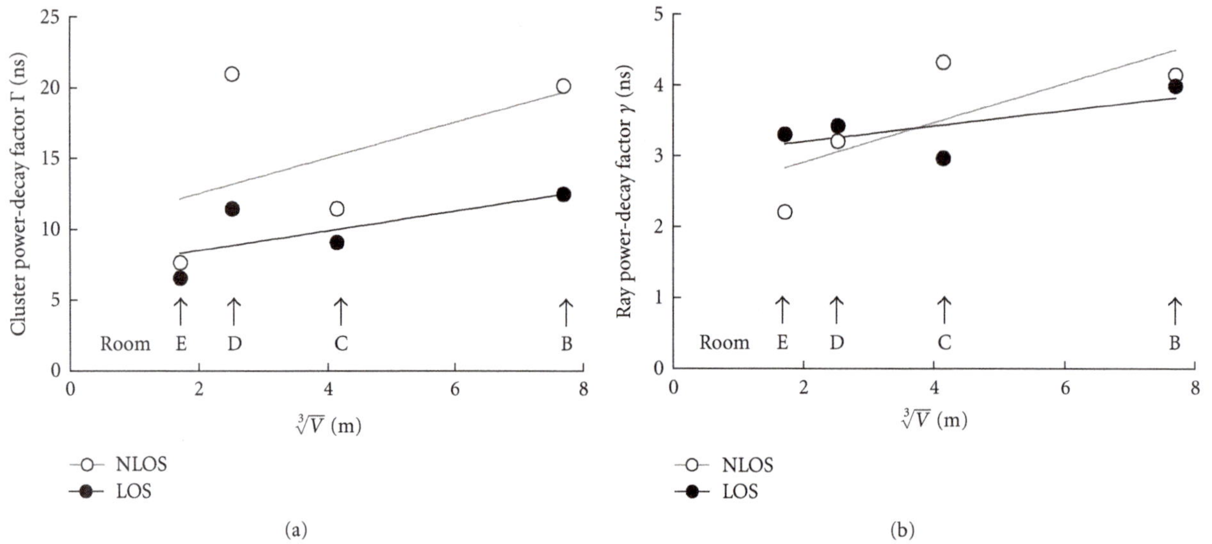

FIGURE 9: Power decay factors against $V^{1/3}$ for (a) the clusters and (b) the rays. The solid lines are the linear fitting.

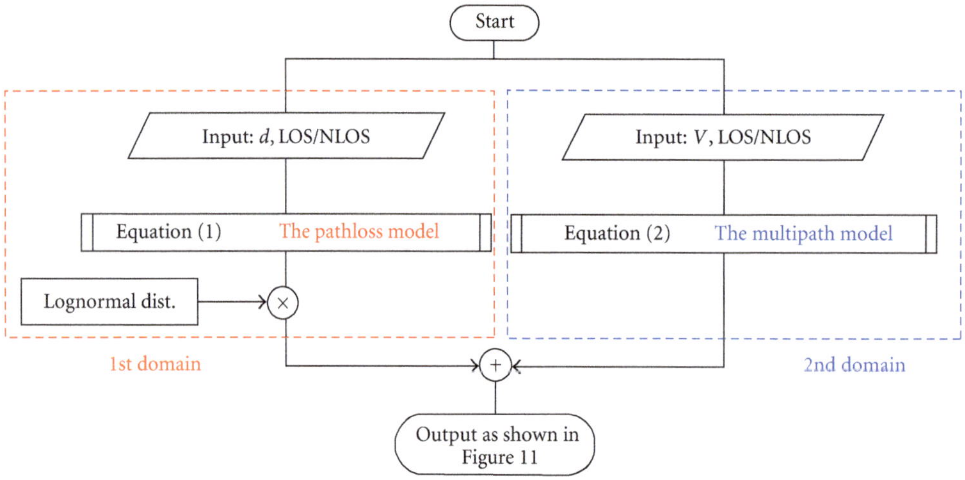

FIGURE 10: The flow chart of our model realization process.

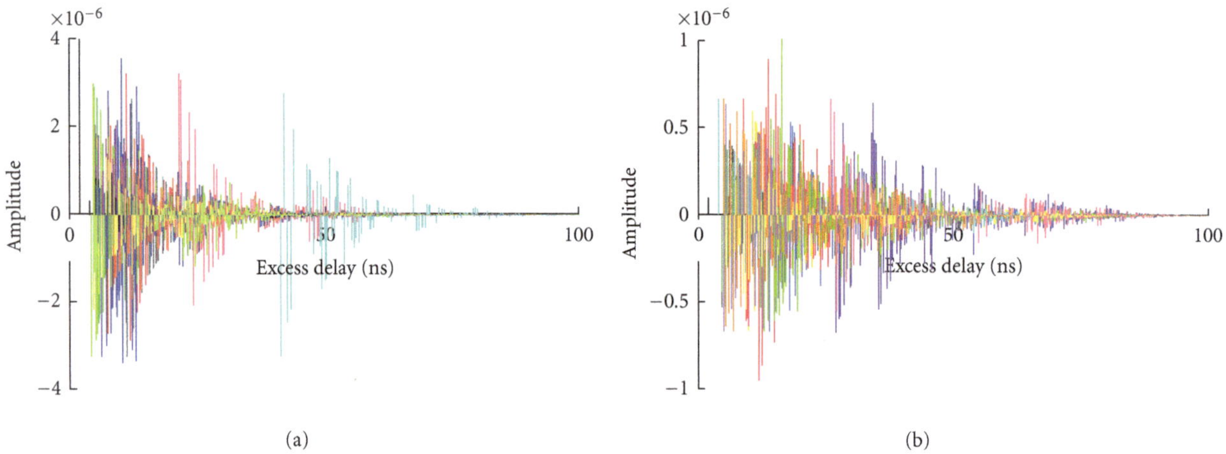

FIGURE 11: Examples of 20 channel response realizations for Room E ($V = 5\,m^3$): (a) LOS and (b) NLOS.

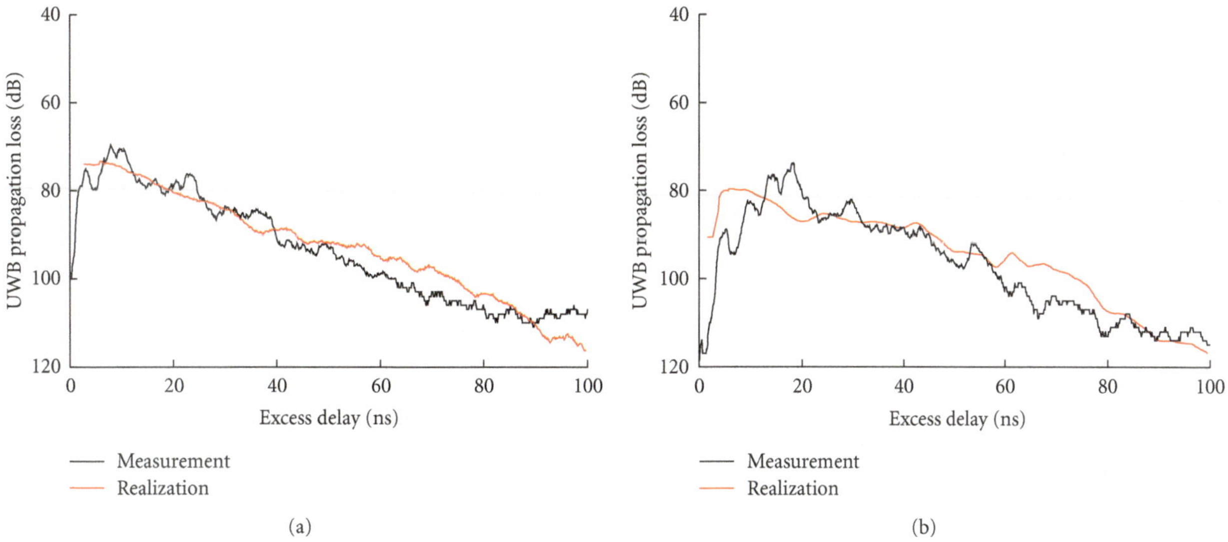

Figure 12: Comparison of the APDPs between 20 realizations and measured delay profiles of Room E (V = 5 m^3): (a) LOS and (b) NLOS.

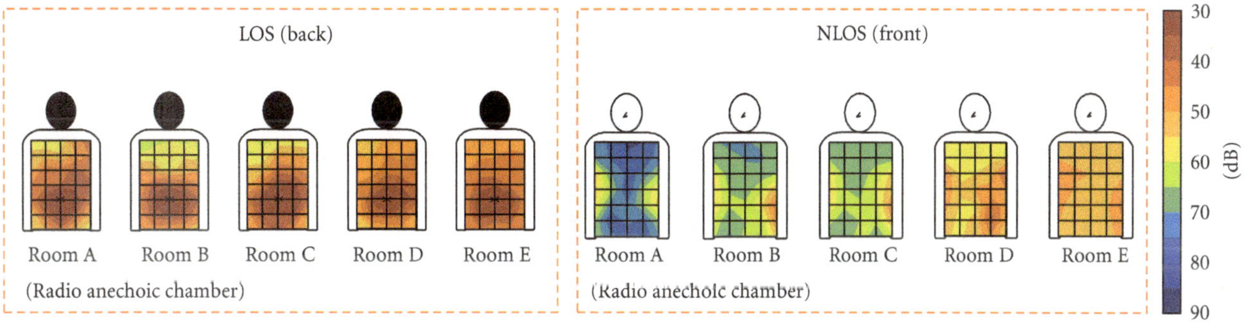

Figure 13: Spatial distribution of UWB propagation losses in the five rooms. The transmitting antenna was placed at point denoted by "*".

Examples of the channel response realizations for LOS and NLOS, assuming d = 200 mm for LOS and 450 mm for NLOS and V = 5 m^3, are presented in Figure 11, where 20 realizations are overwritten. Average power delay profiles (APDPs) for LOS and NLOS were derived from these realizations and compared with the measured data. Moving average was conducted over a 3 ns period for calculating the APDPs. The APDPs derived from the calculated realizations and from the measured delay profiles reasonably agree, as shown in Figure 12. The validity of the proposed composite model was therefore confirmed.

5. Conclusions

In this study, a series of propagation measurements campaign were carried out between onbody antennas in five different rooms. A measured delay profile can be divided into two domains. In the first domain ($0 < t \leq 4$ ns), there is either a direct (for LOS) or diffracted (for NLOS) wave which depends on propagation distance along the perimeter of the body but essentially unrelated to room volume. This domain was modeled with a power decay law against the distance, and its amplitude followed a lognormal distribution. In the second domain ($t > 4$ ns), multipath components are dominant and dependent on room volume. Observations of the second domain indicate that rays generally arrive in clusters. Arrivals of clusters and rays within each cluster were found to be modeled by Poisson processes. As a result, the second domain was modeled by a modified Saleh-Valenzuela model with the use of lognormal distribution rather than Rayleigh distribution for multipath gain coefficients. Finally, the composite model to calculate the UWB onbody channel realizations was obtained by combining the two domains and validated with the use of the measured delay profiles.

Appendices

A. UWB Propagation Loss

Examples of spatial distributions of UWB propagation losses, measured in the same five rooms as those described in Section 2, are shown in Figure 13. The UWB propagation losses were calculated by integrating the power of the losses

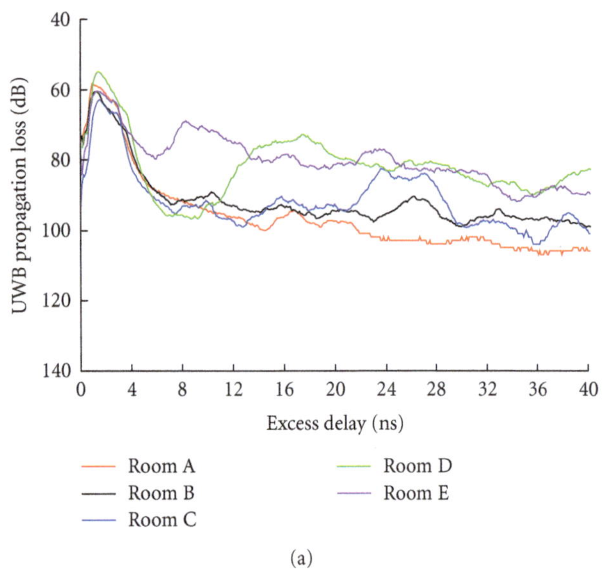

Figure 14: Measured average power delay profiles (averaged over 3 ns): (a) LOS and (b) NLOS.

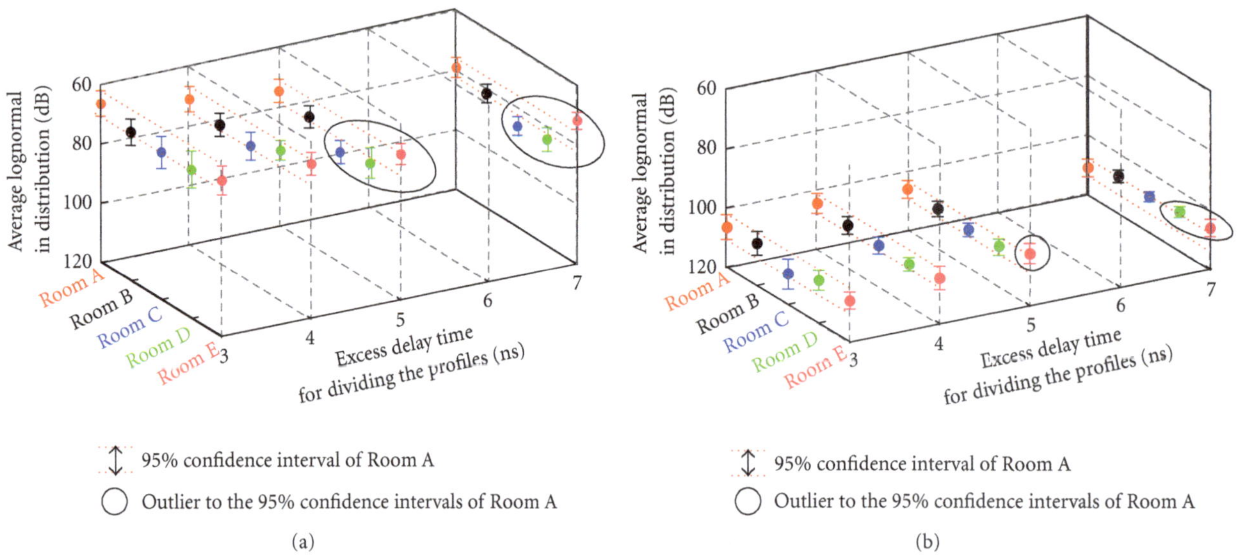

Figure 15: Averages in lognormal distribution fit to the propagation loss for Rooms A to E: (a) LOS and (b) NLOS. The dashed lines are the 95% confidence intervals of Room A (a radio anechoic chamber). The averages encircled by ovals are outliers to the 95% intervals of Room A.

between the feeding points of the antennas over occupied bandwidth:

$$\mathrm{PL}_{\mathrm{dB}} = 10\log\left(\frac{1}{f_H - f_L}\int_{f_L}^{f_H} 10^{\mathrm{PL}_{\mathrm{dB}}(f)/10}\,df\right), \qquad (\mathrm{A}.1)$$

where $\mathrm{PL}_{\mathrm{dB}}(f)$ is the propagation loss in dB measured at frequency f, and f_L and f_H are the lowest and highest frequencies. The propagation losses increased with decreasing the room volume, as shown in Figure 13.

B. Validity of 4 ns for Dividing the Delay Profiles

Figure 14 depicts average power delay profiles (averaged over 3 ns) for LOS and NLOS measured in Rooms A to E.

Curves are almost equal for a period between 0 and approximately 4 ns: the effect of the surrounding environment was insignificant up to 4 ns. Beyond the 4 ns, the propagation loss decreased (the curves move upward) with decreasing room volume. Furthermore, the amplitude distribution was examined to confirm the validity of $t = 4$ ns for dividing the delay profiles. The amplitudes within the measured delay profiles were found to follow lognormal distribution up to an excess delay of 10 ns. The averages in the lognormal distribution up to 3, 4, 5, and 7 ns were estimated for LOS and NLOS, as shown in Figure 15, where the 95% confidence intervals derived of Room A data are plotted by dashed lines. While all the averages up to 4 ns for Rooms B to E fell within the 95% intervals, some (Rooms C, D, and E for LOS and Rooms D and E for NLOS) were outside the intervals, as shown in Figure 15. This fact also ratified the validity of $t = 4$ ns for dividing the profiles.

C. Proposed UWB Propagation Loss Model

Based on a series of propagation measurements conducted in a frequency bandwidth from 3.1 to 10.6 GHz, the authors proposed a UWB propagation loss model [5]:

$$PL_{dB} = PL_{0dB} + 10\left(n_\infty + \frac{n'}{\sqrt[3]{V}}\right)\log\left(\frac{d}{d_0}\right)[\text{dB}], \quad (C.1)$$

where PL_{0dB} is the propagation loss at the reference distance d_0 (= 0.1 m), n_∞ is the propagation loss exponent when the room volume $V = \infty$, and n' is the slope of n against $V^{-1/3}$. The values of PL_{0dB}, n_∞, and n' for LOS and NLOS are given in [5].

Acknowledgments

This study was in part supported by the Japan Society for the Promotion of Science and the Academy of Finland with the Japan-Finland Bilateral Core Program. The authors would like to thank the program participants of the Tokyo Institute of Technology, Japan and Aalto University, Finland, for valuable discussions.

References

[1] T. Zasowski, F. Althaus, M. Stager, A. Wittneben, and G. Troster, "UWB for noninvasive wireless body area networks: channel measurements and results," in *Proceedings of the IEEE Conference on Ultrawideband Systems Technology (UWBST '03)*, pp. 285–289, November 2003.

[2] P. S. Hall and Y. Hao, Eds., *Antennas and Propagation for Body-Centric Wireless Communications*, Artech House, Boston, Mass, USA, 2006.

[3] A. Fort, J. Ryckaert, C. Desset, P. De Doncker, P. Wambacq, and L. Van Biesen, "Ultra-wideband channel model for communication around the human body," *IEEE Journal on Selected Areas in Communications*, vol. 24, no. 4, pp. 927–933, 2006.

[4] A. Sani, A. Alomainy, G. Palikaras et al., "Experimental characterization of UWB on-body radio channel in indoor environment considering different antennas," *IEEE Transactions on Antennas and Propagation*, vol. 58, no. 1, pp. 238–241, 2010.

[5] H. Yamamoto, M. Koiwai, and T. Kobayashi, "Measurements and modeling of ultra-wideband propagation losses around the human body dependent on room volume," *IEICE Transactions on Fundamentals of Electronics, Communications and Computer Sciences*, vol. E93, no. 12, pp. 2624–2633, 2010.

[6] A. F. Molisch, K. Balakrishnan, D. Cassioli et al., "IEEE 802. 15. 4a channel model—final report. Submitted to IEEE 802. 15 WPAN low rate alternative PHY task group 4a (TG4a)," Tech. Rep., IEEE 802. 15, 2004.

[7] C. Roblin, "On the separability of on-body and off-body clusters in the modeling of UWB WBAN channels for various indoor scenarios," in *Proceedings of the 5th European Conference on Antennas and Propagation (EUCAP '11)*, pp. 3148–3152, April 2011.

[8] Skycross, "3.1-10 GHz ultra-wideband antenna," http://www.Skycross.com/Products/PDFs/SMT-3TO10M-A.pdf.

[9] H. Yamamoto and T. Kobayashi, "Effects of feeding cable configurations on propagation measurements between small ultra wideband antennas for WBAN applications," in *Proceedings of the International Workshop on Future Wellness and Medical ICT Systems (FEELIT '08)*, September 2008.

[10] A. A. M. Saleh and R. A. Valenzuela, "A statistical model for indoor multipath propagation," *IEEE Journal on Selected Areas in Communications*, vol. 5, no. 2, pp. 128–137, 1987.

22

Useful Solutions for Plane Wave Diffraction by Dielectric Slabs and Wedges

Gianluca Gennarelli and Giovanni Riccio

D.I.E.I.I., University of Salerno, Via Ponte Don Melillo, Salerno, 84084 Fisciano, Italy

Correspondence should be addressed to Giovanni Riccio, griccio@unisa.it

Academic Editor: Francesco Soldovieri

This work presents an overview of available uniform asymptotic physical optics solutions for evaluating the plane wave diffraction by some canonical geometries of large interest: dielectric slabs and wedges. Such solutions are based on a physical optics approximation of the electric and magnetic equivalent surface currents in the involved scattering integrals. The resulting diffraction coefficients are expressed in terms of the geometrical optics response of the considered structure and the standard transition function of the Uniform Geometrical Theory of Diffraction. Numerical tests and comparisons make evident the effectiveness and reliability of the presented solutions.

1. Introduction

As well-known, the ability to describe and solve electromagnetic scattering problems is highly valued in many areas such as radio planning, remote sensing for monitoring and surveillance of ground, structures and infrastructures, and through-wall building imaging. Numerical techniques represent a possible answer, but they have an inherent drawback: the computation becomes very intensive (if not unmanageable) at high frequencies, where asymptotic methods based on ray-tracing work more efficiently. In this framework, the Geometrical Theory of Diffraction (GTD) [1] and its uniform version (UTD) [2] have received considerable attention because of their peculiarities. When using GTD or UTD, the first step is to resolve the original problem into smaller and simpler components, each related to a canonical geometry, so that the total solution is a superposition of the contributions from each canonical problem. In other words, such ray-based methods allow one to solve a large number of real scattering problems by using the solutions of a relatively small number of simple problems. In addition, they are easy to apply, provide physical insight into the radiation and scattering mechanisms arising from the various parts of the structure and can be combined with numerically rigorous techniques to obtain hybrid methods.

This paper presents a review of Uniform Asymptotic Physical Optics (UAPO) solutions for diffraction problems concerning some typical canonical structures: dielectric slabs (individually considered or forming junctions) and wedges. Examples of application are relevant to (a) through-wall building imaging [3], where the structure can be decomposed into plates forming junctions and corners with other plates and to (b) radar imaging for ground observation, where the presence of ruptures caused by natural disasters can be represented in the form of a wedge-shaped fracture [4].

The starting point for obtaining a UAPO solution is that of considering the scattering integral and using a PO approximation of the electric and magnetic surface currents related to the boundary of the object. A further useful approximation and a uniform asymptotic evaluation of the resulting integrals allow one to obtain the diffraction coefficients in the UTD framework. They result to be expressed in terms of the reflection and transmission coefficients of the structure and the standard transition function of UTD. Note that also the heuristic solutions [5–7] have such characteristics, but they are not derived by means of a rigorous analytical approach. Moreover, the time domain UAPO diffraction coefficients can be determined according to [8] by taking advantage of the UTD-like formulation of their frequency domain counterparts. In light of the above statements, the UAPO solutions have the same effectiveness and ease of handling of those derived in the UTD context and, in addition, they have the inherent advantage of providing the diffracted field from the knowledge of the geometrical optics (GO) field as the heuristic solutions.

The remainder of this paper is organized as follows. Section 2 is devoted to the plane wave diffraction from the edge of a half plane used to model a truncated thin dielectric slab with finite conductivity. The PO surface currents in the scattering integral are equivalent sources originated by the discontinuities of the tangential GO field components across the layer. The UAPO solution presented in Section 2 can be extended to junctions by taking into account the diffraction contributions of the slabs separately. Of course, this useful feature is due to the linearity of the PO scattering integral. A lossless dielectric wedge with obtuse apex angle and arbitrary permittivity is considered in Section 3. Two separate plane wave diffraction problems relevant to the dielectric region and the surrounding space are tackled and solved by using equivalent PO surface currents lying on the internal and external faces of the wedge. Concluding remarks and future investigations are highlighted in Section 4.

2. Dielectric Slabs

The diffraction problem considered in this Section refers to a linearly polarised plane wave impinging on a thin dielectric slab characterised by thickness d, relative complex permittivity ε_r, and relative permeability $\mu_r = 1$ and represented by a lossy penetrable half plane surrounded by free space (see Figure 1). The angles β' and ϕ' fix the incidence direction: the first is a measure of the skewness with respect to the edge ($\beta' = 90°$ corresponds to the normal incidence), and the latter gives the aperture of the edge-fixed plane of incidence with respect to the illuminated face ($\phi' = 0$ corresponds to the grazing incidence). The observation direction is specified by the angles β and ϕ. The $e^{j\omega t}$ time dependence is assumed and suppressed in the following.

The field scattered at the observation point P can be evaluated by using a PO approximation of the electric ($\underline{J}_s^{\mathrm{PO}}$) and magnetic ($\underline{J}_{ms}^{\mathrm{PO}}$) surface currents involved in the well-known integral:

$$\underline{E}^s = -jk_0 \iint_S \left[\left(\underline{\underline{I}} - \hat{R}\hat{R} \right) \zeta_0 \underline{J}_s^{\mathrm{PO}} + \underline{J}_{ms}^{\mathrm{PO}} \times \hat{R} \right] G(\underline{r}, \underline{r}') dS, \tag{1}$$

where ζ_0 is the free-space impedance, k_0 is the free-space propagation constant, $\underline{r} = x\hat{x} + y\hat{y} + z\hat{z} = \underline{\rho} + z\hat{z}$, and $\underline{r}' = x'\hat{x} + z'\hat{z} = \underline{\rho}' + z'\hat{z}$ are the position vectors of P and

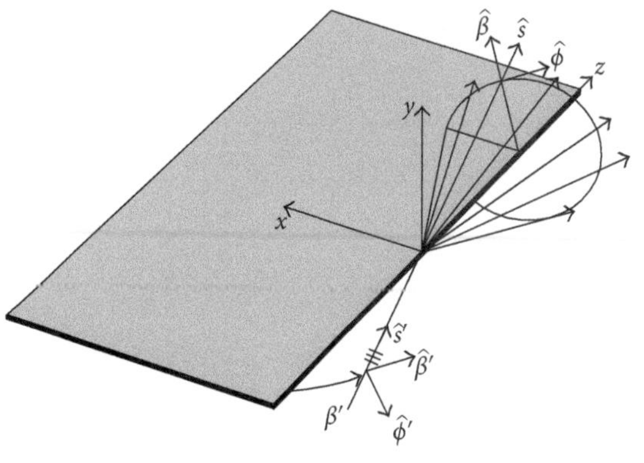

FIGURE 1: Diffraction by a thin dielectric slab.

the source point, respectively, $\hat{R}$ is the unit vector from the source point to P, $\underline{\underline{I}}$ is the 3×3 identity matrix, and

$$\begin{aligned} G(\underline{r}, \underline{r}') &= \frac{e^{-jk_0|\underline{r}-\underline{r}'|}}{4\pi|\underline{r}-\underline{r}'|} = \frac{e^{-jk_0\sqrt{(x-x')^2+y^2+(z-z')^2}}}{4\pi\sqrt{(x-x')^2+y^2+(z-z')^2}} \\ &= \frac{e^{-jk_0\sqrt{|\underline{\rho}-\underline{\rho}'|^2+(z-z')^2}}}{4\pi\sqrt{\left|\underline{\rho}-\underline{\rho}'\right|^2+(z-z')^2}} \end{aligned} \tag{2}$$

is the three-dimensional Green's function. To evaluate the edge diffraction confined to the Keller's cone ($\beta = \beta'$), it is possible to approximate $\hat{R}$ by the unit vector $\hat{s}$ in the diffraction direction, that is, $\hat{R} \cong \hat{s} = \sin\beta'\cos\phi\hat{x} + \sin\beta'\sin\phi\hat{y} + \cos\beta'\hat{z}$. Accordingly, it results that

$$\begin{aligned} \underline{E}^s &= -\frac{jk_0}{4\pi}\left[\left(\underline{\underline{I}} - \hat{s}\hat{s} \right) \zeta_0 \underline{J}_s^* + \underline{J}_{ms}^* \times \hat{s} \right] \\ &\quad \int_0^\infty \int_{-\infty}^\infty e^{jk_0(x'\sin\beta'\cos\phi' - z'\cos\beta')} \\ &\quad \frac{e^{-jk_0\sqrt{|\underline{\rho}-\underline{\rho}'|^2+(z-z')^2}}}{\sqrt{\left|\underline{\rho}-\underline{\rho}'\right|^2+(z-z')^2}} dz'dx' \\ &= \left[\left(\underline{\underline{I}} - \hat{s}\hat{s} \right) \zeta_0 \underline{J}_s^* + \underline{J}_{ms}^* \times \hat{s} \right] I^s. \end{aligned} \tag{3}$$

The expressions of the PO surface currents are obtained in terms of the incident electric field $\underline{E}^i$ by assuming such currents as equivalent sources originated by the discontinuities of the tangential GO field components across the slab, and then

$$\begin{aligned} \zeta_0 \underline{J}_s^* &= (1 - R_\perp - T_\perp) E_\perp^i \cos\theta^i \hat{e}_\perp + (1 + R_\parallel - T_\parallel) E_\parallel^i (\hat{y} \times \hat{e}_\perp), \\ \underline{J}_{ms}^* &= (1 - R_\parallel - T_\parallel) E_\parallel^i \cos\theta^i \hat{e}_\perp - (1 + R_\perp - T_\perp) E_\perp^i (\hat{y} \times \hat{e}_\perp), \end{aligned} \tag{4}$$

where θ^i is the standard incidence angle and $\hat{e}_\perp$ is the unit vector perpendicular to the plane of incidence. The reflection

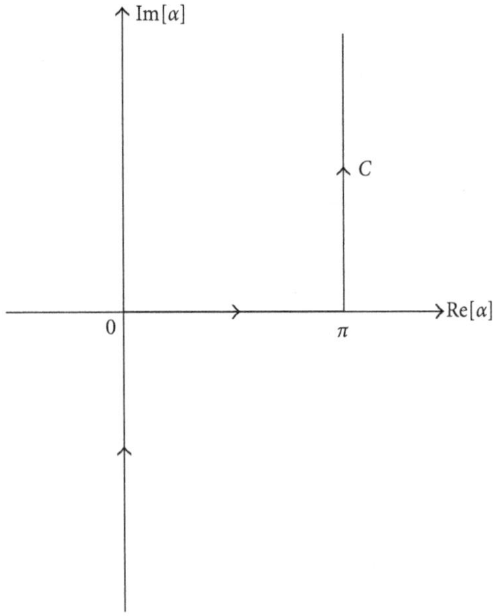

Figure 2: Integration path C.

(R) and transmission (T) coefficients for the parallel ($\|$) and perpendicular ($\perp$) polarisations are given by [9]:

$$R_{\|,\perp} = \frac{\overline{R}_{\|,\perp}\left[1 - e^{-j2\beta_{eq}d/\cos\theta^t} e^{-2\alpha_{eq}d} e^{j2k_0 d \sin\theta^i \tan\theta^t}\right]}{1 - \left(\overline{R}_{\|,\perp}\right)^2 e^{-j2\beta_{eq}d/\cos\theta^t} e^{-2\alpha_{eq}d} e^{j2k_0 d \sin\theta^i \tan\theta^t}}, \quad (5)$$

$$T_{\|,\perp} = \frac{\left[1 - \left(\overline{R}_{\|,\perp}\right)^2\right] e^{-j\beta_{eq}d/\cos\theta^t} e^{-\alpha_{eq}d} e^{jk_0 d \cos(\theta^i - \theta^t)/\cos\theta^t}}{1 - \left(\overline{R}_{\|,\perp}\right)^2 e^{-j2\beta_{eq}d/\cos\theta^t} e^{-2\alpha_{eq}d} e^{j2k_0 d \sin\theta^i \tan\theta^t}}, \quad (6)$$

in which

$$\overline{R}_{\|} = \frac{\varepsilon_r \cos\theta^i - \sqrt{\varepsilon_r - \sin^2\theta^i}}{\varepsilon_r \cos\theta^i + \sqrt{\varepsilon_r - \sin^2\theta^i}}, \qquad \overline{R}_{\perp} = \frac{\cos\theta^i - \sqrt{\varepsilon_r - \sin^2\theta^i}}{\cos\theta^i + \sqrt{\varepsilon_r - \sin^2\theta^i}}. \quad (7)$$

Formulas (5) and (6) account for the propagation of a nonuniform wave in the layer as considered in [10]. The terms β_{eq} and α_{eq} are the equivalent phase and attenuation factors and θ^t is the angle (labelled ψ_2 in [10]) between the direction of propagation in the layer and the direction normal to the surface (direction of attenuation).

According to (4), (3) can be so rewritten in matrix form:

$$\underline{E}^s = \begin{pmatrix} E^s_\beta \\ E^s_\phi \end{pmatrix} = \underline{\underline{M}}_1 \begin{pmatrix} E^i_\| \\ E^i_\perp \end{pmatrix} I^s = \underline{\underline{M}}_1 \underline{\underline{M}}_0 \begin{pmatrix} E^i_{\beta'} \\ E^i_{\phi'} \end{pmatrix} I^s = \underline{\underline{M}} \begin{pmatrix} E^i_{\beta'} \\ E^i_{\phi'} \end{pmatrix} I^s. \quad (8)$$

The matrices $\underline{\underline{M}}_0$ and $\underline{\underline{M}}_1$ involved in the above expression are reported in the appendix and account for the GO response (i.e., reflection and transmission) of the structure and for changes of local coordinate systems. After analytical integrations and representations, the integral I^s can be recast as:

$$I^s = \frac{e^{-jk_0 z \cos\beta'}}{2\sin\beta'} \frac{1}{2\pi j} \int_C \frac{e^{-jk_0\rho \sin\beta' \cos(\alpha \mp \phi)}}{\cos\alpha + \cos\phi'} d\alpha, \quad (9)$$

where C is the integration path in the complex α-plane shown in Figure 2. Then, the Steepest Descent Method is applied to I^s, and a uniform asymptotic evaluation of the integral along the steepest descent path furnishes the diffraction term I^d. As a consequence, the diffracted field $\underline{E}^d$ can be so expressed:

$$\underline{E}^d = \begin{pmatrix} E^d_\beta \\ E^d_\phi \end{pmatrix} = \underline{\underline{M}} \begin{pmatrix} E^i_{\beta'} \\ E^i_{\phi'} \end{pmatrix} I^d = \underline{\underline{D}} \begin{pmatrix} E^i_{\beta'} \\ E^i_{\phi'} \end{pmatrix} \frac{e^{-jk_0 s}}{\sqrt{s}}, \quad (10)$$

where s is the distance along the diffraction direction and the 2×2 diffraction matrix $\underline{\underline{D}}$ is

$$\underline{\underline{D}} = \frac{1}{2\sqrt{2\pi k_0}} \frac{e^{-j\pi/4}}{\sin^2\beta' (\cos\phi + \cos\phi')} F_t\left(2k_0 s \sin^2\beta' \cos^2\left(\frac{\phi \pm \phi'}{2}\right)\right) \underline{\underline{M}}, \quad (11)$$

in which F_t is the UTD transition function [2] and + (−) sign applies when $0 < \phi < \pi$ ($\pi < \phi < 2\pi$).

A set of representative results is reported. They concern a slab characterised by $\varepsilon_r = 4 - j0.23$ and $d = 0.15\lambda_0$ (λ_0 is the free-space wavelength) when a plane wave impinges on it and the observation point moves on a circular path with $\rho = 7\lambda_0$. Figure 3 shows the β-component magnitudes of the GO field and the UAPO diffracted field when $E^i_{\beta'} = 1$, $E^i_{\phi'} = 0$ and $\beta' = 45°$, $\phi' = 60°$. As expected, the GO pattern presents two discontinuities in correspondence of the incidence and reflection shadow boundaries at $\phi = 240°$ and $\phi = 120°$, respectively. The UAPO field contribution is not negligible near such boundaries and guarantees the continuity of the total field across them as shown in Figure 4. The case of normal incidence is considered in the following tests as a useful scenario for comparing the UAPO-based results with those provided by the RF module of COMSOL MULTIPHYSICS [11]. As can be seen in Figures 5 and 6, an excellent agreement is attained, thus confirming the effectiveness of the UAPO solution.

Accounting for the linearity of the PO radiation integral, the UAPO-based approach for the diffraction problem involving one truncated dielectric slab can be extended to junctions by considering the diffraction contributions of the layers separately. Accordingly, in the case of junctions formed by two slabs, it results:

$$\begin{aligned} \underline{E}^s &= -jk_0 \iint_{S_1+S_2} \left[\left(\underline{\underline{I}} - \hat{R}\hat{R}\right)\zeta_0 \underline{J}^{PO}_s + \underline{J}^{PO}_{ms} \times \hat{R}\right] G(\underline{r}, \underline{r}')\, dS \\ &= -jk_0 \iint_{S_1} \left[\left(\underline{\underline{I}} - \hat{R}\hat{R}\right)\zeta_0 \underline{J}^{PO}_{s_1} + \underline{J}^{PO}_{ms_1} \times \hat{R}\right] G(\underline{r}, \underline{r}')\, dS_1 \\ &\quad - jk_0 \int_{S_2} \left[\left(\underline{\underline{I}} - \hat{R}\hat{R}\right)\zeta_0 \underline{J}^{PO}_{s_2} + \underline{J}^{PO}_{ms_2} \times \hat{R}\right] G(\underline{r}, \underline{r}')\, dS_2 \\ &= \underline{E}^s_1 + \underline{E}^s_2 \end{aligned} \quad (12)$$

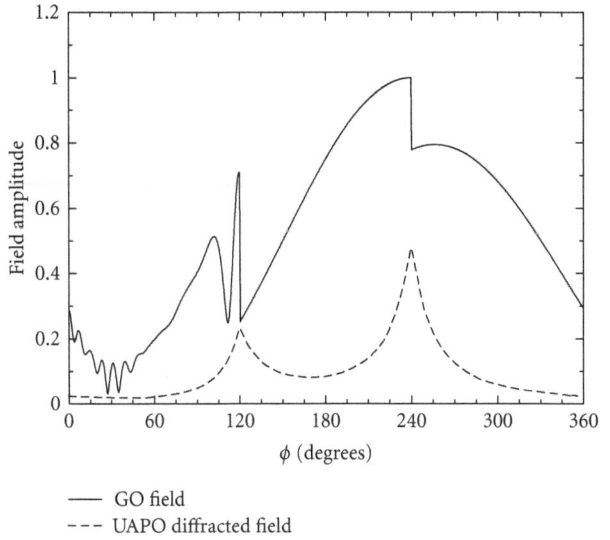

FIGURE 3: Dielectric slab. The β-component magnitudes of the GO field and the UAPO diffracted field at $\rho = 7\lambda_0$ when $\varepsilon_r = 4 - j0.23$, $d = 0.15\lambda_0$, $\beta' = 45^\circ$, $\phi' = 60^\circ$ and $E^i_{\beta'} = 1$, $E^i_{\phi'} = 0$.

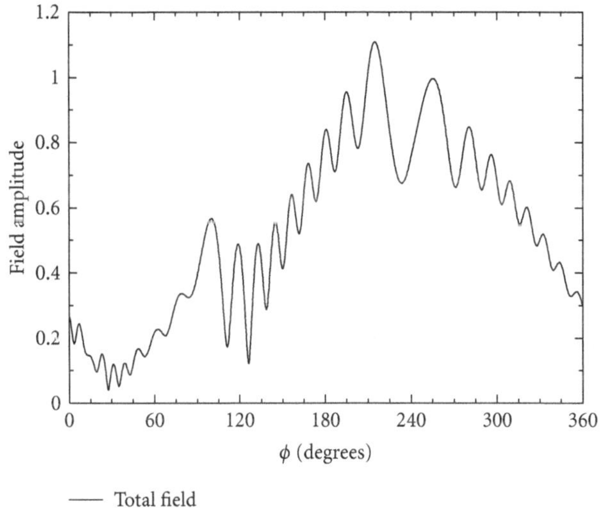

FIGURE 4: Dielectric slab. The β-component magnitude of the total field at $\rho = 7\lambda_0$ when $\varepsilon_r = 4 - j0.23$, $d = 0.15\lambda_0$, $\beta' = 45^\circ$, $\phi' = 60^\circ$ and $E^i_{\beta'} = 1$, $E^i_{\phi'} = 0$.

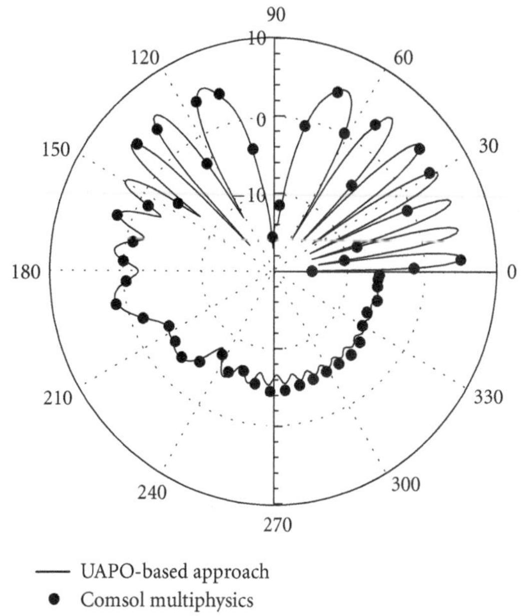

FIGURE 5: Dielectric slab. The relative magnitude of the total field β-component at $\rho = 7\lambda_0$ when $\varepsilon_r = 4 - j0.23$, $d = 0.15\lambda_0$, $\beta' = 90^\circ$, $\phi' = 30^\circ$ and $E^i_{\beta'} = 1$, $E^i_{\phi'} = 0$.

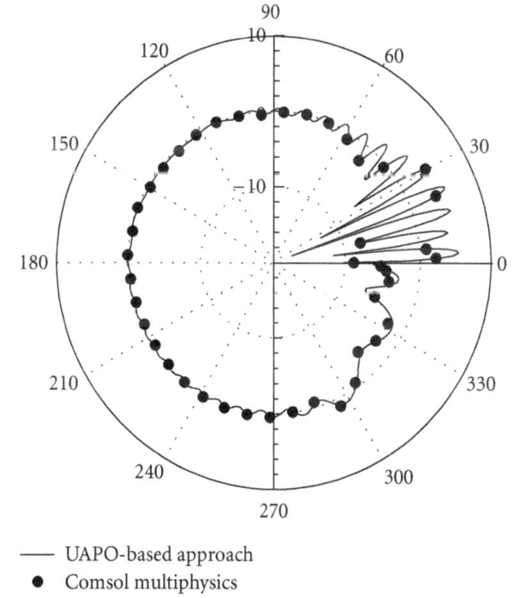

FIGURE 6: Dielectric slab. The relative magnitude of the total field β-component at $\rho = 7\lambda_0$ when $\varepsilon_r = 4 - j0.23$, $d = 0.15\lambda_0$, $\beta' = 90^\circ$, $\phi' = 140^\circ$ and $E^i_{\beta'} = 1$, $E^i_{\phi'} = 0$.

and then $\underline{\underline{D}} = \underline{\underline{D}}_1 + \underline{\underline{D}}_2$, with $\underline{\underline{D}}_1$ given by (11). The diffraction matrix $\underline{\underline{D}}_2$ related to the wave phenomenon originated by the second layer forming the junction can be determined by taking into account that if the external angle of the junction is $n\pi$, a $(n-1)\pi$ rotation of the edge-fixed coordinate system must be considered.

3. Dielectric Wedges

A two-dimensional scattering scenario involving a lossless nonmagnetic dielectric wedge with obtuse apex angle $\alpha = (2-n)\pi$ is sketched in Figure 7. Since $\beta' = 90^\circ$, the incident electric field is resolved in two orthogonal components directed along $\hat{z}$ and $\hat{\phi}'$.

Note that the two-dimensional diffraction problem involving a right-angled dielectric wedge has been tackled and solved by the authors in [12]. In addition, a UAPO solution for the three-dimensional diffraction problem originated by a nonpenetrable arbitrary-angled wedge with finite conductivity can be found in [13].

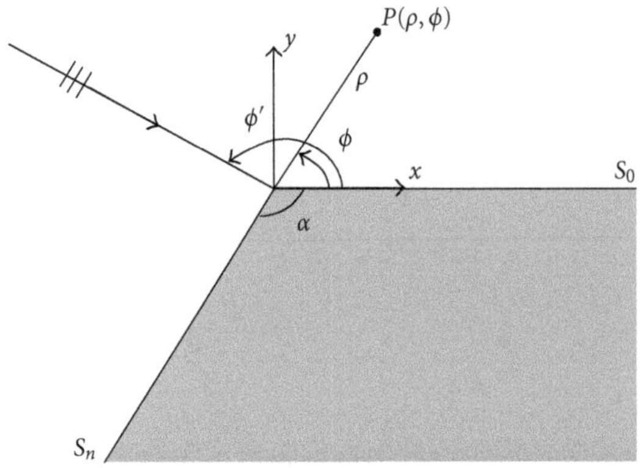

FIGURE 7: Diffraction by an obtuse-angled dielectric wedge.

The methodology adopted for penetrable wedges identifies the inner region (dielectric material) and the surrounding space as separate observation domains and the scattered electric field in each region as originated by electric and magnetic equivalent PO surface currents located on the inner/outer faces of S_0 and S_n. According to [12, 14], the UAPO diffracted field can be so expressed:

$$\begin{pmatrix} E_z^d \\ E_\phi^d \end{pmatrix} = \begin{pmatrix} D_{zz} & 0 \\ 0 & D_{\phi\phi'} \end{pmatrix} \begin{pmatrix} E_z^i \\ E_{\phi'}^i \end{pmatrix} \frac{e^{-jk\rho}}{\sqrt{\rho}}$$
$$= \begin{pmatrix} D_{zz_0} + D_{zz_n} & 0 \\ 0 & D_{\phi\phi'_0} + D_{\phi\phi'_n} \end{pmatrix} \begin{pmatrix} E_z^i \\ E_{\phi'}^i \end{pmatrix} \frac{e^{-jk\rho}}{\sqrt{\rho}}. \quad (13)$$

The analytical expressions of the diffraction matrix elements in (13) depend on the incidence direction and the geometric and electric characteristics of the wedge. In the following, the terms $R_{0,n}$ and $T_{0,n}$ denote the standard reflection and transmission coefficients [10] related to S_0 and S_n. The incidence direction defines two cases: (1) both the faces of the wedge are illuminated; (2) only one face is illuminated.

Case 1 ($(n-1)\pi < \phi' < \pi$). Both the faces of the wedge are illuminated by the impinging wave, and the GO field presents two reflection boundaries in the space surrounding the wedge and two transmission boundaries in the dielectric material.

(i) *Outer region:*

$$D_{zz_0}^{out} = [(1 - R_{0\perp}) \sin\phi' - (1 + R_{0\perp}) \sin\phi] h_0^{out},$$
$$D_{\phi\phi'_0}^{out} = [-(1 - R_{0\parallel}) \sin\phi' + (1 + R_{0\parallel}) \sin\phi] h_0^{out}, \quad (14)$$
$$h_0^{out} = \frac{e^{-j\pi/4}}{2\sqrt{2\pi k_0}} \frac{F_t(2k_0\rho\cos^2((\phi \pm \phi')/2))}{\cos\phi + \cos\phi'},$$

where $+(-)$ sign applies when $0 < \phi < \pi$ ($\pi < \phi < n\pi$), and

$$D_{zz_n}^{out} = [(1-R_{n\perp}) \sin(n\pi-\phi') - (1+R_{n\perp}) \sin(n\pi-\phi)] h_n^{out},$$
$$D_{\phi\phi'_n}^{out} = \left[-\left(1-R_{n\parallel}\right) \sin(n\pi-\phi') + \left(1+R_{n\parallel}\right) \sin(n\pi-\phi)\right] \times h_n^{out},$$
$$h_n^{out} = \frac{e^{-j\pi/4}}{2\sqrt{2\pi k_0}} \frac{F_t(2k_0\rho\cos^2(((n\pi-\phi) \pm (n\pi-\phi'))/2))}{\cos(n\pi-\phi) + \cos(n\pi-\phi')}, \quad (15)$$

in which the sign $+(-)$ holds for $(n-1)\pi < \phi < n\pi$ ($0 < \phi < (n-1)\pi$).

(ii) *Inner region:*

$$D_{zz_0}^{in} = \left[\sqrt{\varepsilon_r} \sin\phi - \sqrt{\varepsilon_r - \cos^2\phi'}\right] T_{0\perp} h_0^{in},$$
$$D_{\phi\phi'_0}^{out} = -\left[\sqrt{\varepsilon_r} \sin\phi - \sqrt{\varepsilon_r - \cos^2\phi'}\right] T_{0\parallel} h_0^{in},$$
$$h_0^{in} = \frac{e^{-j\pi/4}}{2\sqrt{2\pi k}} \frac{F_t(2k\rho\cos^2((\phi - \cos^{-1}(\cos\phi'/\sqrt{\varepsilon_r}))/2))}{\sqrt{\varepsilon_r}\cos\phi + \cos\phi'},$$
$$D_{zz_n}^{in} = \left[\sqrt{\varepsilon_r} \sin(n\pi-\phi) - \sqrt{\varepsilon_r - \cos^2(n\pi-\phi')}\right] T_{n\perp} h_n^{in},$$
$$D_{\phi\phi'_n}^{in} = -\left[\sqrt{\varepsilon_r} \sin(n\pi-\phi) - \sqrt{\varepsilon_r - \cos^2(n\pi-\phi')}\right] T_{n\parallel} h_n^{in}, \quad (16)$$

$$h_n^{in} = \frac{e^{-j\pi/4}}{2\sqrt{2\pi k}} \frac{F_t(2k\rho\cos^2((n\pi-\phi-\cos^{-1}(\cos(n\pi-\phi')/\sqrt{\varepsilon_r}))/2))}{\sqrt{\varepsilon_r}\cos(n\pi-\phi) + \cos(n\pi-\phi')}. \quad (17)$$

Case 2 ($0 < \phi' < (n-1)\pi$). Only the external face of S_0 is illuminated by the incident wave, and two different cases must be separately accounted for.

Subcase 1 ($0 < \phi' < (n-1)\pi$ and $\cos\phi' < \sqrt{\varepsilon_r}|\cos(n\pi)|$). The field transmitted through S_0 does not illuminate the internal face of S_n and therefore $D_{zz_n} = D_{\phi\phi'_n} = 0$. Shadow boundaries related to the incident and reflected fields appear in the external region, and only a transmission shadow boundary is present in the wedge. The expressions of D_{zz_0} and $D_{\phi\phi'_0}$ are equal to those reported for the Case 1.

Subcase 2 ($0 < \phi' < (n-1)\pi$ and $\cos\phi' > \sqrt{\varepsilon_r}|\cos(n\pi)|$). The field transmitted through S_0 penetrates in the dielectric region and illuminates the internal face of S_n. It is subsequently transmitted through S_n with coefficient T_{0n} and reflected with coefficient R_{0n}. Accordingly, shadow boundaries related so the incident, reflected, and transmitted-transmitted fields are in the space surrounding the wedge, whereas only the shadow boundary associated to the transmitted-reflected field exists in the dielectric material.

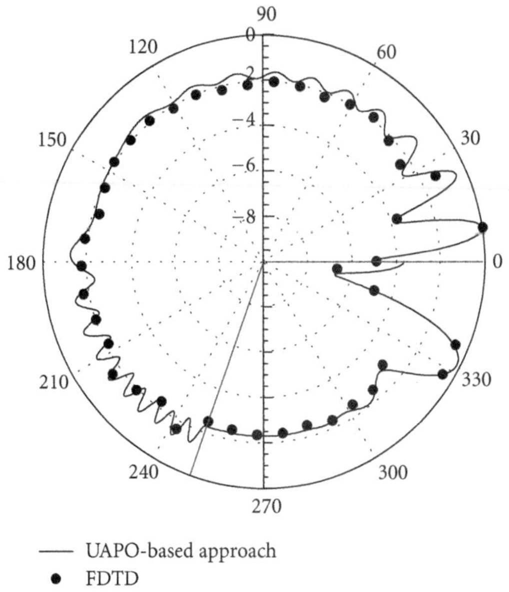

FIGURE 8: Dielectric wedge. The relative magnitude of the total field z-component at $\rho = 5\lambda_0$ when $\alpha = 110°$, $\varepsilon_r = 1.25$, and $\phi' = 160°$.

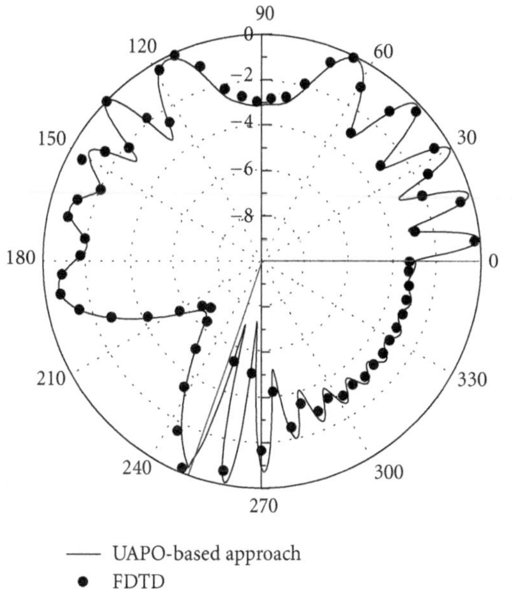

FIGURE 9: Dielectric wedge. The relative magnitude of the total field z-component at $\rho = 5\lambda_0$ when $\alpha = 110°$, $\varepsilon_r = 1.25$, and $\phi' = 30°$.

(i) *Outer region:*

$$D_{zz_n}^{\text{out}} = -\left[\cos\theta_n^t + \sin(n\pi - \phi)\right] T_{0\perp} T_{0n\perp} h_n^{\text{out}}, \tag{18}$$

$$D_{\phi\phi'_n}^{\text{out}} = \left[\cos\theta_n^t + \sin(n\pi - \phi)\right] T_{0\parallel} T_{0n\parallel} h_n^{\text{out}}, \tag{19}$$

$$h_n^{\text{out}} = \frac{e^{-j\pi/4}}{2\sqrt{2\pi k_0}} \frac{F_t\left(2k_0\rho\cos^2\left(\left(\cos^{-1}(-\eta) \pm (\phi + (1-n)\pi)\right)/2\right)\right)}{\cos(n\pi - \phi) + \eta}, \tag{20}$$

where

$$\theta_n^t = \sin^{-1}\left(\sqrt{\varepsilon_r}\sin\left((2-n)\pi - \sin^{-1}(\cos\phi'/\sqrt{\varepsilon_r})\right)\right), \tag{21}$$

$$\eta = \cos(n\pi)\cos\phi' + \sin(n\pi)\sqrt{\varepsilon_r - \cos^2\phi'}, \tag{22}$$

and the $+(-)$ sign holds when $(n-1)\pi < \phi < n\pi$ $(0 < \phi < (n-1)\pi)$. The expressions of $D_{zz_0}^{\text{out}}$ and $D_{\phi\phi'_0}^{\text{out}}$ are equal to those presented for the Case 1. Note that the total reflection occurs on S_n when $\cos\phi' \leq |\cos(n\pi) + \sin(n\pi)\sqrt{\varepsilon_r - 1}|$. In such a case it is appropriate to set $D_{zz_n}^{\text{out}} = D_{\phi\phi'_n}^{\text{out}} = 0$, since the surface waves are not taken into account in the UAPO-based approach.

(ii) *Inner region:*

$$\begin{aligned} D_{zz_n}^{\text{in}} &= \left[(1 - R_{0n\perp})\cos(n\pi + \theta_0^t) + (1 + R_{0n\perp})\sin(n\pi - \phi)\right] \\ &\quad \times \sqrt{\varepsilon_r} T_{0\perp} h_n^{\text{in}}, \\ D_{\phi\phi'_n}^{\text{in}} &= -\left[\left(1 - R_{0n\parallel}\right)\cos(n\pi + \theta_0^t) + \left(1 + R_{0n\parallel}\right)\right. \\ &\quad \left.\sin(n\pi - \phi)\right]\sqrt{\varepsilon_r} T_{0\parallel} h_n^{\text{in}}, \end{aligned} \tag{23}$$

$$h_n^{\text{in}} = \frac{e^{-j\pi/4}}{2\sqrt{2\pi k}} \frac{F_t\left(2k\rho\cos^2\left(\left(\cos^{-1}(-\eta/\sqrt{\varepsilon_r}) - (\phi + (1-n)\pi)\right)/2\right)\right)}{\sqrt{\varepsilon_r}\cos(n\pi - \phi) + \eta}, \tag{24}$$

in which $\theta_0^t = \sin^{-1}(\cos\phi'/\sqrt{\varepsilon_r})$. The expressions of $D_{zz_0}^{\text{in}}$ and $D_{\phi\phi'_0}^{\text{in}}$ are equal to those presented for the Case 1.

Figures 8 and 9 show comparisons with data provided by an "ad hoc" developed code implementing the Finite-Difference Time-Domain (FDTD) method [15]. They refer to a wedge characterised by $\alpha = 110°$ and $\varepsilon_r = 1.25$ when the observation point moves along a circular path with $\rho = 5\lambda_0$. The good agreements assess the effectiveness of the presented UAPO solution also when considering dielectric wedges.

Note that the case corresponding to $\pi < \phi' < n\pi$ is not reported here because it can be tackled and solved in a similar way.

4. Conclusions

UAPO solutions have been presented in the UTD context for evaluating the field diffracted by penetrable dielectric slabs and wedges. They are in closed form, simple and easy to handle and yield total field levels in good agreement with data obtained via numerical tools. In addition, the UTD-like formulation of UAPO diffraction coefficients facilitates the analytical evaluation of the time domain counterparts. These characteristics encourage the use of UAPO solutions for diffraction problems of interest in many application areas. Future research activities will be directed towards three-dimensional scenarios involving penetrable wedges with finite conductivity.

Appendix

$$\underline{\underline{M}}_0 = \frac{1}{\sqrt{1-\sin^2\beta'\sin^2\phi'}}\begin{pmatrix}\cos\beta'\sin\phi' & \cos\phi' \\ -\cos\phi' & \cos\beta'\sin\phi'\end{pmatrix},$$

$$\underline{\underline{M}}_1 = \underline{\underline{M}}_7\left[\underline{\underline{M}}_2\underline{\underline{M}}_4\underline{\underline{M}}_5 + \underline{\underline{M}}_3\underline{\underline{M}}_4\underline{\underline{M}}_6\right],$$

$$\underline{\underline{M}}_7 = \begin{pmatrix}\cos\beta'\cos\phi & \cos\beta'\sin\phi & -\sin\beta' \\ -\sin\phi & \cos\phi & 0\end{pmatrix},$$

$$\underline{\underline{M}}_2 = \begin{pmatrix}1-\sin^2\beta'\cos^2\phi & -\sin\beta'\cos\beta'\cos\phi \\ -\sin^2\beta'\sin\phi\cos\phi & -\sin\beta'\cos\beta'\sin\phi \\ -\sin\beta'\cos\beta'\cos\phi & \sin^2\beta'\end{pmatrix},$$

$$\underline{\underline{M}}_3 = \begin{pmatrix}0 & -\sin\beta'\sin\phi \\ -\cos\beta' & \sin\beta'\cos\phi \\ \sin\beta'\sin\phi & 0\end{pmatrix},$$

$$\underline{\underline{M}}_4 = \frac{1}{\sqrt{1-\sin^2\beta'\sin^2\phi'}}\begin{pmatrix}-\cos\beta' & -\sin\beta'\cos\phi' \\ -\sin\beta'\cos\phi' & \cos\beta'\end{pmatrix},$$

$$\underline{\underline{M}}_5 = \begin{pmatrix}0 & (1-R_\perp - T_\perp)\cos\theta^i \\ 1+R_\parallel - T_\parallel & 0\end{pmatrix},$$

$$\underline{\underline{M}}_6 = \begin{pmatrix}(1-R_\parallel - T_\parallel)\cos\theta^i & 0 \\ 0 & -(1+R_\perp - T_\perp)\end{pmatrix}. \tag{A.1}$$

References

[1] J. B. Keller, "Geometrical theory of diffraction," *Journal of the Optical Society of America*, vol. 52, pp. 116–130, 1962.

[2] R. G. Kouyoumjian and P. H. Pathak, "A uniform geometrical theory of diffraction for an edge in a perfectly conducting surface," *Proceedings of the IEEE*, vol. 62, no. 11, pp. 1448–1461, 1974.

[3] P. C. Chang, R. J. Burkholder, J. L. Volakis, R. J. Marhefka, and Y. Bayram, "High-frequency em characterization of through-wall building imaging," *IEEE Transactions on Geoscience and Remote Sensing*, vol. 47, no. 5, Article ID 4810144, pp. 1375–1387, 2009.

[4] A. I. Kozlov, L. Lighart, and A. I. Logvin, "Radar reflection from surfaces with ruptures," in *Proceedings of 13th International Conference on Microwaves, Radar and Wireless Communications (MIKON '00)*, vol. 1, pp. 347–350, 2000.

[5] W. D. Burnside and K. W. Burgener, "High frequency scattering by a thin lossless dielectric slab," *IEEE Transactions on Antennas and Propagation*, vol. 31, no. 1, pp. 104–110, 1983.

[6] R. J. Leubbers, "Finite conductivity uniform GTD versus knife edge diffraction in prediction of propagation path loss," *IEEE Transactions on Antennas and Propagation*, vol. 32, no. 1, pp. 70–76, 1984.

[7] P. Bernardi, R. Cicchetti, and O. Testa, "A three-dimensional UTD heuristic diffraction coefficient for complex penetrable wedges," *IEEE Transactions on Antennas and Propagation*, vol. 50, no. 2, pp. 217–224, 2002.

[8] T. W. Veruttipong, "Time domain version of the uniform GTD," *IEEE Transactions on Antennas and Propagation*, vol. 38, no. 11, pp. 1757–1764, 1990.

[9] F. Ferrara, C. Gennarelli, G. Gennarelli, M. Migliozzi, and G. Riccio, "Scattering by truncated lossy layers: A UAPO-based approach," *Electromagnetics*, vol. 27, no. 7, pp. 443–456, 2007.

[10] C. A. Balanis, *Advanced Engineering Electromagnetics*, Wiley, New York, NY, USA, 1989.

[11] COMSOL Multiphysics, v. 3.5a, Users' guide, RF Module, 2008.

[12] G. Gennarelli and G. Riccio, "A uniform asymptotic solution for the diffraction by a right-angled dielectric wedge," *IEEE Transactions on Antennas and Propagation*, vol. 59, no. 3, Article ID 5677594, pp. 898–903, 2011.

[13] G. Gennarelli and G. Riccio, "A UAPO-based model for propagation prediction in microcellular environments," *Progress In Electromagnetics Research B*, no. 17, pp. 101–116, 2009.

[14] G. Gennarelli and G. Riccio, "Plane-wave diffraction by an obtuse-angled dielectric wedge," *Journal of the Optical Society of America A*, vol. 28, no. 4, pp. 627–632, 2011.

[15] A. Taflove and S. Hagness, *Computational Electrodynamics: The Finite Difference Time Domain Method*, Artech House, London, UK, 2000.

Planar Printed Shorted Monopole Antenna with Coupled Feed for LTE/WWAN Mobile Handset Applications

Do-Gu Kang and Youngje Sung

Department of Electronic Engineering, Kyonggi University Yeongtong-gu, Suwon-si, Gyeonggi-do, Republic of Korea

Correspondence should be addressed to Youngje Sung; yjsung@kgu.ac.kr

Academic Editor: Antonio Faraone

A shorted monopole antenna with coupled feed for LTE/WWAN mobile handset applications is described. The basic resonance of the shorted monopole combines with the resonance formed by the coupling between the coupling strip and the feeding pad to cover the LTE700, GSM850, and GSM900 bands. Both the feeding pad and the coupling strip operate with the shorting strip as a loop antenna. The resonance of the loop antenna and the harmonics of the shorted monopole combine to cover the GSM1800, GSM1900, UMTS, and LTE2300 bands. A stable and omnidirectional radiation pattern with reasonable gain has been observed over the operating bandwidth.

1. Introduction

Technological advancements and the diversity of needs in wireless communication are leading to the development of antennas for smaller, lighter, and more multifunctional mobile handsets [1]. Consequently, planar monopoles [2–4] enabling pentaband WWAN (wireless wide area network) operation have been developed. The WWAN band covers the GSM850 (824–894 MHz), GSM900 (880–960 MHz), GSM1800 (1710–1880 MHz), GSM1900 (1850–1990 MHz), and UMTS (1920–2170 MHz) bands. These developed planar monopoles are, respectively, a monopole with coupled strips [2], a CPW-fed LI-shaped monopole [3], and a coupled-fed shorted monopole [4]. Recently, coupled-fed shorted monopoles covering not only the WWAN band but also the LTE band have been developed. The LTE band covers LTE700 (698–798 MHz), LTE2300 (2300–2400 MHz), and LTE2500 (2500–2690 MHz) [5, 6]. These monopoles operating in the WWAN/LTE bands are, respectively, a monopole with a tuning stub [5] and a monopole with a chip inductor-loaded shorted strip [6].

In this paper, a shorted monopole antenna with coupled feed for LTE/WWAN bands is proposed. The base resonance of the shorted monopole combines with the resonance formed by the coupling between the feeding pad and the coupling strip to include the LTE700, GSM850, and GSM900 bands. The loop antenna is formed by both the coupled feed structure and the shorting strip. The loop resonance and a harmonic element combine to cover the GSM1800, GSM1900, UMTS, and LTE2300 bands. Therefore, the proposed structure covers the LTE700, LTE2300, GSM850, GSM900, GSM1800, GSM1900, and UMTS bands.

2. Antenna Geometry

Figure 1 shows the proposed antenna which consists of a long radiator (AB), a coupling strip (BC), a feeding pad, and a shorting strip. l_1 denotes the total length of the long radiator. l_2 denotes the length of the part generating the coupled feed, and g denotes the gap between the feeding pad and the coupling strip; these are the two parameters that affect the coupled feed. l_3 denotes the length of the feeding pad. The antenna covers an area of $18 \times 49\ \text{mm}^2$ on the upper surface of an FR4 substrate with a 0.6 mm thick. The total area of the substrate is $118 \times 49\ \text{mm}^2$; the ground (GND) found on the lower surface of the substrate covers an area of $100 \times 49\ \text{mm}^2$.

3. Simulated and Measured Results

Figure 2 shows the simulated and measured reflection coefficients for the proposed antenna and reference 1. The

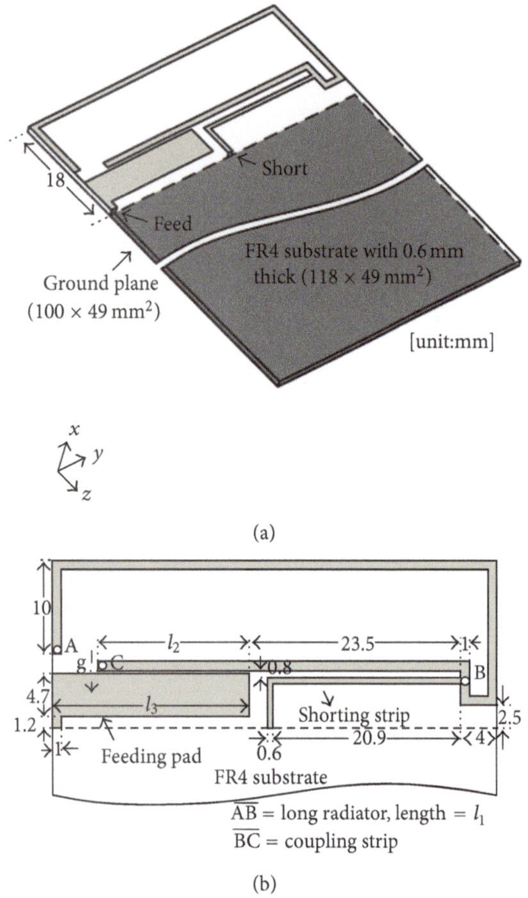

FIGURE 1: (a) 3D geometry of the proposed antenna. (b) The antenna geometry as seen from above.

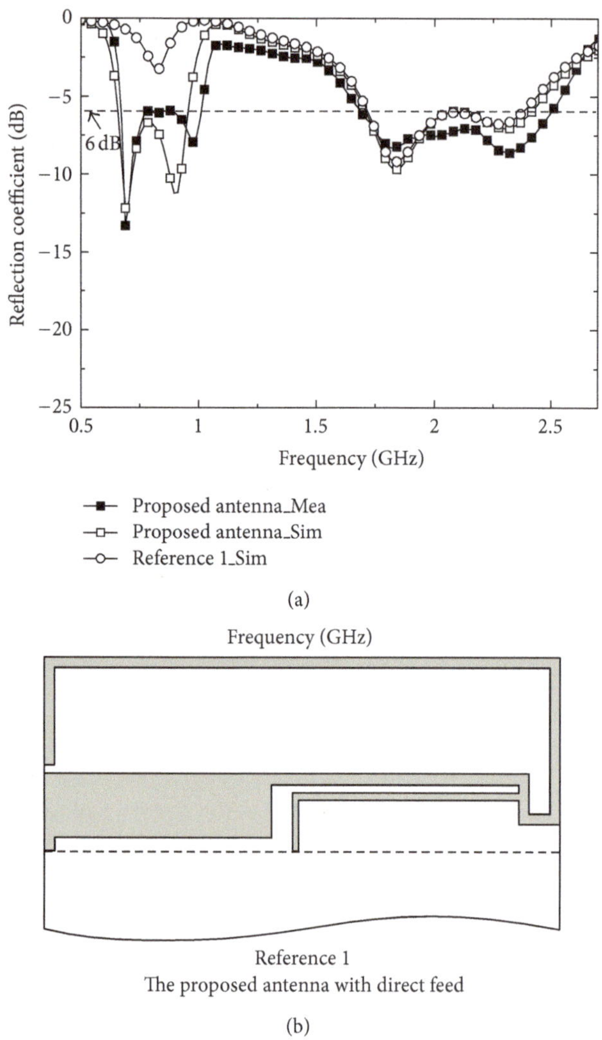

FIGURE 2: Simulated and measured reflection coefficients for the proposed antenna and reference 1.

parameters for the proposed antenna were set to l_1 = 78 mm, l_2 = 16.5 mm, g = 0.3 mm, and l_3 = 21.5 mm. The prototype of the proposed antenna is based on the foregoing optimized parameters and is fabricated as shown in Figure 3. The simulation was carried out through HFSS. The simulated results were nearly identical to the measured results. The proposed antenna has four resonant frequencies (f_1, f_2, f_3, and f_4). The low band covers the LTE700, GSM850, and GSM900 bands, and the high band covers the GSM1800, GSM1900, UMTS, and LTE2300 bands. All the bands operated properly with a −6 dB bandwidth which is a widely used value for internal WWAN antennas in practical mobile phone application.

If the proposed antenna's result is compared to reference 1's (the proposed antenna with direct feed) result, the effect of the coupled feed can be observed. The matching characteristics of the base resonance were improved by the coupled feed that increases the electrical length. Therefore, the resonance frequency decreased. The resonance combined with the second resonance that was formed by the coupled feed. The two resonances form dual resonance. Therefore, the proposed antenna achieves a low wideband through the coupled feed.

FIGURE 3: Fabricated prototype of the proposed antenna.

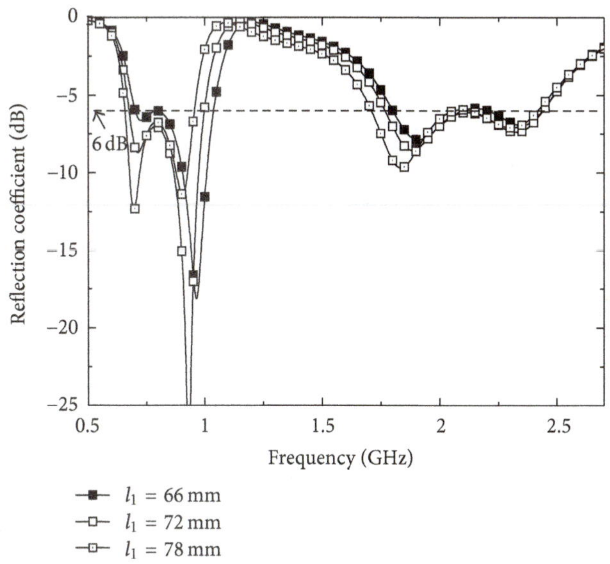

FIGURE 4: Simulated reflection coefficients relative to changes in the total length of the long radiator l_1.

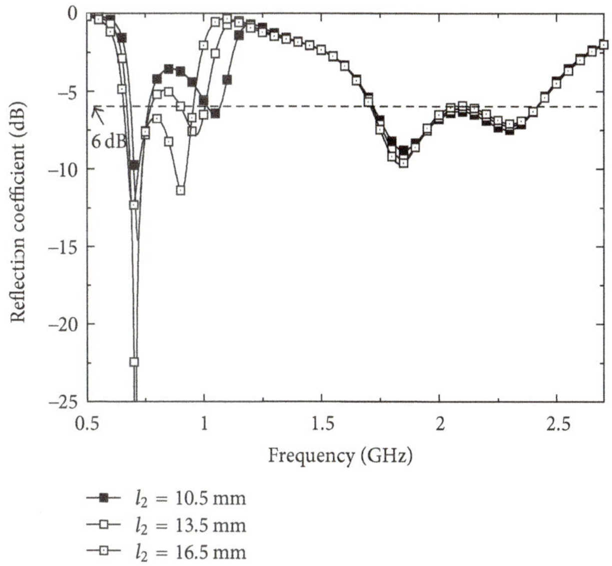

FIGURE 5: Simulated reflection coefficients relative to changes in the length of the part generating the coupled feed l_2.

Figure 4 shows the simulated reflection coefficients relative to the changes in the total length l_1 of the long radiator. The base resonance f_1 is formed by the long radiator's length which was a resonance length of $\lambda/4$ at f_1 (0.69 GHz). Further, f_3 decreased, which proves that f_3 is a harmonic of f_1. As l_1 increased, the lower edge of the high band decreased from 1.8 to 1.71 GHz. Thus, the antenna covers the GSM1800 band. Therefore, considering the high band, l_1 was selected as 78 mm.

Figure 5 shows the simulated reflection coefficients relative to the changes in length l_2 of the part generating the coupled feed l_2. Because l_2 affects the coupled feed, the

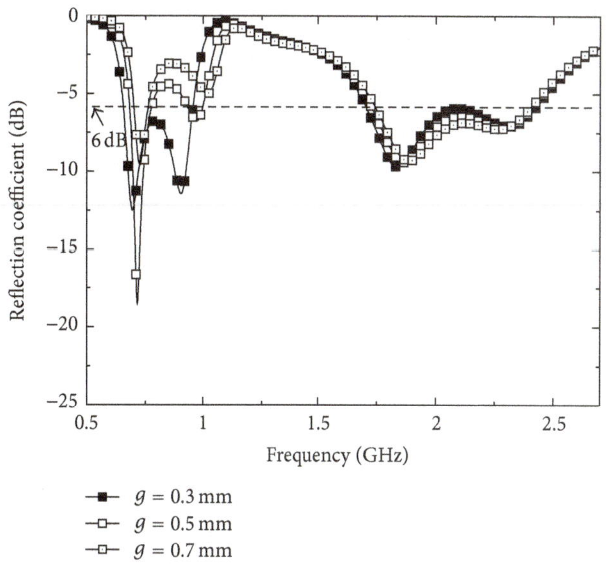

FIGURE 6: Simulated reflection coefficients relative to changes in the gap between feeding pad and coupling strip g.

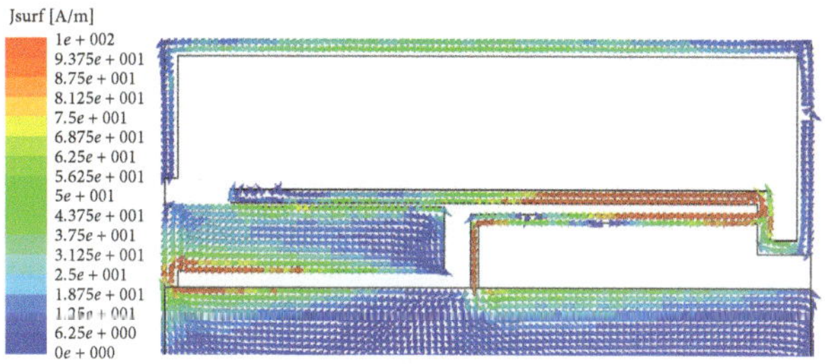

FIGURE 7: Surface current distribution simulated on the proposed antenna's radiator and GND at f_4 = 2.3 GHz.

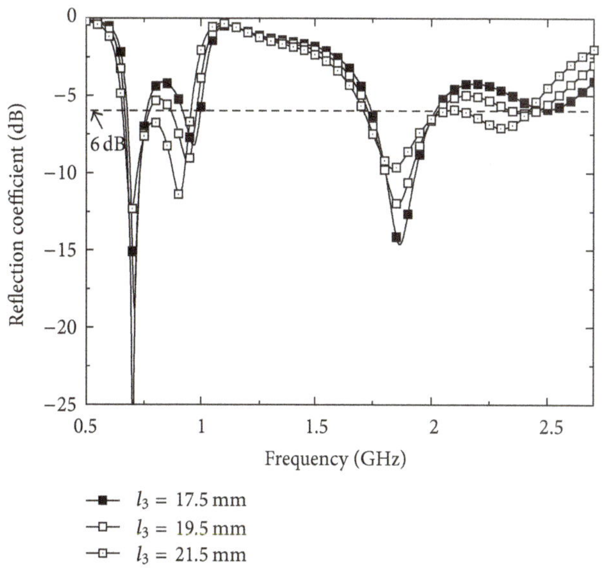

FIGURE 8: Simulated reflection coefficients relative to changes in the length of the feeding pad l_3.

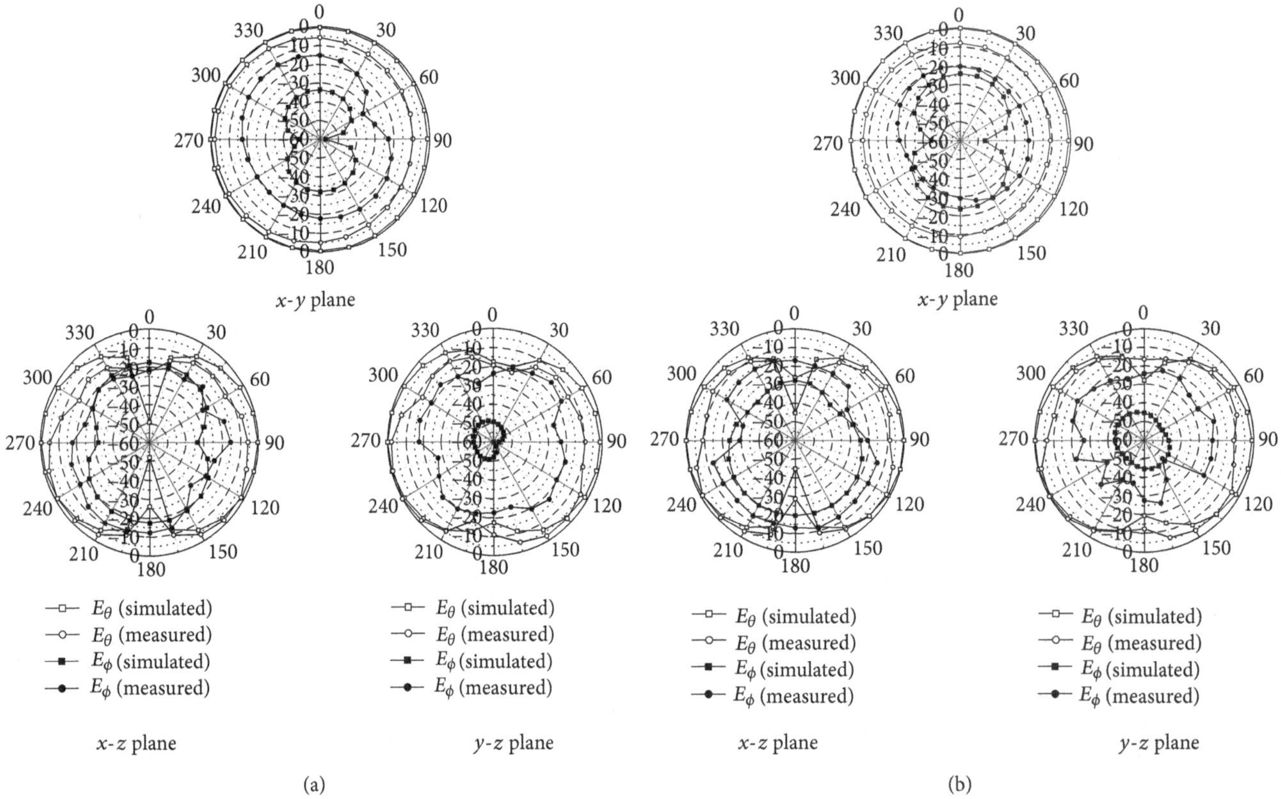

FIGURE 9: Simulated and measured 2D radiation patterns of proposed antenna: (a) $f_1 = 0.69$ GHz, (b) $f_2 = 0.9$ GHz.

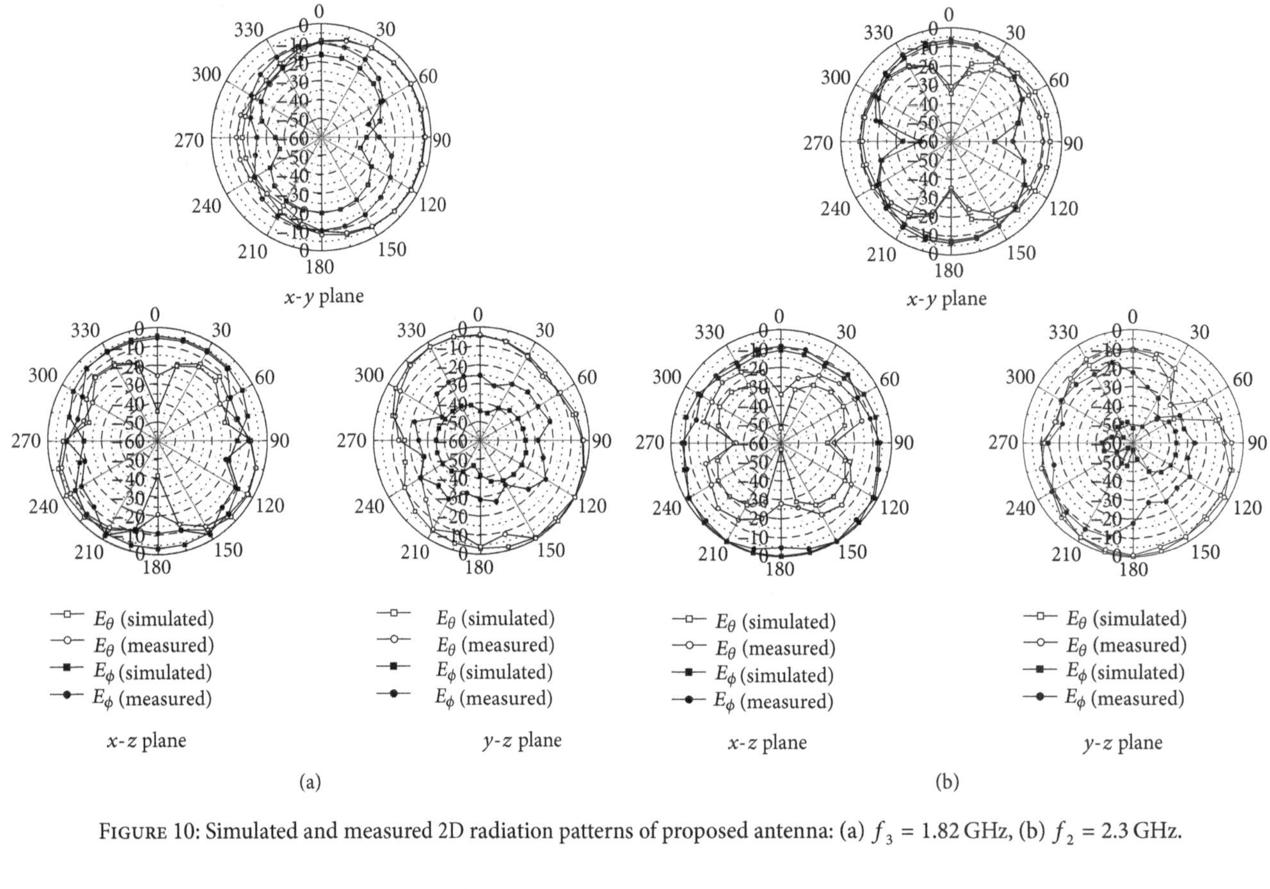

FIGURE 10: Simulated and measured 2D radiation patterns of proposed antenna: (a) $f_3 = 1.82$ GHz, (b) $f_2 = 2.3$ GHz.

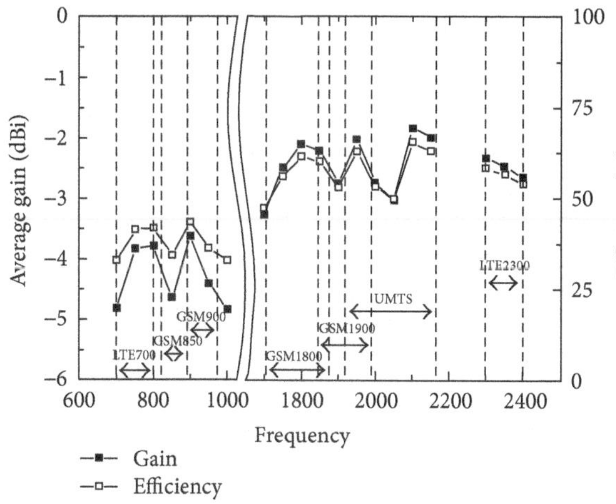

FIGURE 11: Measured average gain and efficiency of the proposed antenna.

resonance formed by the coupled feed was affected as l_2 changed. The base resonance formed by the long radiator further combined with the resonance formed by the coupled feed as l_2 increased. l_2 was selected as 16.5 mm to allow the antenna to have low wideband characteristics.

Figure 6 shows the simulated reflection coefficients relative to the changes in gap g between the feeding pad and the coupling strip. As g decreased, the matching of f_2 improved in the state the matching between f_1 and f_2 improved. Thus, considering the low band, g was selected as 0.3 mm.

Figure 7 shows the surface current distribution simulated on the proposed antenna's radiator and GND at $f_4 = 2.3$ GHz. The red area indicates a strong field, whereas the blue area indicates a weak field. The maximum current is located at the beginning of the feeding pad, and the null of the current is located at the end. The coupled gap was electrically connected. The maximum current on the coupling strip generated the maximum current on the shorting strip's end. The null is located between the maximum current on the shorting strip's end and the maximum current on the shorting strip's beginning. Through these findings, the path from the feeding pad to the coupled gap was combined with the path from the coupling strip to the shorting strip, which was the path of the loop operating at the λ resonant length. Therefore, f_4 was the resonant frequency of the loop antenna.

Figure 8 shows the simulated reflection coefficients relative to the changes in length l_3 of the feeding pad. As l_3 increased, the matching characteristics between f_1 and f_2 improved with a decrease of f_2. Because l_3 affected the length of the coupled feed, its effect was analogous to the effect of l_2 and g in the low band. Further, the matching characteristics between f_3 and f_4 improved with a decrease of f_4. To take into consideration the effect of both the high band and low band, l_3 was set to 21.5 mm. When a parameter study of l_3 was compared with other parameter studies, it showed that f_4 was relatively more transferred by l_3. This is because l_3 resides in the loop path.

Figures 9 and 10 show the simulated and measured 2D radiation patterns of the proposed antenna at $f_1 = 0.69$ GHz, $f_2 = 0.9$ GHz, $f_3 = 1.82$ GHz, and $f_4 = 2.3$ GHz. The radiation pattern was measured on the x- y, x- z, and y- z planes, on the basis of the direction of the antenna placement shown in Figure 1(a). The measured co-pols closely matched the simulated co-pols; the measured cross-pols were higher than the simulated cross-pols. The reason for this is that the GND was extended by the cabling used to measure the radiation pattern. Dipole-like radiation patterns with omnidirectional radiation in the azimuthal plane (x- y plane) are observed at two resonances of the low band in Figure 9. Conversely, more nulls and changes occurred in the radiation pattern observed for the two high band resonances in Figure 10. The obtained radiation patterns were analogous to the radiation patterns of WWAN mobile phone antennas [7, 8]. In addition, stable radiation patterns were obtained in both the low band and the high band. Figure 11 shows the measured average gain and efficiency of the proposed antenna. At the LTE700/GSM850/GSM900 band, the average gain varies from −4.8 to −3.6 dBi. At the GSM1800/GSM1900/UMTS band, it varies from −3.2 to −1.8 dBi, while it varies from −2.6 to −2.3 dBi at the LTE2300 band.

4. Conclusions

In this work, a shorted monopole antenna with a coupled feed for LTE/WWAN mobile handset applications was proposed. The base resonance of the shorted monopole and the resonance formed by the coupled feed structure combine to form a low wideband. The coupled feed structure and the shorting strip generate the loop antenna. The loop resonance combines with the harmonic element, which forms a high wideband. Therefore, the proposed structure is suitable for seven-band mobile handset applications, covering the LTE700, GSM850, GSM900, GSM1800, GSM1900, UMTS, and LTE2300 bands. Good radiation characteristics for frequencies over the operating bands were observed.

Acknowledgment

This research was supported by Basic Science Research Program through the National Research Foundation of Korea (NRF) funded by the Ministry of Education, Science, and Technology (2010-0021222).

References

[1] P. Ciais, R. Staraj, G. Kossiavas, and C. Luxey, "Design of an internal quad-band antenna for mobile phones," *IEEE Microwave and Wireless Components Letters*, vol. 14, no. 4, pp. 148–150, 2004.

[2] H. W. Liu and C. F. Yang, "Miniature multiband monopole antenna for WWAN operation in laptop computer," *Electronics Letters*, vol. 46, no. 1, pp. 21–23, 2010.

[3] Y. Jee and Y. M. Seo, "Triple-band CPW-fed compact monopole antennas for GSM/PCS/DCS/WCDMA applications," *Electronics Letters*, vol. 45, no. 9, pp. 446–448, 2009.

[4] C.-T. Lee and S.-W. Su, "Compact and printed, standalone wireless wide area network antenna incorporating a conductive wire at low cost," *Microwave and Optical Technology Letters*, vol. 54, no. 1, pp. 109–113, 2012.

[5] S. C. Chen and K. L. Wong, "Small-size 11-band LTE/WWAN/WLAN internal mobile phone antenna," *Microwave and Optical Technology Letters*, vol. 52, no. 11, pp. 2603–2608, 2010.

[6] S. C. Chen and K. L. Wong, "Wideband monopole antenna coupled with a chip-inductor-loaded shorted strip for LTE/WWAN mobile handset," *Microwave and Optical Technology Letters*, vol. 53, no. 6, pp. 1293–1298, 2011.

[7] K. L. Wong and C. H. Huang, "Compact multiband PIFA with a coupling feed for internal mobile phone antenna," *Microwave and Optical Technology Letters*, vol. 50, no. 10, pp. 2487–2491, 2008.

[8] B. N. Kim, S. O. Park, J. K. Oh, and G. Y. Koo, "Wideband built-in antenna with new crossed c-shaped coupling feed for future mobile phone application," *IEEE Antennas and Wireless Propagation Letters*, vol. 9, pp. 572–575, 2010.

Electromagnetic Reciprocal Cloak with Only Axial Material Parameter Spatially Variant

Jing Jing Yang, Ming Huang, Fu Chun Mao, Guang Hui Cai, and Yao Zhong Lan

School of Information Science and Engineering, Yunnan University, Kunming 650091, China

Correspondence should be addressed to Ming Huang, huangming@ynu.edu.cn

Academic Editor: Tat Yeo

Reciprocal cloak is an intriguing metamaterial device, in which a hidden antenna or a sensor can receive electromagnetic irradiation from the outside but its presence will not be detected. Based on transformation optics, a cylindrical electromagnetic reciprocal cloak with only axial parameter varying with radius is designed and validated by full wave simulation. When two dispersive reciprocal cloaks are put together, they do not interfere with each other. Our work demonstrates the electromagnetic compatibility (EMC) ability of the reciprocal cloak which is very important in multi antenna and sensor design.

1. Introduction

In the past decade, a fascinating term "metamaterials" has attracted a great deal of attentions since it provides many possibilities to explore unknown physical phenomena and fabricate novel devices [1–6] such as perfect lens, sensors, and antennae. Recently, theory of transformation optics generalized by Pendry et al. and Leonhardt [7, 8] built up a bridge between the function of metamaterials and material parameter distribution and provided a complete prescription for controlling and guiding electromagnetic wave as well [9–17]. One of the most popular and intriguing topic in this field is perhaps the "invisible cloak" [9–12], which makes the incident wave detour around the cloaked region without disturbance, thus an arbitrary object might be hidden inside; but the object has to be "blind," since no outside electromagnetic waves can reach within the cloaked space [18].

To overcome this limitation, some novel cloaking schemes such as open cloak and external cloak were developed. The open cloak [19–21] strategy provides an approach to open a window on the surface of a cloak, so that exchanging information and matter with the outside can be achieved. The external cloak [22, 23] is capable of hiding arbitrary objects at a distance outside the cloaking shell and thus share information with the surrounding. It is based on the idea of scattering cancelation, that is, the scattered field of an object is cancelled by its antiobject located nearby. Besides, scattering of a target object can also be cancelled or minimized by coating with a layer of well-designed materials. This has been firstly considered by Chew and Kerker [24]. Latterly, Alù and Engheta [25] showed that it is possible to reduce the scattering cross-section of spherical and cylindrical objects using lossless plasmonic or metamaterial covers. In the long wavelength limit, Gao et al. [26, 27] developed isotropic permittivity and permeability tensors of small coated sphere and cylinder. They demonstrated that total scattering width can be dramatically reduced by suitable adjustment of the dielectric anisotropy of the shell. In our previous work, we proposed a cloaking scheme named as reciprocal cloak [28] based on the theory of transformation optics. We show that arbitrary object can be cloaked when coated with a complementary media. Since electromagnetic waves penetrate into the cloaked space, the object hidden inside can exchange information with the outside, which may lead to many potential applications in electromagnetic engineering. However, practical fabrication of the reciprocal cloak is not so far achievable for the material parameters which are highly anisotropic and even singular. In this paper, we establish a flexible transformation procedure to design a cylindrical reciprocal cloak with only axial material parameter varying with radius, which makes it possible to be realized by layered metamaterials [29, 30]. EMC ability of the cloak is validated by full wave simulation. The impact

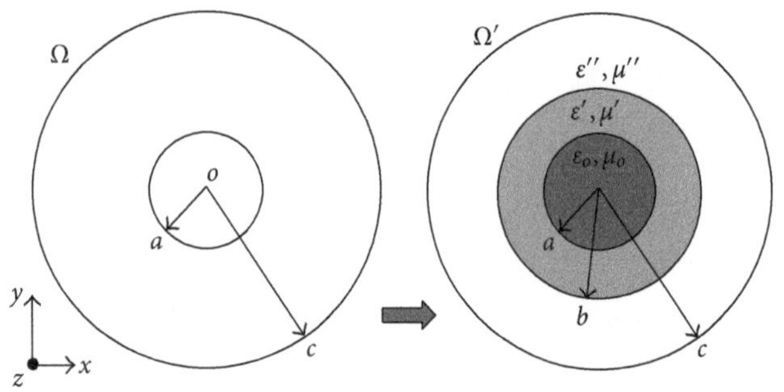

FIGURE 1: Schematic diagrams for describing the coordinate transformation of the reciprocal cloak.

of metamaterial loss and material parameter deviation on cloaking effect is also analyzed in detail.

2. Method and Simulation Model

Figure 1 shows the schematic diagram for the construction of the cylindrical reciprocal cloak. A point O in the virtual space (Ω) is transformed into a circle with a radius of b in the physical space (Ω'), while keeping the inner and outer boundary unchanged. So the whole transformation space is divided into three regions, that is, the cloaked object ($r' < a$), the complementary media ($a < r' < b$), and the recovering layer ($b < r'' < c$). The working principle of the cloak can be described in two steps. First, the object is optically cancelled by the complementary media. Then, correct optical path in the canceled space is restored by the recovering layer. In the complementary media layer, we suppose the transformation function between the physical space and the virtual space is in the form of $r = f(r')$, where $f(r')$ is a continuous function of r', and it satisfies $f(b) = \delta$, $f(a) = a$. Here, δ is a constant that approximates zero. To avoid singularity, we set it to be 0.002 in this paper. According to the optical transformation theory and the form invariance of Maxwell's equations, relative permittivity ε' and permeability μ' of the complementary media can be written as

$$\frac{\varepsilon'_r}{\varepsilon_{or}} = \frac{\mu'_r}{\mu_{or}} = \frac{f(r')}{r' df(r')/dr'}, \quad (1a)$$

$$\frac{\varepsilon'_\varphi}{\varepsilon_{o\varphi}} = \frac{\mu'_\theta}{\mu_{o\varphi}} = \frac{r' df(r')/dr'}{f(r')}, \quad (1b)$$

$$\frac{\varepsilon'_z}{\varepsilon_{oz}} = \frac{\mu'_z}{\mu_{oz}} = \frac{f(r') df(r')/dr'}{r'}, \quad (1c)$$

where ε_o and μ_o are the relative permittivity and permeability of the cloaked object. We can notice that if we set (1b) as a constant, (1a) will be fixed. Thus, we suppose that

$$\frac{\varepsilon'_\varphi}{\varepsilon_{o\varphi}} = \frac{\mu'_\varphi}{\mu_{o\varphi}} = \frac{r' df(r')/dr'}{f(r')} = m'_0. \quad (2)$$

A solution of this differential equation is given by

$$r = f(r') = m'_1 r'^{m'_0}, \quad (3)$$

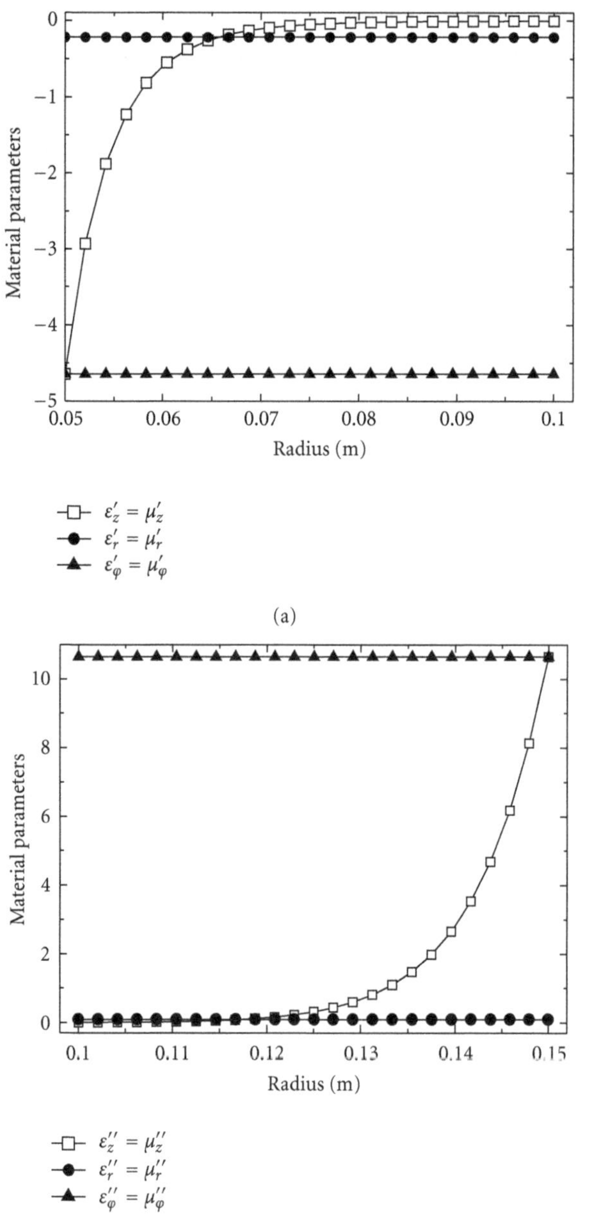

FIGURE 2: Constitutive parameters distribution along radial direction of the reciprocal cloak. (a) The complementary layer ($a < r' < b$). (b) The recovering layer ($b < r'' < c$).

where $m'_0 = \log_{(b/a)}(\delta/a)$, $m'_1 = a^{(1-m'_0)}$. Then, material parameters of the complementary media layer can be obtained as follows:

$$\frac{\varepsilon'_r}{\varepsilon_{or}} = \frac{\mu'_r}{\mu_{or}} = \frac{1}{m'_0},$$
$$\frac{\varepsilon'_\varphi}{\varepsilon_{o\varphi}} = \frac{\mu'_\varphi}{\mu_{o\varphi}} = m'_0, \quad (4)$$
$$\frac{\varepsilon'_z}{\varepsilon_{oz}} = \frac{\mu'_z}{\mu_{oz}} = m'_0 \left(\frac{r'}{a}\right)^{2(m'_0-1)}.$$

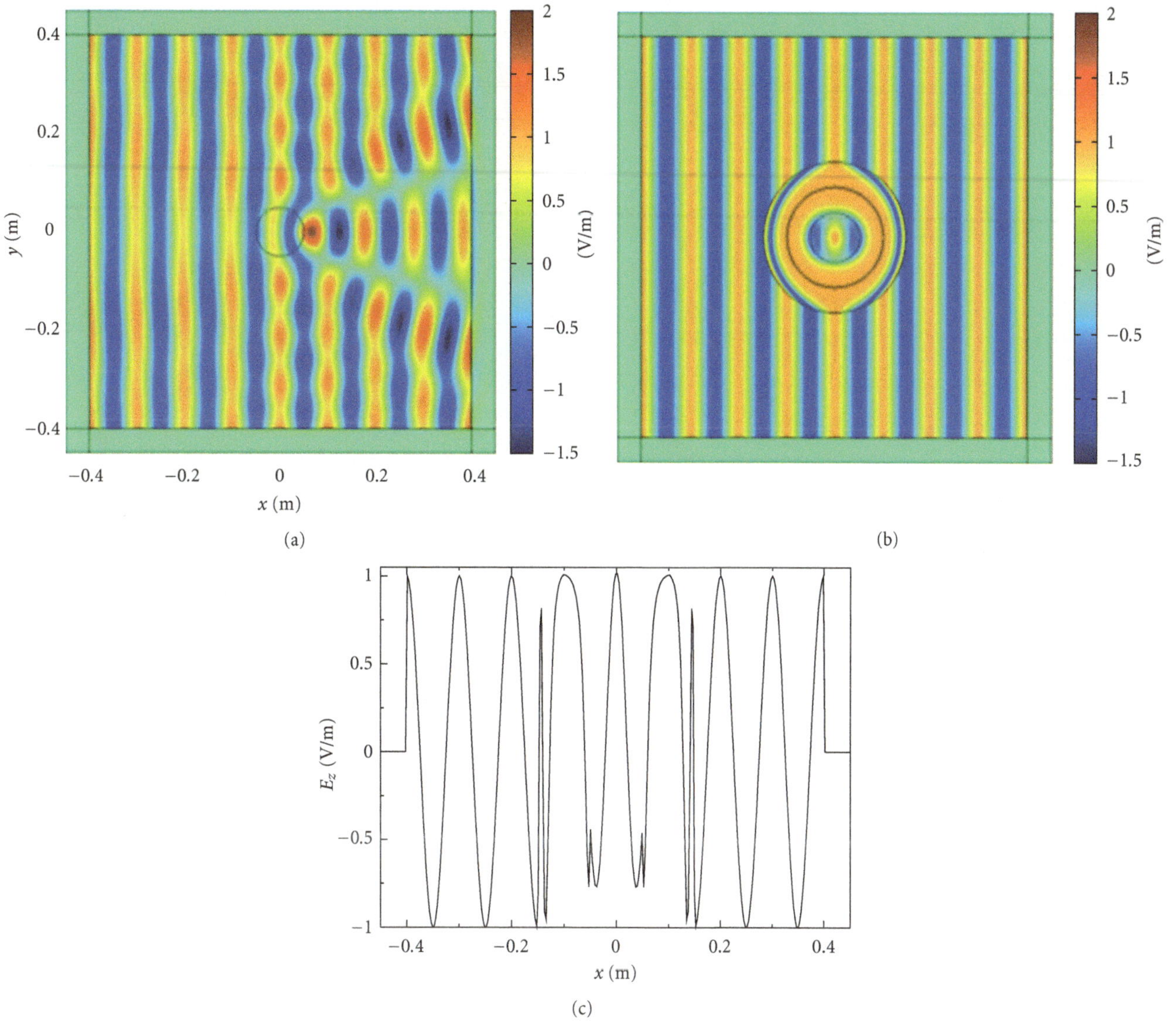

FIGURE 3: Electric field (E_z) distributions in the computational domain. (a) Scattering pattern of a dielectric object with $\varepsilon_{zo} = 2$, $\mu_{ro} = 1$, $\mu_{\varphi o} = 1$. (b) The object is hidden by the reciprocal cloak. (c) Electric field distribution along x axis.

In the recovering layer, we suppose the coordinate transformation equation is $r = f(r'')$, where $f(r'')$ satisfies $f(b) = \delta$, $f(c) = c$. Similarly, according to the preceding procedure, relative permittivity and permeability of the recovering layer can be obtained as

$$\varepsilon_r'' = \mu_r'' = \frac{1}{m_0''},$$
$$\varepsilon_\varphi'' = \mu_\varphi'' = m_0'', \qquad (5)$$
$$\varepsilon_z'' = \mu_z'' = m_0''\left(\frac{r''}{c}\right)^{2(m_0''-1)},$$

where $m_0'' = \log_{(b/c)}(\delta/c)$. With the geometry parameters of $a = 0.05$ m, $b = 0.1$ m, $c = 0.15$ m, constitutive parameters of the reciprocal cloak can be calculated through (4) and (5), as shown in Figure 2. It is seen that the complementary layer is a double negative shell, while all components in the recovering layer are positive. Moreover, both r and φ components in the complementary layer and the recovering layer are constants; only z component is dependent on the radius of the reciprocal cloak.

3. Results and Discussions

In this section, performance of the reciprocal cloak is simulated based on the commercia software COMSOL Multiphysics. We consider the case of transverse-electric (TE) polarization, and only ε_z, μ_r, μ_φ components of the material parameters are required. The incident wave with electric field polarized along z axis is impinging onto the cloak. Frequency of the incident wave is set to be 3 GHz. In the simulation, the whole computational domain is surrounded by a perfectly matched layer that absorbs the outward traveling waves.

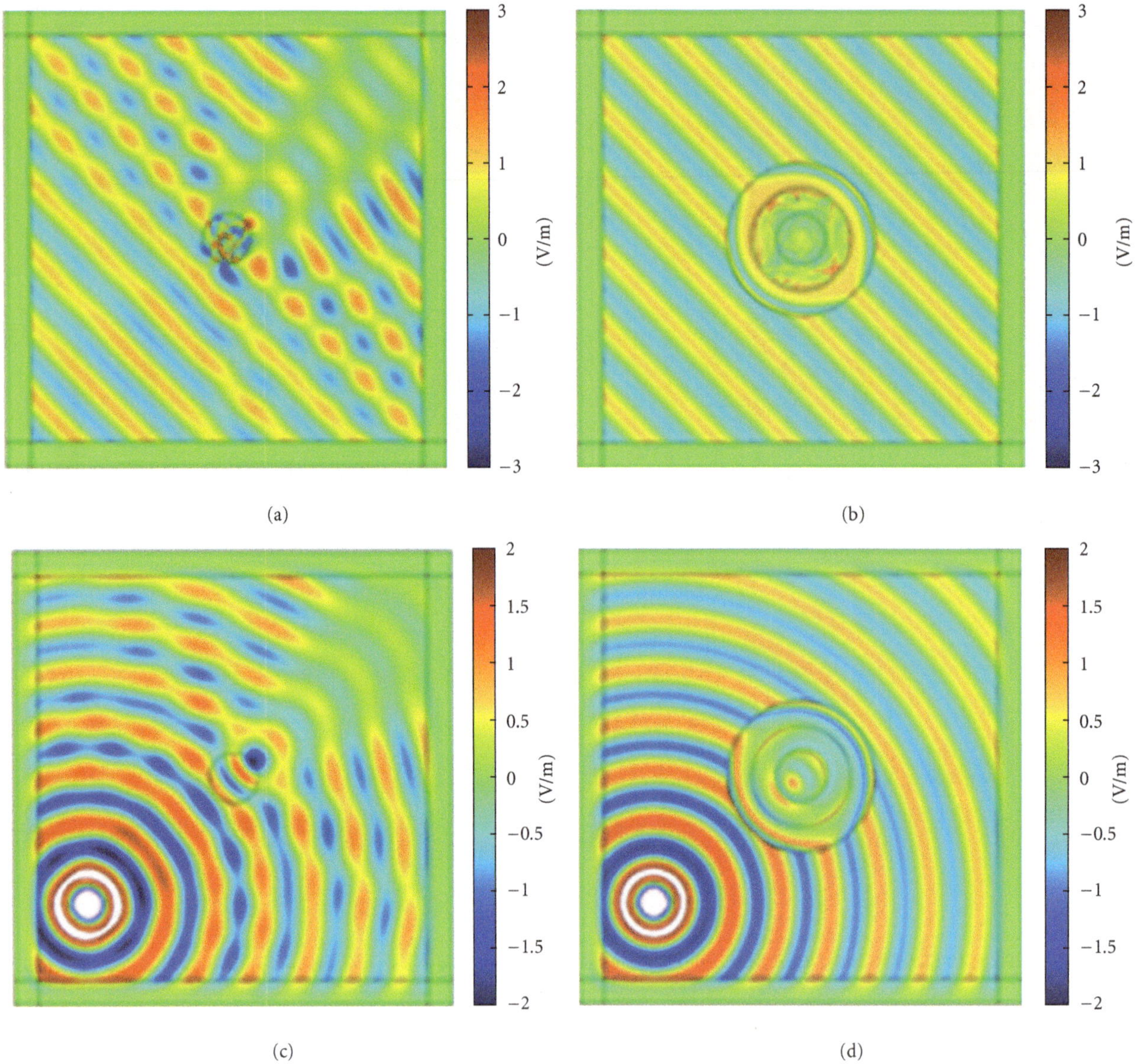

FIGURE 4: Electric field (E_z) distributions in the computational domain. (a) Scattering pattern of the object with anisotropic permeability ($\mu_{ro} = -1$, $\mu_{\varphi o} = 2$). The plane wave is incident at an angle of 45°. (b) The object in (a) is hidden by the reciprocal cloak. (c) Scattering pattern of an object with linearly changing permittivity ($\varepsilon_{zo} = 2r + 3$). The current source with amplitude of 10^{-3}A is located at (−0.3 m, −0.25 m). (d) The object in (c) is hidden by the reciprocal cloak.

Figure 3(a) shows the scattering pattern of a dielectric circular cylinder with material parameters of $\varepsilon_{zo} = 2$, $\mu_{ro} = 1$, $\mu_{\varphi o} = 1$. In order to make it invisible, we put it inside a reciprocal cloak, as shown in Figure 3(b). The incident plane wave with unit amplitude propagates from left to right along x axis. The perfect recovered wavefront demonstrates the effectiveness of the cloak. Figure 3(c) plots the electric field distribution along x axis of the computational domain. We can clearly observe that the electric field intensity inside the cloak is equal to that outside. It indicates that the object enclosed inside the reciprocal cloak can receive information from the outside without distortion, while its presence cannot be detected. It is worth noting that arbitrary object with isotropic or anisotropic material parameters can be cloaked as long as the complementary media layer is "custom-made" according to (4), and the performance of the cloak is independent on the direction of the incident wave and the excitation source. For example, in Figures 4(a) and 4(b), we demonstrate the cloaking of the object with anisotropic permeability $\mu_{ro} = -1$, $\mu_{\varphi o} = 2$. The plane wave is incident at an angel of 45°. Besides, the object with linear changing permittivity, such as $\varepsilon_{zo} = 2r + 3$, can also be cloaked under cylindrical wave irradiation, as shown in Figures 4(c) and 4(d).

Since electromagnetic wave penetrates into the cloaking region, the hidden object such as an antenna or a sensor can receive information from the outside. This indicates that the reciprocal cloak may have potential application for

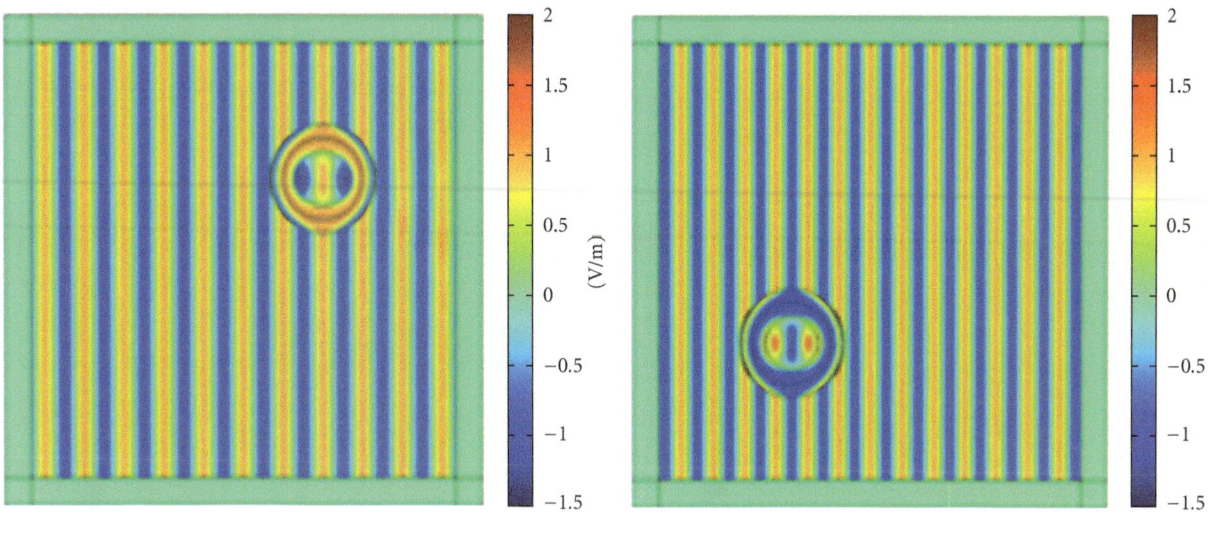

(a) (b)

FIGURE 5: (a) Cloak A with working frequency of 4 GHz. (b) Cloak B with working frequency of 5 GHz.

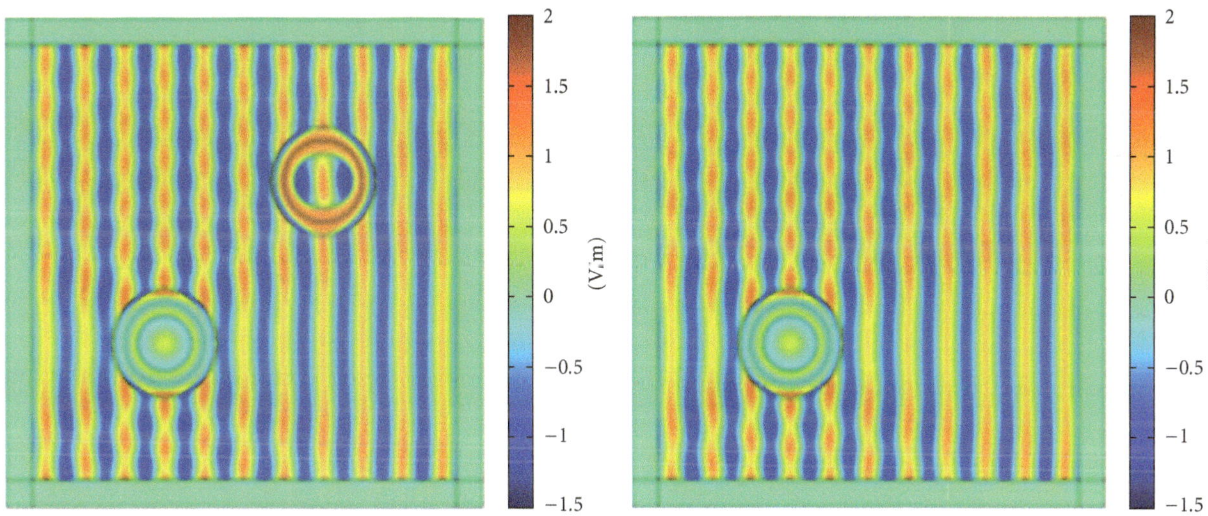

(a) (b)

FIGURE 6: (a) Electric field distribution in the computational domain of the cloaking system. (b) Cloak B at disabled state.

enhancing the electromagnetic compatibility of an electronic device. To demonstrate the EMC application of the cloak, we simulate a cloaking system, which is consisted of two close-ly neighboring reciprocal cloaks. In order to assess the frequency behavior of the cloaking system, the Drude dispersion model has been employed for the double-negative shell of the cloak. In (6), the quantity ω_p is the plasma frequency, and it is chosen such that negative permittivity and permeability are obtained at the respective design frequency. The parameter Γ is the collision frequency representing the losses in the material:

$$\varepsilon(\mu) = 1 - \frac{\omega_p^2}{\omega(\omega + i\Gamma)}. \tag{6}$$

Figures 5(a) and 5(b) show the simulation results of two reciprocal cloaks with working frequency of 4 GHz (Cloak A) and 5 GHz (Cloak B), respectively. Now we put the two cloaks together and set the frequency of the impinging plane wave to be 4 GHz which is the working frequency of cloak A. Electric field distribution of the cloaking system is shown in Figure 6(a). As a comparison, electric field distribution in the computational domain of Cloak B at the disabled state is also simulated and shown in Figure 6(b). It is seen that field patterns in the two figures are the same. That means when Cloak A is at the working state, it will be invisible and it does not influence the field pattern of Cloak B. Contrarily, when Cloak B is at the working state, it will be invisible and it does not influence field pattern of Cloak A. Therefore, the two

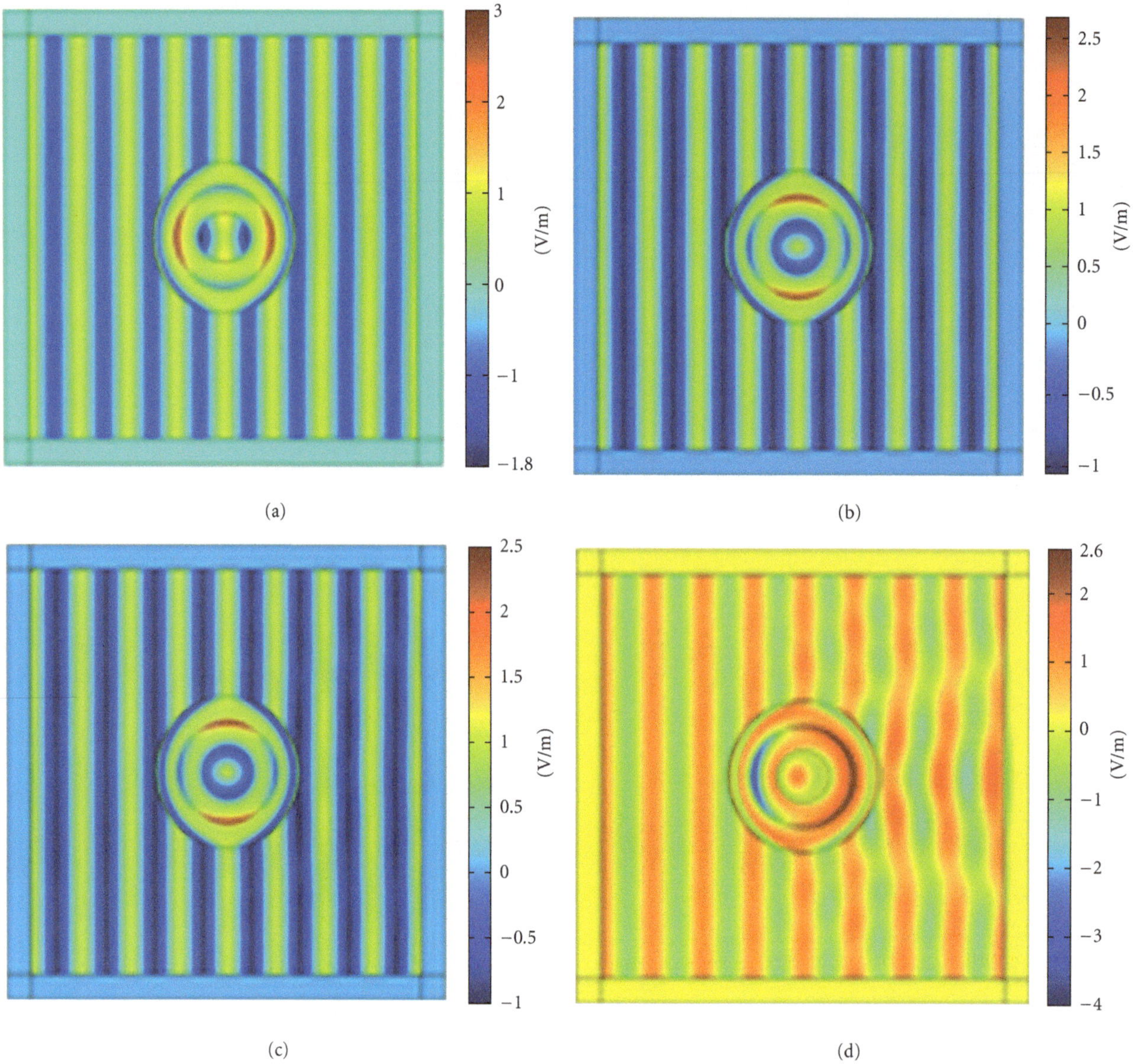

FIGURE 7: The electric field (E_z) distributions in the vicinity of the cloak with loss tangents of 0.001 (a), 0.01 (b), 0.015 (c), and 0.1 (d), respectively.

closely neighboring reciprocal cloaks do not interfere with each other. If the reciprocal cloaks were applied to protect an antenna array, mutual interference could be avoided.

In practical applications, metamaterials are always lossy. Thus, in what follows, we investigate the influence of loss tangent on the performance of the cloak, and the material parameter deviation is also considered. Electric field distribution in the vicinity of the reciprocal cloak with electric and magnetic loss tangents of 0.001, 0.01, 0.015, and 0.1 are displayed in Figures 7(a), 7(b), 7(c), and 7(d). Here, material parameters of the cloaked object are set to be $\varepsilon_{zo} = 2$, $\mu_{ro} = 1$, $\mu_{\varphi o} = 1$. From Figures 7(a)–7(c), we can observe that field distribution is basically undisturbed when loss tangent is less then 0.015; but when the loss tangents of metamaterials is 0.1 or more than that, it deteriorates the performance of the reciprocal cloak especially in the forward scattering region, as shown in Figure 7(d). Therefore, loss tangents of metamaterials being less than or equal to 0.015 is acceptable.

To investigate the influence of material parameter deviation on cloaking effect, we multiply transverse constitutive material parameters (μ_r, μ_φ) for both the complementary media layer and recovering layer by coefficient δ_1, and axial constitutive material parameter (ε_z) by coefficient δ_2. This represents the slight changes in the transverse and axial material parameters, respectively. When there is a slight change in the axial parameter, the following three cases are considered: (i) the transverse parameters are kept invariant, (ii) the impedance ($Z = \sqrt{\mu_\varphi/\varepsilon_z}$ and $\sqrt{\mu_r/\varepsilon_z}$) is kept invariant, (iii) the refractive index ($n = \sqrt{\mu_\varphi \varepsilon_z}$ and $\sqrt{\mu_r \varepsilon_z}$) is kept invariant. The corresponding changes in the transverse parameters can

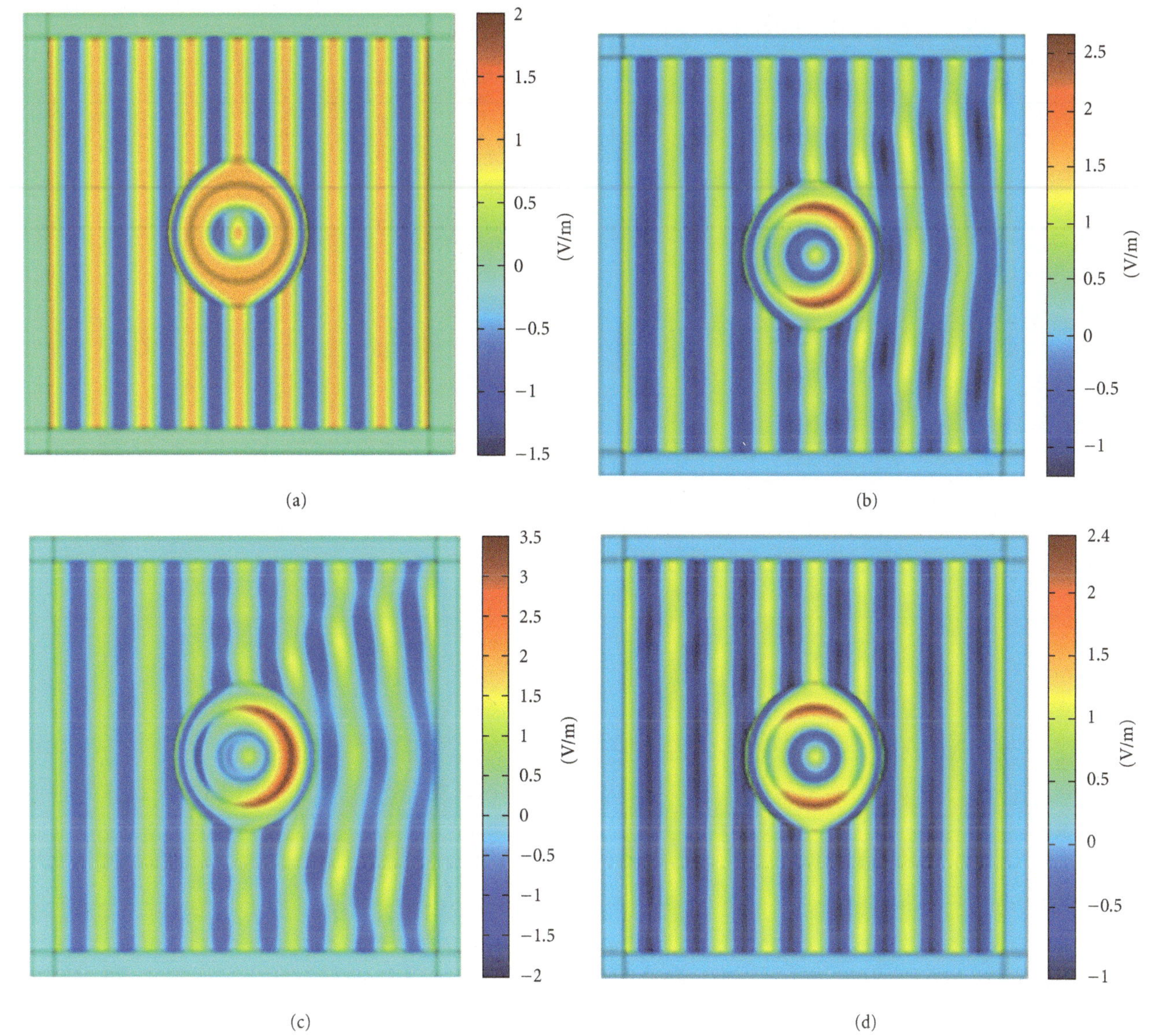

FIGURE 8: Electric field (E_z) distributions in the vicinity of the reciprocal cloak without perturbation (a), with negative perturbation ($\delta_2 = 0.9$) while transverse parameters are invariant ($\delta_1 = 1$) (b), impedances are invariant ($\delta_1 = 0.9$) (c), and refraction index is invariant ($\delta_1 = 1/0.9$) (d).

be easily derived. In the simulation, the negative perturbation ($\delta_2 = 0.9$) and positive perturbation ($\delta_2 = 1.1$) in the axial parameter are taken into consideration, and the material parameters of circular dielectric object to be cloaked is $\varepsilon_{zo} = 2$, $\mu_{ro} = 1$, $\mu_{\varphi o} = 1$. Figures 8(b)–8(d) and Figures 9(b)–9(d) show the corresponding electric field distributions in the vicinity of the reciprocal cloak. Simulation results for the cloak without perturbation are also simulated for comparisons, as shown in Figures 8(a) and 9(a). It is clear that whether the transverse parameters change or not, the performance of the reciprocal cloak will be affected by the positive or negative perturbation in the axial parameters (as shown in Figures 8(b)-8(c), and Figures 9(b)-9(c)). In practice, if we want to keep the cloaking effects of the reciprocal cloak while at the same time minimizing its scattering field, it is the best choice to keep the refractive index invariant (shown in Figures 8(d) and 9(d)).

4. Conclusions

In conclusion, a cylindrical electromagnetic reciprocal cloak with only axial material parameter spatially variant is proposed and designed by means of transformation optics. The EMC ability of the cloak is demonstrated based on full wave simulation. Besides, it is found that the reciprocal cloak is robust against the perturbation in the axial material parameters when the refractive index is kept invariant. The simplicity of the material parameters would put a step further towards the practical realization the reciprocal cloak, which shows some advantages in opening up an avenue for

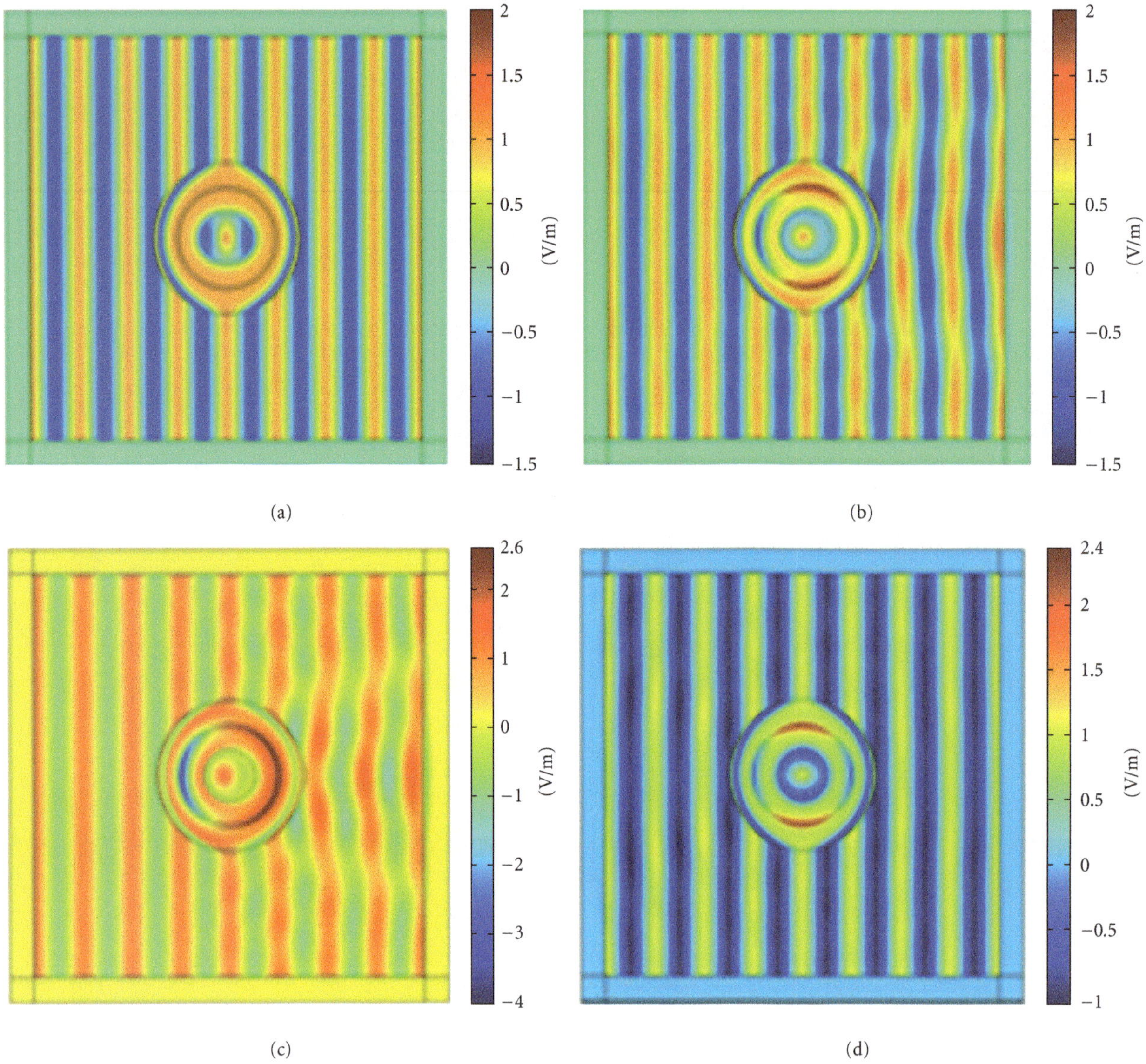

FIGURE 9: Electric field (E_Z) distributions in the vicinity of the reciprocal cloak without perturbation (a), with positive perturbation (δ_2 = 1.1) while transverse parameters are invariant (δ_1 = 1) (b), impedances are invariant (δ_1 = 1.1) (c), and refraction index is invariant (δ_1 = 1/1.1) (d).

designing antennae and sensors with perfect electromagnetic compatibility.

Acknowledgments

This work was supported by the National Natural Science Foundation of China (Grant no. 60861002, 61161007, 61162004), the Natural Science Foundation of Yunnan Province (Grant no. 2011FB018), and NSFC-YN (Grant no. U1037603).

References

[1] V. G. Veselago, "The electrodynamics of substances with simultaneously negative values of ε and μ," *Soviet Physics Uspekhi*, vol. 10, no. 4, pp. 509–514, 1968.

[2] J. B. Pendry, "Negative refraction makes a perfect lens," *Physical Review Letters*, vol. 85, no. 18, pp. 3966–3969, 2000.

[3] J. J. Yang, M. Huang, and J. Sun, "Double negative metamaterial sensor based on microring resonator," *IEEE Sensors Journal*, vol. 11, no. 10, pp. 2254–2259, 2011.

[4] M. Huang, J. Yang, S. Jun, S. Mu, and Y. Lan, "Simulation and analysis of a metamaterial sensor based on a microring resonator," *Sensors*, vol. 11, no. 6, pp. 5886–5899, 2011.

[5] M. Huang, J. Yang, J. Sun, J. Shi, and J. Peng, "Modelling and analysis of ω-shaped double negative material-assisted microwave sensor," *Journal of Infrared, Millimeter, and Terahertz Waves*, vol. 30, no. 11, pp. 1131–1138, 2009.

[6] A. Azarbar and J. Ghalibafan, "A compact low-permittivity dual-layer EBG structure for mutual coupling reduction," *International Journal of Antennas and Propagation*, vol. 2011, Article ID 237454, 6 pages, 2011.

[7] J. B. Pendry, D. Schurig, and D. R. Smith, "Controlling electromagnetic fields," *Science*, vol. 312, no. 5781, pp. 1780–1782, 2006.
[8] U. Leonhardt, "Optical conformal mapping," *Science*, vol. 312, no. 5781, pp. 1777–1780, 2006.
[9] D. Schurig, J. J. Mock, B. J. Justice et al., "Metamaterial electromagnetic cloak at microwave frequencies," *Science*, vol. 314, no. 5801, pp. 977–980, 2006.
[10] X. Liu, C. Li, K. Yao et al., "Experimental verification of broadband invisibility using a cloak based on inductor-capacitor networks," *Applied Physics Letters*, vol. 95, no. 19, Article ID 191107, 2009.
[11] P. Alitalo and S. Tretyakov, "Electromagnetic cloaking with metamaterials," *Materials Today*, vol. 12, no. 3, pp. 22–29, 2009.
[12] U. Leonhardt and T. Philbin, *Geometry and Light: The Science of Invisibility*, Dover Publications, Mineola, NY, USA, 2010.
[13] U. Leonhardt and T. G. Philbin, "General relativity in electrical engineering," *New Journal of Physics*, vol. 8, article no. 247, 2006.
[14] J. Yang, M. Huang, C. Yang, Z. Xiao, and J. Peng, "Metamaterial electromagnetic concentrators with arbitrary geometries," *Optics Express*, vol. 17, no. 22, pp. 19656–19661, 2009.
[15] J. J. Yang, T. H. Li, M. Huang, and M. Cheng, "Transparent device with homogeneous material parameters," *Applied Physics A*, vol. 104, no. 2, pp. 733–737, 2011.
[16] W. X. Jiang, J. Y. Chin, and T. J. Cui, "Anisotropic metamaterial devices," *Materials Today*, vol. 12, no. 12, pp. 26–33, 2009.
[17] H. Chen, C. T. Chan, and P. Sheng, "Transformation optics and metamaterials," *Nature Materials*, vol. 9, no. 5, pp. 387–396, 2010.
[18] A. E. Dubinov and L. A. Mytareva, "Invisible cloaking of material bodies using the wave flow method," *Physics-Uspekhi*, vol. 53, no. 5, pp. 455–479, 2010.
[19] H. Ma, S. Qu, Z. Xu, and J. Wang, "The open cloak," *Applied Physics Letters*, vol. 94, no. 10, Article ID 103501, 2009.
[20] T. C. Han, C. W. Qiu, and X. H. Tang, "Creating rigorous open cloaks," *Journal of Electromagnetic Waves and Applications*, vol. 24, no. 13, pp. 1839–1847, 2010.
[21] T. Han, C. Qiu, and X. Tang, "The general two-dimensional open-closed cloak with tunable inherent discontinuity and directional communication," *Applied Physics Letters*, vol. 97, no. 12, Article ID 124104, 2010.
[22] Y. Lai, H. Chen, Z. Q. Zhang, and C. T. Chan, "Complementary media invisibility cloak that cloaks objects at a distance outside the cloaking shell," *Physical Review Letters*, vol. 102, no. 9, Article ID 093901, 2009.
[23] C. Yang, J. Yang, M. Huang, Z. Xiao, and J. Peng, "An external cloak with arbitrary cross section based on complementary medium and coordinate transformation," *Optics Express*, vol. 19, no. 2, pp. 1147–1157, 2011.
[24] H. Chew and M. Kerker, "Abnormally low electromagnetic scattering cross sections," *Journal of the Optical Society of America*, vol. 66, no. 5, pp. 445–449, 1976.
[25] A. Alù and N. Engheta, "Achieving transparency with plasmonic and metamaterial coatings," *Physical Review E*, vol. 72, no. 1, Article ID 016623, 2005.
[26] L. Gao, T. H. Fung, K. W. Yu, and C. W. Qiu, "Electromagnetic transparency by coated spheres with radial anisotropy," *Physical Review E*, vol. 78, no. 4, Article ID 046609, 2008.
[27] Y. Ni, L. Gao, and C. W. Qiu, "Achieving invisibility of homogeneous cylindrically anisotropic cylinders," *Plasmonics*, vol. 5, no. 3, pp. 251–258, 2010.
[28] J. J. Yang, M. Huang, C. F. Yang, and J. Yu, "Reciprocal invisibility cloak based on complementary media," *European Physical Journal D*, vol. 61, no. 3, pp. 731–736, 2011.
[29] B. Wood, J. B. Pendry, and D. P. Tsai, "Directed subwavelength imaging using a layered metal-dielectric system," *Physical Review B*, vol. 74, no. 11, Article ID 115116, 2006.
[30] Y. Huang, Y. Feng, and T. Jiang, "Electromagnetic cloaking by layered structure of homogeneous isotropic materials," *Optics Express*, vol. 15, no. 18, pp. 11133–11141, 2007.

25

Reconfigured and Notched Tapered Slot UWB Antenna for Cognitive Radio Applications

Tamer Aboufoul, Akram Alomainy, and Clive Parini

School of Electronic Engineering and Computer Science, Queen Mary University of London, Mile End Road, London E1 4NS, UK

Correspondence should be addressed to Tamer Aboufoul, ta302@eecs.qmul.ac.uk

Academic Editor: Hala A. Elsadek

A compact reconfigurable and notched ultra-wideband (UWB) tapered slot antenna (TSA) is presented. The antenna reconfiguration operation principle relies on 2 mechanisms: in the first mechanism a resonator parasitic microstrip line electrically coupled to the TSA is used to notch the TSA at a specific frequency and the second mechanism relies on changing the input impedance matching of the antenna by means of changing the length of a stub line extended from an additional tiny partial ground on the back side of the antenna. The reflection coefficient, radiation patterns, and gain simulations and measurements for the proposed antenna are presented to verify the design concepts featuring a very satisfactory performance. Total efficiency simulations and measurements are also presented to highlight the filtering performance of the reconfigured antenna. When the antenna was reconfigured from the UWB to work into multiple frequency bands, the radiation patterns were still the same and the total peak gain has slightly improved compared to the UWB case. In addition, when the antenna operated in the notched mode, the gain has significantly dropped at the notch frequency. The simplicity and flexibility of the proposed multimode antenna make it a good candidate for future cognitive radio front ends.

1. Introduction

Reconfigurable UWB antennas can be deployed in cognitive radio (CR) front ends by reconfiguring UWB antenna to work into multiple predefined frequency bands. When the antenna is in the UWB mode, it can be used for sensing the spectrum and when it is operating in the reconfigured modes, the antenna can be used for communication purpose [1–4]. Several designs of UWB antenna with band rejection characteristics have also targeted the cognitive radio applications [5, 6]. The proposed reconfigurable antenna here is mixing both approaches (e.g., reconfiguration and band rejection in the same antenna design). The reference antenna which will be reconfigured is presented in [7]. This antenna features wide operating bandwidth, very simple and compact structure, ease of fabrication, good radiation patterns over the entire bandwidth, and good time domain performance. In the proposed antenna design, reconfigurability is introduced by employing stubs and changing the electrical length of the stubs by using multiple switches. These stubs are connected to an additional tiny partial ground printed on the back side of the tapered slot antenna (TSA). When the stub is disconnected from the ground it can be used as a parasitic resonator to notch the UWB antenna at specific unwanted frequency. In Section 2, the antenna design is presented and the operation of the antenna is briefly discussed. In Section 3 we show the simulated and measured results for the performance of the antenna. In Section 4 we briefly discuss the applicability of this new design in cognitive radio applications. Finally, conclusions are drawn and presented in Section 5.

2. Antenna Design and Operation

Figure 1 shows the geometry of the proposed reconfigured UWB TSA antenna. The proposed antenna substrate is a Taconic board with thickness of 1.52 mm, relative dielectric constant of 3, and loss tangent of 0.003. The width of the antenna is 16 mm ($0.28\lambda_g$) and the length of the antenna is 27 mm ($0.47\lambda_g$) where λ_g is the guided wavelength at 3 GHz. On the back side of the substrate, a small partial ground (12 mm $\times$ 1 mm) is printed and there is a stub

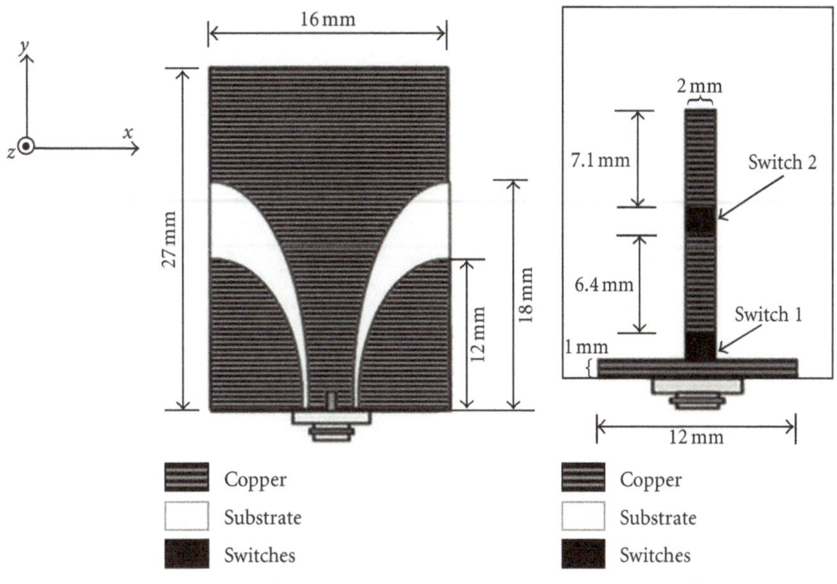

(a) (b)

FIGURE 1: Geometry of the reconfigured tapered slot antenna: (a) top view, (b) bottom view.

(a) (b)

FIGURE 2: Antenna prototype: (a) top view, (b) bottom view.

connected to this ground by switch 1. By changing the length of the stub, we can change the input impedance matching of the antenna to work in predefined frequency bands. The length of the ground stub is determined from circuit analysis using ADS [8]. When the stub length is 8.4 mm, the antenna should be working between 3 and 4 GHz. When the stub length increases to 17.5 mm when both switches 1 and 2 are ON, the frequency band of operation will change to 4.5–5.5 GHz. When the ground stub is not connected to the partial ground (e.g., switch 1 is OFF and switch 2 is ON), the stub will act as a parasitic element electrically coupled to the TSA. The length of the parasitic resonator determines the notch frequency and if we want to notch the upper band of WLAN at 5.8 GHz, the length of the resonator should be around 15 mm which is equal to $\lambda_g/2$ where λ_g is the guided wavelength at 5.8 GHz. Full UWB operation without any notches can be achieved if we can kill the resonance coupling of the parasitic element (e.g., the total length of the parasitic resonator is changed when switch 2 is OFF); consequently, it cannot support the resonating currents and no notch occurs at 5.8 GHz. The operation modes of the antenna for different switch states are summarized in Table 1.

TABLE 1: Operation modes of the antenna.

Switch 1	Switch 2	
	ON	OFF
ON	Band II	Band I
OFF	Notched-UWB	UWB

A prototype antenna was fabricated to verify the performance. Figure 2 shows the proposed prototype antenna. RF switches used in this prototype are realized as metal pads with dimensions 2 mm × 2 mm. Although this model is ideal, it gives a good approximation for the commercial PIN diode switches.

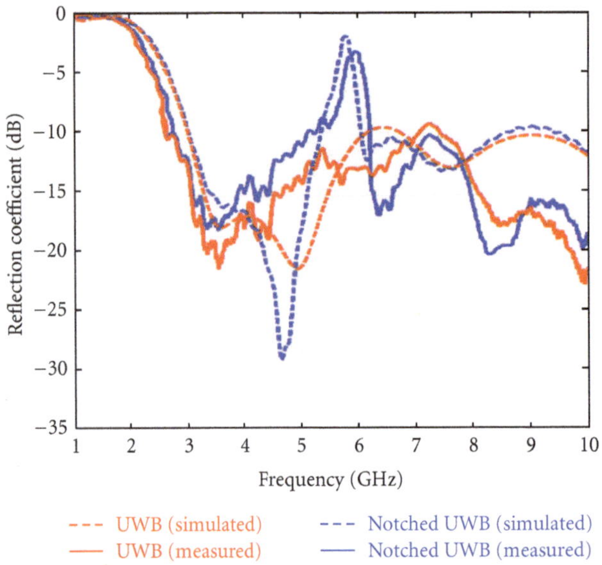

FIGURE 3: Simulated (dashed lines) and measured (solid lines) reflection coefficient curves when antenna is in UWB mode and in the notched UWB mode.

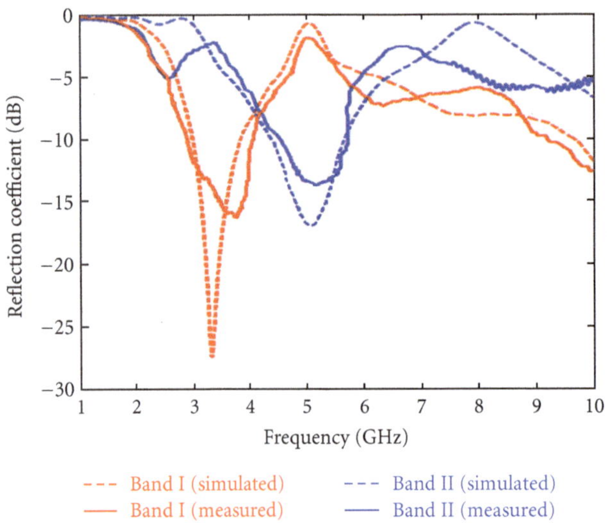

FIGURE 4: Simulated (dashed lines) and measured (solid lines) reflection coefficient curves when antenna is reconfigured in Band I: 3-4 GHz and Band II: 4.5–5.5 GHz.

3. Simulation and Measurements Results

3.1. Reflection Coefficient. All the simulations in this paper are performed using the CST microwave studio package which utilizes the finite integration technique for electromagnetic computation [9]. Figure 3 shows the simulated results for the reflection coefficient of the antenna when it is in the UWB mode and when it is notched around 5.8 GHz. Figure 4 depicts the simulated reflection coefficient when the antenna is configured to work into Band I and Band II. In general, a good agreement between simulated and measured results has been achieved. Small frequency shift (<200 MHz) between the simulated and measured results has occurred; this can be accounted for by fabrication and material tolerances.

3.2. Surface Current. Figures 5(a) and 5(b) show the surface current distributions for the notched UWB antenna at (3.4 GHz) in the passband and at the stopband (5.8 GHz) respectively. At the stop band, the surface currents are condensed around the parasitic resonator and a strong coupling occurs (the direction of current flow on the resonator is opposite to the flow of the current on the nearby slots edges). Eventually, the surface current is largely reduced along the edges of the 2 slots and no radiation occurs, while in the pass band frequency (3.4 GHz) the current flows smoothly on the edges of the slots without any perturbation by the resonator and hence radiation is as normal. Figure 6(a) shows the surface current distributions for the UWB antenna when it is operating in Band I at (3.4 GHz) in the passband, and Figure 6(b) shows the surface current distributions for the

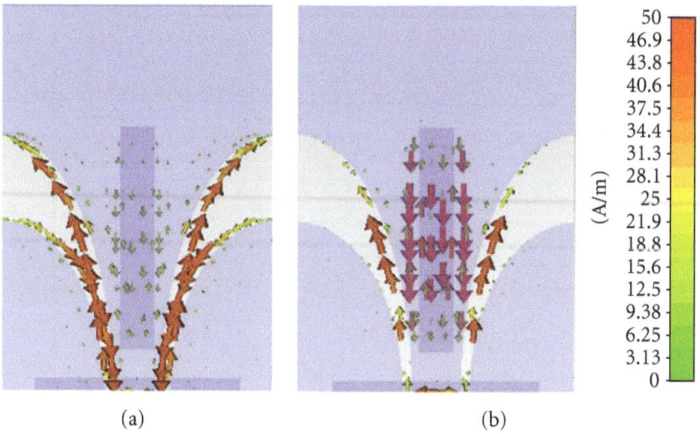

FIGURE 5: Simulated current distribution of the notched UWB antenna at (a) 3.4 GHz (normal operation frequency) and (b) 5.8 GHz (notched frequency).

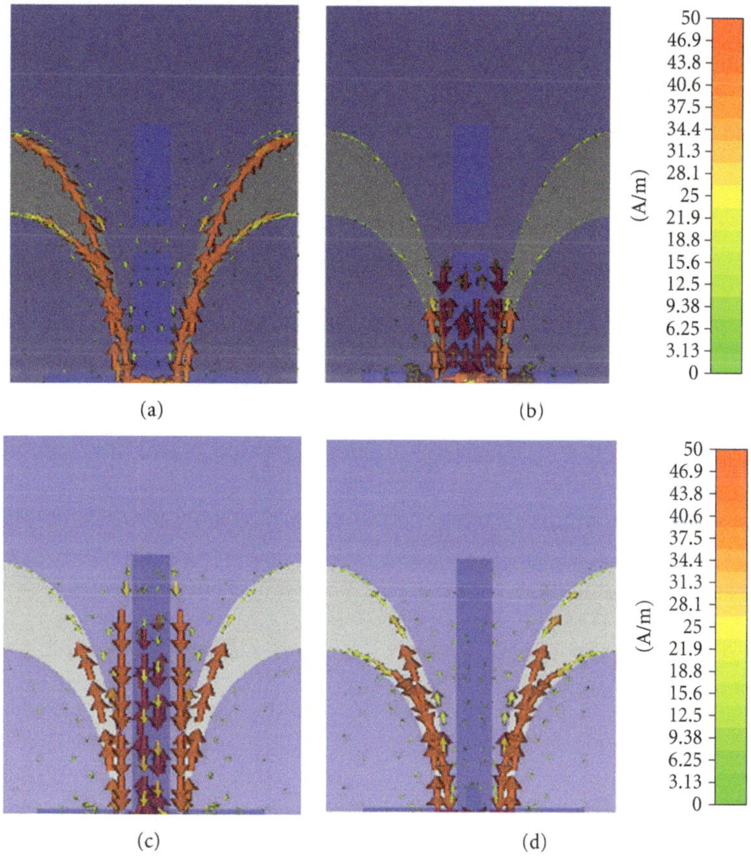

FIGURE 6: Simulated current distribution of the reconfigured UWB antenna in Band I: (a) 3.4 GHz, (b) 5 GHz, Band II: (c) 3.4 GHz, (d) 5 GHz.

UWB antenna when it is operating in the same band but at 5 GHz which is out of band frequency, respectively. Figures 6(c) and 6(d) show the surface current distributions for the UWB antenna when it is operating in Band II at (3.4 GHz) out of band frequency and Band II at (5 GHz) which is in the passband, respectively. At the out of band frequency, the stub which is connected to the partial ground will change the input impedance of the antenna and the surface current will be more dominant around the stub while the surface current along the slots of the antenna is reduced. The effect of the stubs on matching the input impedance of the antenna can also be highlighted from Figure 7 which depicts the input impedance of the UWB antenna when it is reconfigured to work in Band I. For the UWB operation, the antenna input resistance (R_i) should be around 50 Ω and the input reactance (X_i) should be close to 0 Ω for the whole operating bandwidth. After reconfiguring the antenna to work into Band I, R_i is close to 50 Ω and X_i is around 0 Ω only at

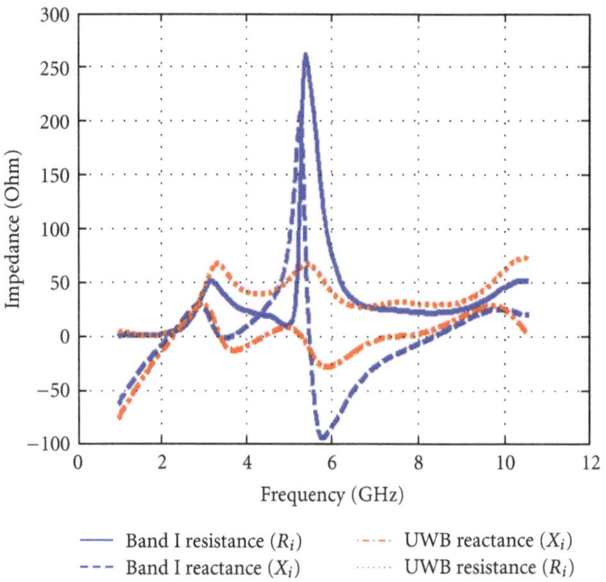

FIGURE 7: Simulated input impedance curves for the UWB antenna when the antenna is configured to operate in Band I.

TABLE 2: Simulated (S) and measured (M) realized gain.

Frequency (GHz)	Gain (dBi)							
	UWB		Notched UWB		Band I		Band II	
	S	M	S	M	S	M	S	M
3.4	1.9	2.4	1.9	2.5	2.1	3	0.8	0.7
5	2.9	3	3.0	3	−6.4	−4	3.7	3.5
5.8	3.6	3	−7	−4	1.9	2	3.8	3.7

TABLE 3: Measured total efficiency.

Frequency (GHz)	Total efficiency (%)			
	UWB	Notched UWB	Band I	Band II
3.4	87	88	90	40
5	90	87	30	91
5.8	85	45	55	88

frequencies within the operating bandwidth (e.g., 3.4 GHz) and their values largely deviate from the nominal values outside the passband.

3.3. Radiation Patterns and Gain. The simulated and measured normalized radiation patterns at 3.4 and 5 GHz for the UWB mode operation are plotted in Figures 8(a) and 8(b), respectively. The simulated and measured normalized radiation patterns when the antenna is configured in Band I at 3.4 and in Band II at 5 GHz are plotted in Figures 9(a) and 9(b), respectively. The *H*-plane patterns are omnidirectional and the *E*-plane patterns look like number "8" like a traditional monopole.

For radiation patterns measurements, the quality of the far field anechoic chamber is critically important. Slight misalignment uncertainties in the antenna under test (AUT) positioning can lead to discrepancies between simulated and measured radiation patterns. Moreover, the effects of the connectors and cables connected to the AUT were not simulated in the computational domain in spite of their capability to add some reflection in the chamber and deteriorate the perfect line-of-sight (LOS) assumption and add some ripples on the radiation patterns.

It is worth mentioning here that the radiation patterns did not change when we reconfigure from the UWB operation to the reconfigured cases. Table 2 summarizes a comparison of the simulated and measured realized peak gain between the UWB case, the notched UWB, and the 2 reconfigured cases. At 5.8 GHz (notch frequency), the UWB notched antenna showed hugely reduced broadside gain compared to the UWB case and the reconfigured cases show slight increase in peak gain for passband frequencies compared to the UWB case.

3.4. Efficiency. Radiation efficiency is the ratio of input power that ends up as radiated power to the net power accepted by the antenna and the total efficiency is the ratio of the total power radiated from the antenna to the total net power applied at the antenna input terminal. Total efficiency results usually give better indication than return loss results to highlight the filtering performance of the antenna; this is because the antenna total efficiency not only accounts for the input mismatch of the antenna but also accounts for the ohmic losses in the antenna. Figure 10 depicts the simulated values of the total efficiency, when the antenna is operating in the UWB mode and the notched UWB mode and when the antenna is configured to work in Band I and Band II. The total efficiency was measured using the modified wheeler cap methods as in [10–12]. Table 3 shows the measured total efficiency when the antenna is operating in the UWB mode and the notched UWB mode and when the antenna is configured to work in Band I and Band II. The measured values agree well with the simulated ones. From Figure 10 and Table 3 we can see how poor the total efficiency gets at

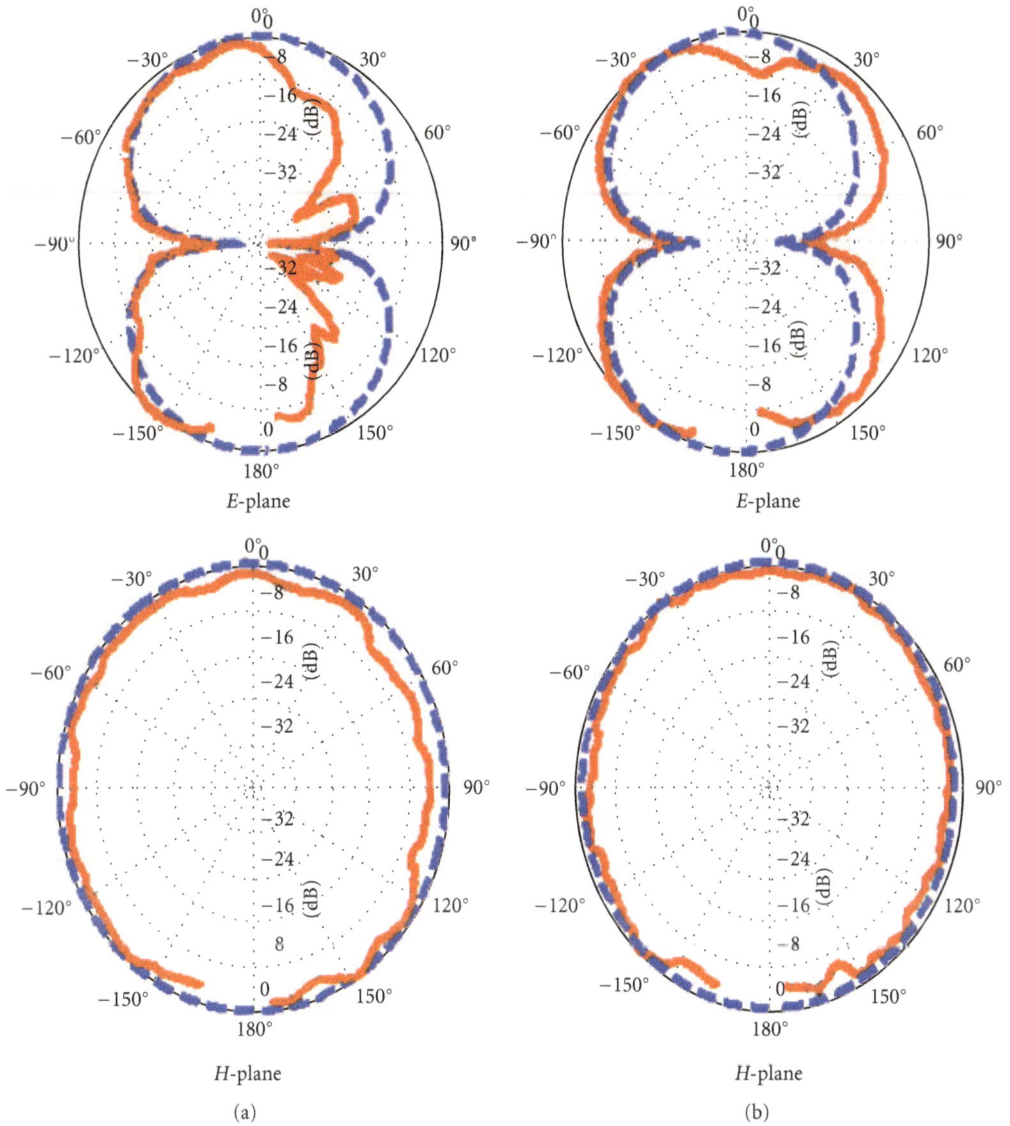

FIGURE 8: Simulated (dashed blue line) and measured (solid red line) normalised E-plane (X-Y) and H-plane (X-Z) radiation patterns, when antenna is in the UWB mode: (a) 3.4 GHz and (b) 5 GHz.

the notch frequency in UWB notched case or at out-of-band frequencies in the reconfigured bands I and II.

4. Applicability to Cognitive Radio

The previously discussed antenna can be easily integrated into cognitive radio; when the antenna is operating in the UWB mode, it can sense the whole spectrum to find the spectrum holes and then from sensing information it will decide to switch to work in one of the available predefined bands (e.g., Band I or II). Usually sensing places severe requirements on sensitivity, linearity, and dynamic range of the cognitive radio RF front end. This is because the RF signal presented at the antenna of a cognitive radio includes signals from close or separated transmitters and from transmitters operating at different power levels [13]. Detection of weak signals might frequently be performed in the presence of very strong signals. Thus, to relax the strict requirements placed on the RF analogue circuits, it would be beneficial from a system point of view if we can mitigate an interferer before it saturate our RF front end. This can be done by using the UWB notched mode during sensing. It is worth mentioning here that not only this notched UWB mode can be good for sensing, but also it can be helpful in communication between cognitive secondary users who can still communicate more efficiently using the notched UWB channel (e.g., more allowed transmitted power and longer transmission range with higher data rate) without affecting or being affected by a specific legacy primary user who is currently operating at the notched frequency band.

5. Conclusion

A reconfigurable UWB tapered slot antenna (TSA) has been demonstrated and reconfigurability is introduced by

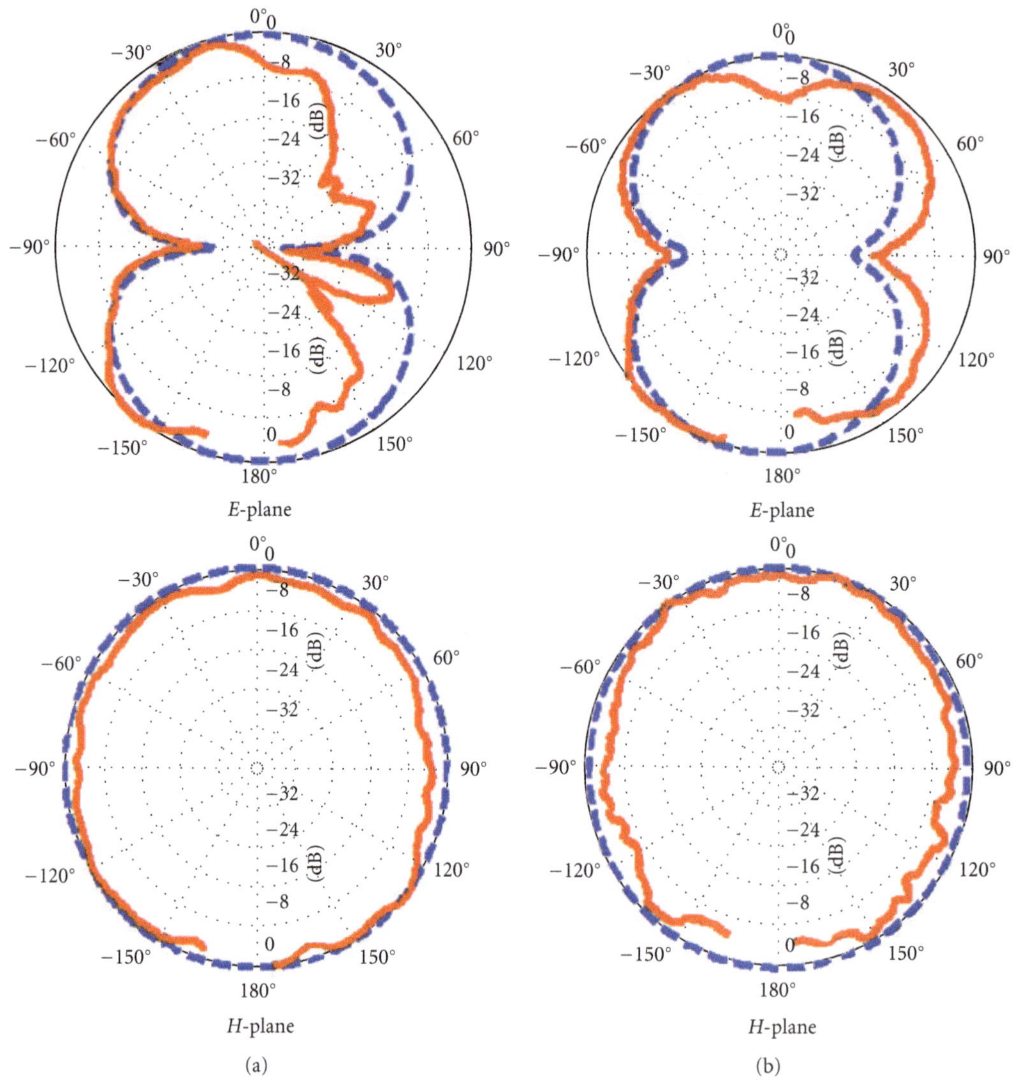

FIGURE 9: Simulated (dashed blue line) and measured (solid red line) normalised E-plane (X-Y) and H-plane (X-Z) radiation patterns, when the antenna is configured at: (a) 3.4 GHz in Band I and (b) 5 GHz in Band II.

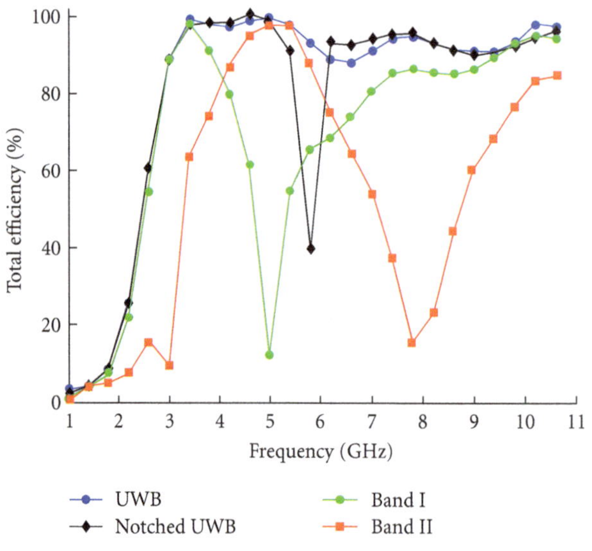

FIGURE 10: Simulated total efficiency when the antenna is operating in different modes.

varying the length of a stub connected to an additional tiny partial ground on the back of the antenna. The proposed topology is versatile in terms of the availability of different reconfiguration bands and the antenna is very compact and simple. Another interesting feature of the antenna is that this ground stub can be used as a parasitic element electrically coupled to the TSA antenna to notch the UWB operation at a specific frequency. The simulated and measured reflection coefficient, radiation patterns, and gain for the UWB mode, the notched UWB mode, and the reconfigured Bands I and II were presented and they all showed a very good performance. Applications requiring frequency band switching such as cognitive radio could benefit from the proposed reconfigurable TSA antenna as we can use the UWB mode or the notched UWB mode in sensing and the reconfigured bands for communication purpose.

References

[1] J. R. Kelly, P. S. Hall, and P. Gardner, "Integrated wide-narrow band antenna for switched operation," in *Proceedings of the 3rd European Conference on Antennas and Propagation (EuCAP '09)*, pp. 3757–3760, March 2009.

[2] T. Aboufoul and A. Alomainy, "Reconfigurable printed UWB circular disc monopole antenna," in *Proceedings of the Antennas & Propagation Conference (LAPC '11)*, pp. 1–4, Loughborough, UK, November 2011.

[3] T. Aboufoul and A. Alomainy, "Single-element reconfigurable planar ultra wideband antenna for cognitive radio front end," in *Proceedings of the 4th International Conference on Cognitive Radio and Advanced Spectrum Management (CogART '11)*, Barcelona, Spain, October 2011.

[4] T. Aboufoul, A. Alomainy, and C. Parini, "Reconfiguring UWB monopole antenna for cognitive radio applications using GaAs FET switches," *IEEE Antennas and Wireless Propagation Letters*, vol. 11, pp. 392–394, 2012.

[5] X. Artiga, J. Perruisseau-Carrier, P. Pardo-Carrera, I. Llamas-Garro, and Z. Brito-Brito, "Halved Vivaldi antenna with reconfigurable band rejection," *IEEE Antennas and Wireless Propagation Letters*, vol. 10, pp. 56–58, 2011.

[6] J. Perruisseau-Carrier, P. Pardo-Carrera, and P. Miskovsky, "Modeling, design and characterization of a very wideband slot antenna with reconfigurable band rejection," *IEEE Transactions on Antennas and Propagation*, vol. 58, no. 7, pp. 2218–2226, 2010.

[7] A. Rahman and Y. Hao, "A novel tapered slot CPW-fed antenna for ultra-wideband applications and its on/off-body performance," in *Proceedings of the IEEE International Workshop on Antenna Technology (iWAT '07)*, pp. 503–506, Cambridge, UK, March 2007.

[8] Advanced Design System (ADS), "Agilent Technologies," Santa Clara, Calif, USA, 2005, http://www.agilent.com.

[9] CST-Microwave Studio, "User's Manual," 2011.

[10] H. A. Wheeler, "The radian sphere around a small antenna," in *Proceedings American Institute of Electrical Engineers (IRE '59)*, vol. 47, pp. 1325–1331, August 1959.

[11] H. G. Schantz, "Measurement of UWB antenna efficiency," in *Proceedings of the IEEE VTS 53rd Vehicular Technology Conference (VTS SPRING '01)*, vol. 2, pp. 1189–1191, May 2001.

[12] M.-C. Huynh, *Wideband compact antennas for wireless communication [Ph.D. dissertation]*, Department of Electrical and Computer Engineering, Virginia Polytechnic Institute, 2004.

[13] D. Cabric, S. M. Mishra, and R. W. Brodersen, "Implementation issues in spectrum sensing for cognitive radios," in *Proceedings of the Conference Record of the 38th Asilomar Conference on Signals, Systems and Computers*, vol. 1, pp. 772–776, November 2004.

26

Novel Design of Electromagnetic Bandgap Using Fractal Geometry

Huynh Nguyen Bao Phuong,[1] Dao Ngoc Chien,[2] and Tran Minh Tuan[3]

[1] *School of Electronics and Telecommunications, Hanoi University of Science and Technology, Hanoi 10000, Vietnam*
[2] *Department of High Technology, Ministry of Science and Technology, Hanoi 10000, Vietnam*
[3] *National Institute of Information and Communications Strategy, Ministry of Information and Communications, Hanoi 10000, Vietnam*

Correspondence should be addressed to Huynh Nguyen Bao Phuong; phuong.huynhnguyenbao@hust.edu.vn

Academic Editor: Eric Lheurette

A novel electromagnetic bandgap (EBG) structural design based on Fractal geometry is presented in this paper. These Fractals, which are the Sierpinski triangles, are arranged to repeat each 60° to produce the hexagonal unit cells. By changing the gap between two adjacent Sierpinski triangles inside EBG unit cell, we can produce two EBG structures separately that have broadband and dual bandgap. By using the suspending microtrip method, two arrays 3 × 4 of EBG unit cells are utilized to investigate the bandgap of the EBG structures. The EBG operation bandwidth of the broadband structure is about 87% and of the dual-band structure is about 40% and 35% at the center bandgap frequencies, respectively. Moreover, a comparison between the broadband EBG and the conventional mushroom-like EBG has been done. Experimental results of the proposed design show good agreement in comparison with simulation results.

1. Introduction

Recently, an intensive interest in metamaterials, such as high impedance surface (HIS) composed of metallic or dielectric unit cells, can be observed due to their unique characteristics in antennas and microwave circuits applications [1]. In general, HIS has two unique properties included of electromagnetic bandgap (EBG) and artificial magnetic conductor (AMC). One of the most important properties of EBG structure is to prohibit the propagation of the surface wave [2, 3]. Practical applications of EBG structure usually have difficulty in accommodating its physical size, because the period of EBG lattices has to be a half-wavelength at the bandgap frequency. This problem had not been solved until the mushroom-like EBG was proposed by Sievenpiper et al. [4]. Then, several of novel EBG structures were presented such as spiral EBG, uniplanar compact EBG (UC-EBG), and fork-like EBG. These structures have several advantages, such as compact size low loss. However, due to their resonant property, all of those structures are narrowband which makes them not practical for use in broadband applications. It is a real challenging task to create an EBG with wide bandwidth.

Several broadband EBG structures were found in the literature. Typically, there have been two approaches aiming to obtain EBG structures with wider bandwidth: the use of EBGs with via-holes [5] (with the inconveniences of complex and expensive manufacturing process) and the adoption of multilayered FSSs over a metallic ground plane or multiperiod mushroom-like structure [6] (which yields less compact designs and is rather expensive). Recent research efforts focus on the development of planar EBG that does not need vias and that can be integrated antenna to enhance the gain and reduce the backward radiation and increasing efficiency [7–9].

As [10], several of Fractal shapes can be used for antennas includes: (1) the von Koch curve, (2) the Minkowski curve, (3) the Hilbert curve, (4) the Fractal tree, (5) the Sierpinski (gasket and carpet) Fractals, and (6) the Cantor set. These geometries also can be used in order to create EBG structures. However, only the Fractal shapes that are used for multiband antennas are able to form broader bandgap EBG. Actually, when an EBG structure has multiband, the broadband can be done by overlapping the different single-band to each other. In literature, Sierpinski (gasket and carpet) structures are widely used to create multi-band antenna [10]. From the idea of this design that is produced a hexagonal EBG with Fractal geometry, due to the characteristic of an equilateral triangle, Sierpinski gasket is chosen as a metal patch on the top surface in order to create the hexagonal EBG.

In this paper, the advantage of the novel hexagonal EBG structure was found by investigating in case of changing the gap between two adjacent Sierpinski triangles inside EBG unit cell. As a result, two structures are proposed; in which one structure introduces broad bandwidth and the others introduce dual bandgap. The design of EBG structures involving the mode-2 Sierpinski Gasket triangles (see Figure 1) and without via-holes or multilayer substrate is presented. The proposed EBG structures were simulated and measured using the method of suspended microstrip, which is recommended in [11] and a comparison between the measured results and simulated results has been carried out.

The rest of this paper is organized as follows. In Section 2, the detailed designs of the novel EBG structures are presented. By using the suspending microstrip method, the scatter parameters of the proposed EBG structure based on Sierpinski triangle that is formed by different steps, are characterized and simulated in Section 3. Besides, the bandgap of the conventional EBG structure is also determined for verifying the broader bandwidth of the proposed EBG structure. Next, the measurement of the proposed EBG structures in fabrication and discussions are also utilized in Section 4, while the conclusion is provided in Section 5.

2. Hexagonal EBGs Design

The resonance frequency and the bandwidth of an EBG structure depend on the unit-cell geometry together with substrate's relative dielectric permittivity and thickness. Each unit cell implements a distributed parallel network having one or more resonant frequencies. The resonance frequency of a parallel circuit is defined as follows:

$$f_C = \frac{1}{2\pi\sqrt{LC}}. \tag{1}$$

The bandwidth is also an important consideration and is given by

$$BW = \frac{1}{\eta}\sqrt{\frac{L}{C}}, \tag{2}$$

where η is the free space impedance, which is $120\,\pi$.

The designed unit cell geometry exhibits six symmetry planes, which makes it polarization-angle independent. The proposed EBG structures are shown in Figure 2, which are printed on an FR4 dielectric slab with dielectric constant of 4.4 and thickness of 1.6 mm and loss tangent of 0.02. Dark

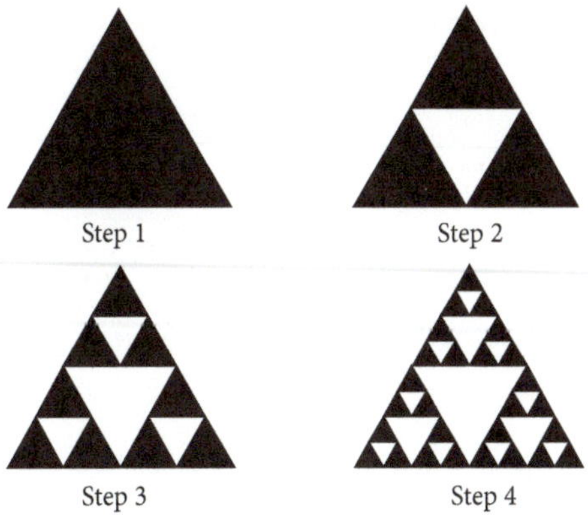

FIGURE 1: Four steps to form a mode-2 Sierpinski gasket structure.

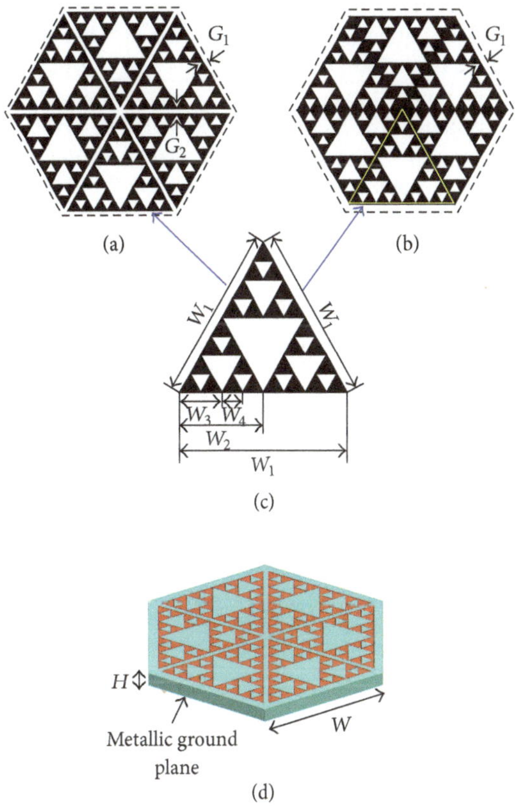

FIGURE 2: Unit cell geometries: (a) BEBG, (b) DEBG, (c) detailed Sierpinski gasket triangles, and (d) slide view of BEBG. Detailed dimensions: $W_4 = W_1/8$; $W_3 = W_1/4$; $W_2 = W_1/2$; $W_1 = W - 2(G_2 + G_1)/\sqrt{3}$, $G_2 = 0.5$ mm, $G_1 = 1$ mm.

parts in this figure represent the metallic periodic structure, which is etched on a dielectric substrate. Unit cell dimension is $W = 10$ mm, and the metallization thickness is 18 μm.

The Sierpinski gasket is a well-known Fractal. The way it can be constructed and its main properties can be found in [12–14]. In [15], it is shown that the Sierpinski gasket is a special case of a wider class of Fractals that can be derived from the well-known Pascal's triangle. This class of Fractals can be derived in the following way. Consider an equiangular triangular grid whose rows shall be labelled by $n = 1, 2, 3, \ldots$. Each row contains n nodes, and to each node a number

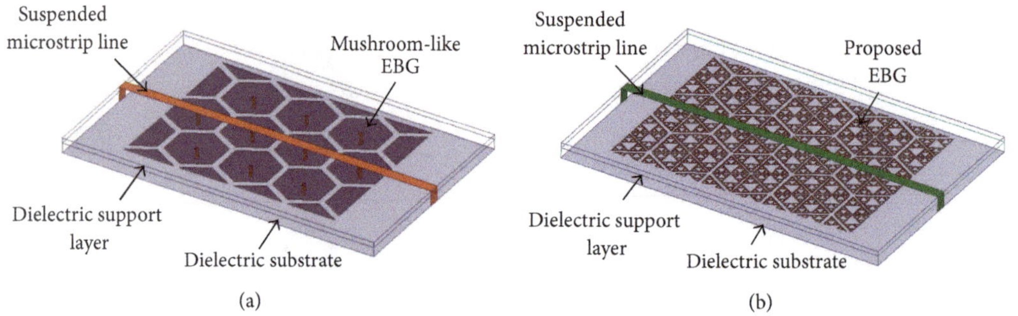

FIGURE 3: Geometry of 3 × 4 EBG arrays: (a) mushroom-like EBG array, and (b) proposed EBG array.

is attached. This number is the coefficient of the binomial expansion of $(x + y)^{n-1}$. Now delete from this grid those nodes that are attached to numbers that are divisible by p, where p is a prime number. The result is a self-similar Fractal that will be referred to as the mod - p Sierpinski gasket [13].

In this paper, The Sierpinski gasket triangle is formed by the following process (see Figure 2). Firstly, create an equilateral triangle patch with length edge of W_1. The second step of this Fractal EBG is constructed by etching out the center equilateral triangle of length edge W_2 inside the patch and subtracting it from the patch. The next stage is achieved by subtracting additional three equilateral triangles of length edge W_3 of the center triangles that follows the same technique adopted to realize the mode-2 Sierpinski gasket geometry (see Figure 2(c)).

From this design, the Sierpinski gasket triangles are arranged to repeat each 60°. The aim of this work allows modifying the resonance frequencies and the EBG operation bandwidth by changing the width and the gap G_1 between the hexagonal lattices. Moreover, by changing the gap G_2 between Sierpinski triangle units, two EBG structures are formed separately. The first one, which is called broadband EBG structure (BEBG), has a broader bandgap with the value of G_2 greater than 0 mm. The second one, which is called dual-band EBG structure (DEBG), has dual bandgap when the value of G_2 is set at 0 mm. These structures are shown in Figures 2(a) and 2(b), respectively.

3. The Bandgap Characteristics

In this section, the bandgap characteristics of two hexagonal EBG structures, which are formed by the Sierpinski triangle at $G_2 = 0$ and $G_2 = 0.5$ mm, are investigated. Besides, the bandgap of the conventional mushroom-like EBG is also determined in comparison with the proposed EBG structures. In order to analyze the bandgap properties of these EBG structures with finite unit number, an experiment concerning transmission through the above structures has been carried out. A 3 × 4 conventional EBG structure and proposed EBG structures have been simulated using the method of suspended microstrip, which is proposed by Fan to measure the bandgap characteristic of the EBGs, as shown in Figure 3 [11]. The operating frequency is set at 7 GHz, and the 50 Ω microstrip line is placed on a dielectric support layer with the thickness of 0.5 mm. The bandgap bandwidth will be defined with S_{11} above −5 dB and S_{21} below −30 dB at the same time.

3.1. EBG Structures at Different Steps. The hexagonal EBG structures with Sierpinski triangles at different steps are investigated in case the value of W is fixed at 10 mm while the value of G_2 is set at 0 mm and 0.5 mm. The simulation results of the EBG structures based on the Sierpinski triangle at step 1, step 2, and step 3 are shown in Figures 4, 5, and 6, respectively. In case of step 1, as can be seen in Figure 4(a), the EBG structure introduces a bandgap, which is ranging from 5.07 GHz to 7.58 GHz, as G_2 is equal to 0.5 mm. However, the bandgap was not found in case of G_2 value is equal to 0 mm (see Figure 4(b)). As can be seen from Figure 5(a), the EBG structure based on the Sierpinski triangle at step 2 has a lower bandgap than the one in the case of step 1. The bandwidth of this bandgap covers from 4.22 to 6.88 GHz. Two bandgaps are defined in Figure 5(b) when the value of G_2 is set at 0 mm. The lower bandgap covers from 2.25 GHz to 2.96 GHz, and the higher one spans from 4.14 GHz to 5.34 GHz. In this case, both bandgaps are defined not clearly, because the ranges of frequencies of S_{11}, which are up to −5 dB, are not flat. As can be obversed from Figure 6(a), the bandgap of EBG structure in step 3 is larger than the ones in the case of step 2 as G_2 is set at 0.5 mm. This bandgap spans the frequencies from 4.32 to 7.92 GHz. Moreover, when G_2 is equal to 0 mm the bandgaps are defined easier than the one in the case of step 2. From Figure 6(b), the curve of S11 is quite flat in two bandgaps. The first bandgap spans the frequencies from 2.15 to 3.02 GHz; the second one has the bandwidth that is covered from 3.81 GHz to 5.20 GHz. Next, the bandgap properties of the EBG structures at step 4 are considered by investigating the effect of parameters such as the size of unit cell W, and the gap between two adjacent unit cells G_1 to the bandgaps in both cases of DEBG and BEBG.

3.2. Broadband EBG (BEBG). In general, in order to obtain a wider EBG operation bandwidth, it is necessary to increase L and reduce C. Equivalent inductance L can be increased using a thicker dielectric substrate and also including in the geometry narrow and long strips (lines). Equivalent capacitance C can be reduced by reducing substrate's relative dielectric permittivity and increasing the gap between the metallization edge and the unit cell edge (and so the gap

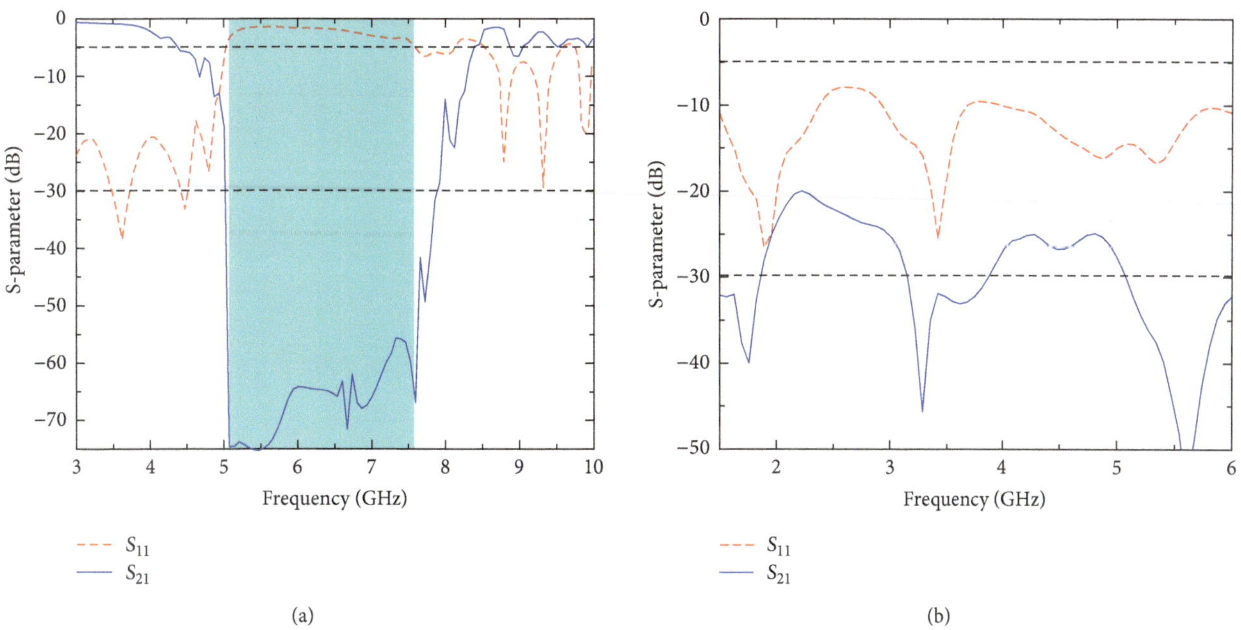

FIGURE 4: Operation bandwidth of hexagonal EBG structure based on Sierpinski triangle at step 1: (a) $G_2 = 0.5$ mm and (b) $G_2 = 0$ mm.

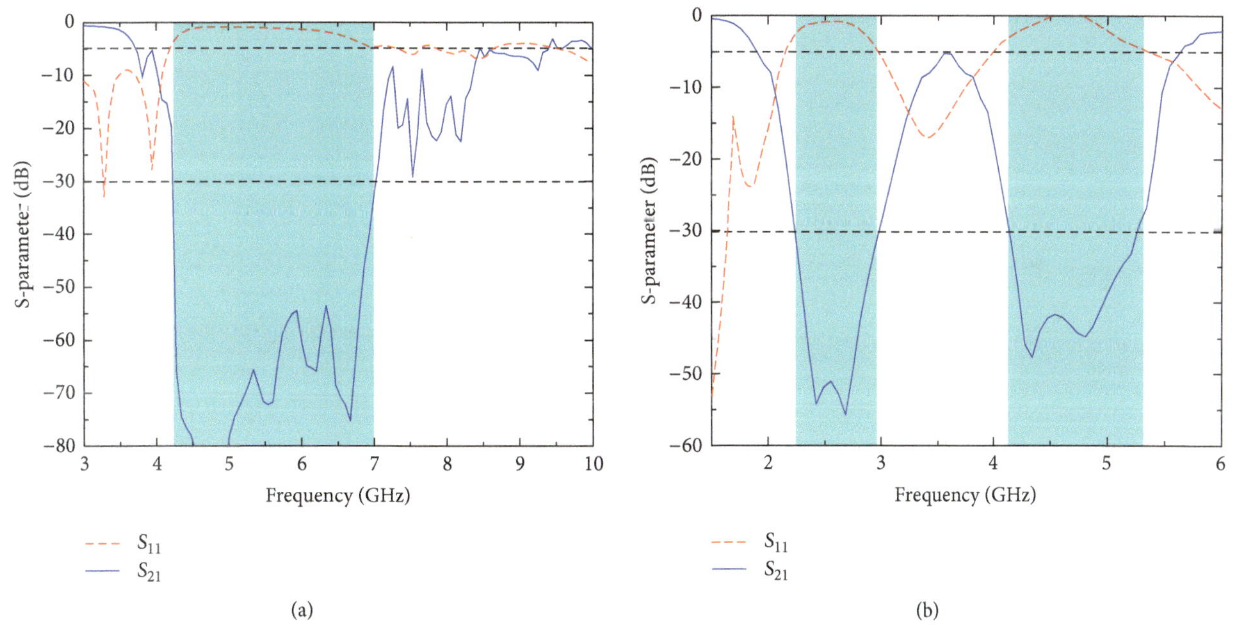

FIGURE 5: Operation bandwidth of hexagonal EBG structure based on Sierpinski triangle at step 2: (a) $G_2 = 0.5$ mm and (b) $G_2 = 0$ mm.

between adjacent unit cells). Moreover, the thickness of substrate has to be adopted in order to obtain both compact size and broad EBG operation bandwidth.

In the EBG design procedure, if the dielectric material and its thickness have been chosen in this analysis, the total equivalent inductance L cannot be altered. Therefore, only capacitance C can be changed. Regarding the total equivalent C of the unit cell, the contribution due to the gap between adjacent EBG cells (coplanar C) varies as the distance between unit cells reduces (which increases C) and also the distance between the Sierpinski triangles in the EBG unit cells (which decreases coplanar C). More important, in addition, the contribution due to the capacitance between unit-cell metallization and the unit-cell ground plane (parallel plates) is reduced for smaller unit-cell size (W).

Finally, the wider EBG operation bandwidth for smaller unit-cell size (W) of this novel design is due to more significantly decreased C values of the parallel equivalent circuit. Moreover, the EBG operation bandwidth is also increased by reducing the relative dielectric permittivity for a given unit-cell size.

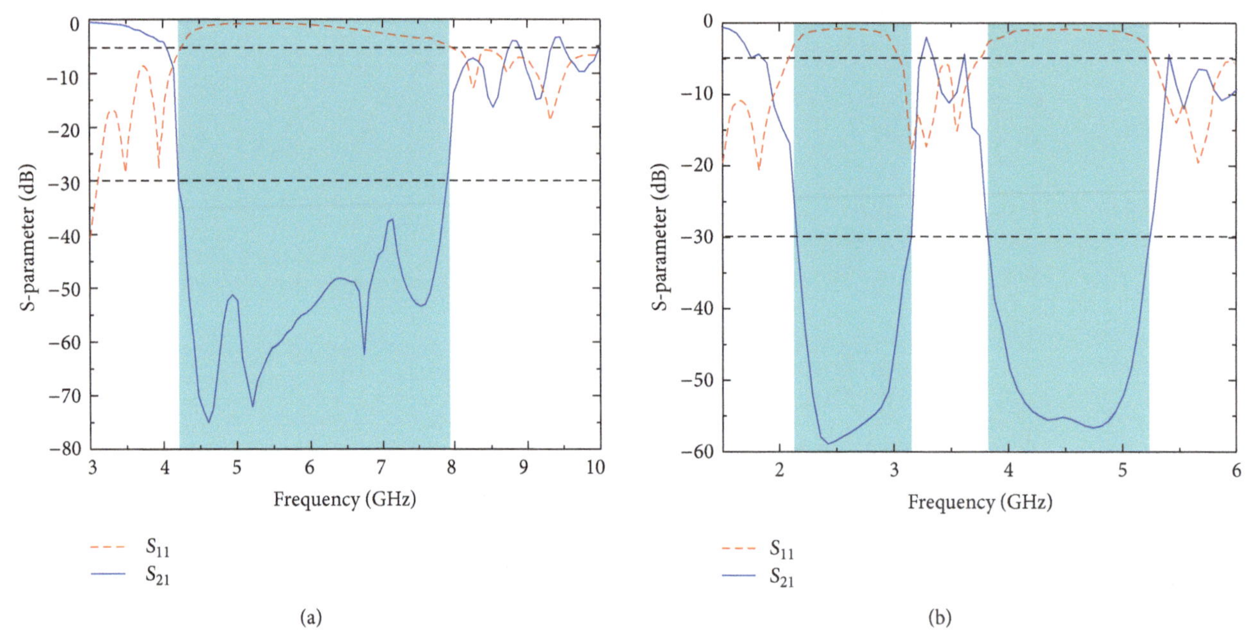

FIGURE 6: Operation bandwidth of Hexagonal EBG structure based on Sierpinski triangle at step 3: (a) $G_2 = 0.5$ mm and (b) $G_2 = 0$ mm.

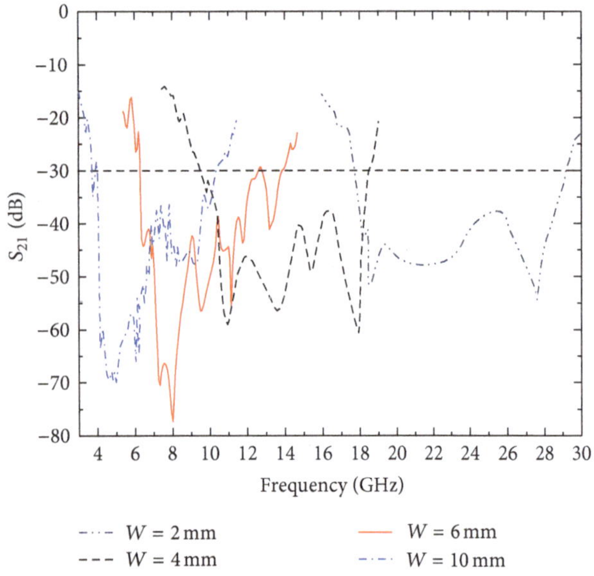

FIGURE 7: EBG operation bandwidth variations versus EBG unit-cell size (W).

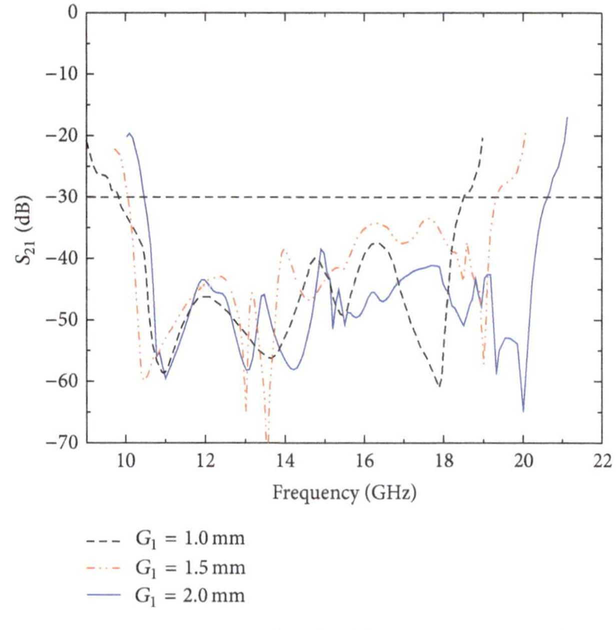

FIGURE 8: BEBG operation bandwidth variations versus the gap between adjacent EBG cells (G_1) while the unit-cell (W) is fixed at 4 mm.

In order to analyses the effect of the parameters of dimension on the total equivalent capacitance C, the unit-cell (W) and the gap between adjacent EBG units (G_1) are investigated while other parameters are constant. The unit-cell size (W) is very important parameter to take effect on the bandgap, which affects the equivalent capacitance of the LC resonant circuit of the EBG structure. Firstly, we investigate the bandgap of the EBG structure with different values of W. The unit cell size is increased from 2 mm to 10 mm while other parameters of the EBG structure are unchanged. The bandgap moves to lower frequency region with patch size increasing, and the bandgap becomes narrow. From Figure 7, we can see that the equivalent capacitance diminished when the patch size increases, so the resonant frequency is decreased. Since the bandgap width is proportional to L/C, the bandgap becomes narrow. The detail of this is shown in Table 1.

Moreover, the effect of distance between adjacent EBG cells (G_1) on the width of the bandgap is also investigated, and the simulated results are shown in Figure 8. As can

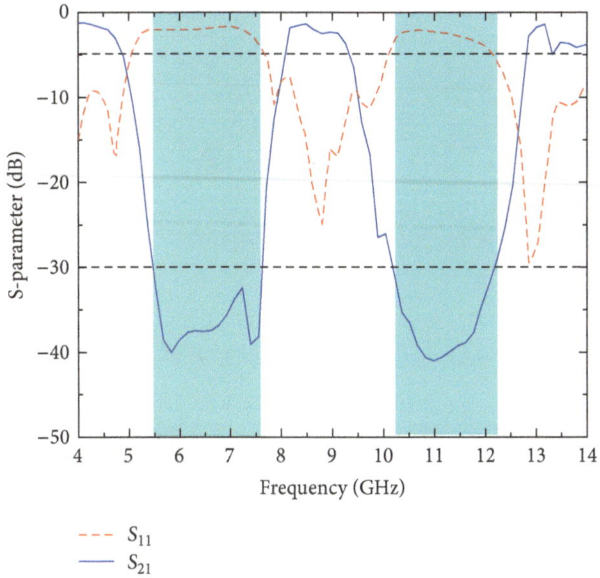

FIGURE 9: DEBG operation bandwidth with G_2 is equal to 0 mm and W is equal to 4 mm while the other parameters are fixed.

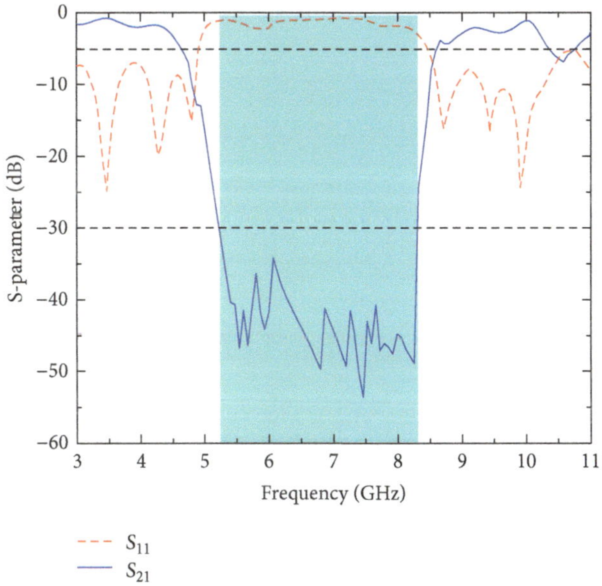

FIGURE 10: Simulated results of scatter parameters of conventional mushroom-like EBG structure (W = 10 mm).

be observed from Figure 10, the EBG operation bandwidth becomes wider by increasing the gap G_1 between adjacent EBG cells and the bandgap moves to the higher frequency area. It is shown that the coplanar C is increased with the higher values of G_1.

3.3. Dual-Band EBG (DEBG). In this section, the gap G_2 between two adjacent Sierpinski triangles inside the EBG unit cell is also investigated while other parameters are constant. From Figure 9, it can be observes that two bandgaps are determined by transmission coefficient. The unique characteristic

TABLE 1: Parametric analysis of EBG unit cell.

Reference	Unit cell size (mm)	Thickness (mm)	ε_r	Bandwidth (%)	Resonance frequency (GHz)
This paper	2 (λ/6.4)	1.6 (λ/8)	4.4	49	23.50
	4 (λ/5.2)	1.6 (λ/13)	4.4	65	14.05
	6 (λ/4.9)	1.6 (λ/5)	4.4	76	10.06
	10 (λ/4.0)	1.6 (λ/4)	4.4	87	7.20

TABLE 2: Relative operation bandwidth of EBG structure for various W while G_2 is Fixed at 0 mm and 0.5 mm.

Unit cell size W (mm)	G_2 = 0.5 mm Bandgap (GHz)	G_2 = 0 mm Bandgap 1 (GHz)	G_2 = 0 mm Bandgap 2 (GHz)
2	17.74–29.25	10.77–14.47	18.80–21.95
4	9.48–18.62	5.51–7.73	10.20–12.33
6	6.23–13.88	3.72–5.07	6.72–8.53
10	4.05–10.35	2.06–3.08	3.71–5.26

TABLE 3: Operation bandwidth of proposed EBG structures with W = 10 mm.

Methods	BEBG (G_2 = 0.5 mm) Bandgap (GHz) *(bandwidth (%))*	DEBG (G_2 = 0 mm) Bandgap 1 (GHz) *(bandwidth (%))*	DEBG (G_2 = 0 mm) Bandgap 2 (GHz) *(bandwidth (%))*
Simulated	4.05–10.35 *(87)*	2.06–3.08 *(40)*	3.71–5.26 *(35)*
Measured	4.15–10.00 *(83)*	2.17–2.97 *(32)*	3.58–5.32 *(39)*

of this design that is the structure can be transformed the bandwidth of broadband into dual bandgap when G_2 is equal to zero. When the unit cell size W is set at 4 mm, the first bandgap spans from 5.51 GHz to 7.73 GHz. The bandwidth of the second one is spreading from 10.20 GHz to 12.18 GHz. This can be indicated more clearly in Table 2.

It can be observed from Table 2 that these bandgaps move forward to the lower frequency regions. It is clear that when the value of G_2 is up to zero, two above bandgaps are moving to overlap each other and the broader bandgap can be obtained.

3.4. Conventional Mushroom-Like EBG. In order to verify the bandgap characteristic of the broadband EBG structure (BEBG), an array of 3 × 4 unit cells of hexagonal mushroom-like EBG is investigated (see Figure 3(a)). This EBG is also constructed on a dielectric substrate with a relative permittivity of 4.4 and thickness of 1.6 mm. The unit cell dimension is fixed at 10 mm. The simulated result of transmission coefficient is shown in Figure 10. The bandgap is center at 6.77 GHz, and the bandgap bandwidth is from 5.22 GHz to 8.32 GHz, achieve the bandgap bandwidth about 46%. As can be observed from Table 1, the bandgap bandwidth of

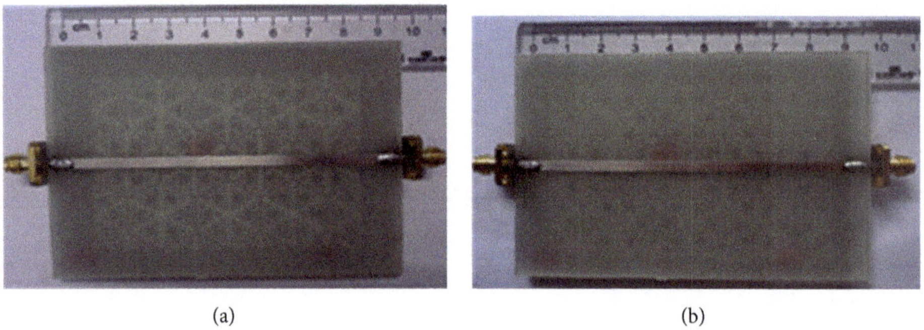

FIGURE 11: Photos of the proposed structures with suspended microstrip: (a) array of 3 × 4 DEBG cells and (b) array of 3 × 4 BEBG cells.

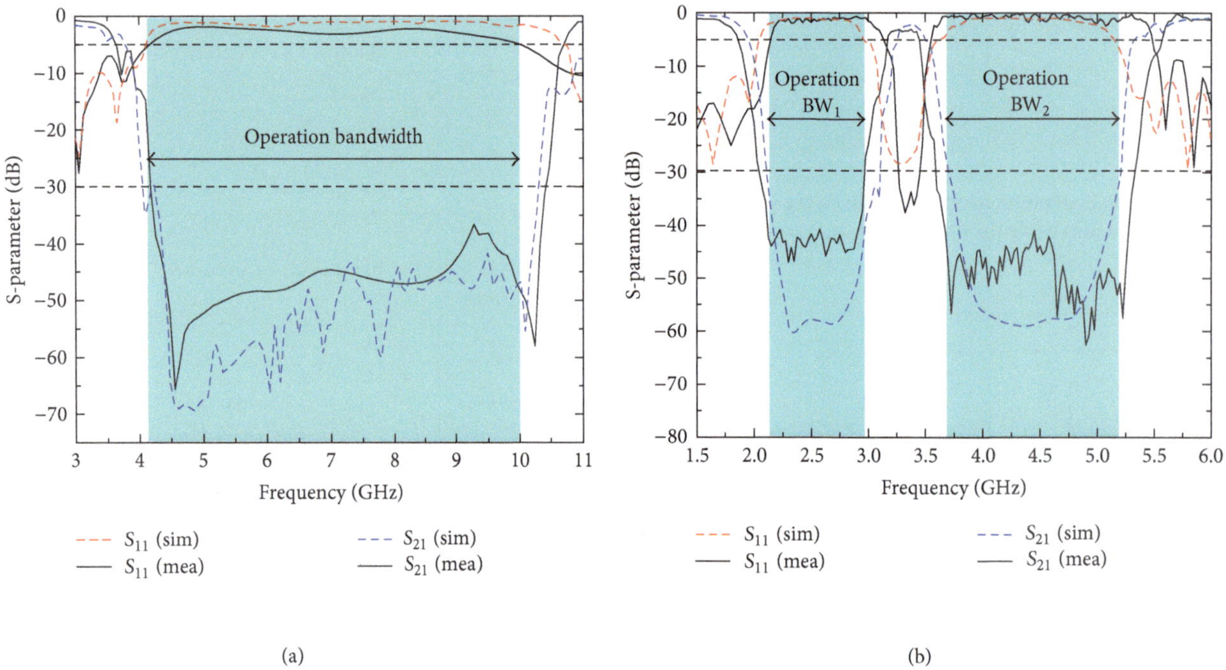

FIGURE 12: EBG operation bandwidth (W = 10 mm): (a) BEBG and (b) DEBG.

the BEBG structure at W = 10 mm is larger than that of the conventional EBG at the same dimension of the unit cell.

4. Results and Discussions

Both schematics of proposed EBG structures were simulated and taken into comparison with the measurement of the fabrication. The microstrip lines are soldered with SMA connector to measure the scatter parameters. The measured results of scatter parameters of the arrays are performed by Anritsu 37369D network analyzer. Photos of the proposed EBG structures are shown in Figure 11. The simulated and the measured results of the parameters of BEBG structure and DEBG structure are shown in Figures 12(a) and 12(b), respectively. It is very efficient to determine the bandgap by the transmission curves. From the measurement of S_{11} with the criteria of −5 dB, the operation bandwidth can be determined easily when the transmission coefficient S_{21} is below −30 dB. As can be observed from Figure 12(a), the BEBG introduces a bandgap between the frequencies 4.15 GHz–10 GHz with the criteria of −30 dB, from that the electromagnetic wave cannot propagate. The bandgap bandwidth is about 83%. The reason is that the surface impedance of the EBG structure becomes very high within the bandgap. From Figure 12(b), two distinct bandgaps of DEBG structure are defined in which the first bandgap is ranging from 2.17 to 2.97 GHz and the bandwidth of the second bandgap is spreading from 3.58 to 5.32 GHz. The bandgap bandwidths of the DEBG structure are 32% and 39%, respectively. It can be observed from Table 3 that the measured results show good agreement with the simulated results.

5. Conclusion

A novel EBG design is presented using Sierpinski gasket triangles, which are arranged to repeat 60° to form the hexagonal EBG unit cell. Two EBG structures, which have broadband and dual bandgap, are proposed by setting the value of the gap between two adjacent Sierpinski inside the unit cell is equal to 0 mm and greater than 0 mm, respectively.

The suspending microstrip method is used to simulate the scatter coefficients of the EBG structure. The bandwidth of BEBG structure is much larger than that of the conventional EBG. The results show a good agreement between simulations and measurements. Due to several advantages of the structures such as using inexpensive dielectric substrate FR4 and planar structure, the EBG structures are promising for low profile and low cost antennas in broadband or multiband applications.

References

[1] F. R. Yang, K. P. Ma, M. Yongxi Qian, and T. Itoh, "A uni-planar compact photonic-bandgap (UC-PBG) structure and its applications for microwave circuits," *IEEE Transactions on Microwave Theory and Techniques*, vol. 47, no. 8, pp. 1509–1514, 1999.

[2] J. D. Joannopoulos, R. D. Meade, and J. N. Winn, *Photonic Crystals*, Princeton University Press, Princeton, NJ, USA, 1995.

[3] E. Yablonovitch, "Photonic band-gap structure," *Journal of the Optical Society of America B*, vol. 10, no. 2, pp. 283–295, 1993.

[4] D. Sievenpiper, L. Zhang, R. F. Jimenez Broas, N. G. Alexöpolous, and E. Yablonovitch, "High-impedance electromagnetic surfaces with a forbidden frequency band," *IEEE Transactions on Microwave Theory and Techniques*, vol. 47, no. 11, pp. 2059–2074, 1999.

[5] A. Stark, S. Prorok, and A. F. Jacob, "Broadband EBG structures with compact unit cell," in *Proceedings of the 38th European Microwave Conference, EuMC 2008*, pp. 698–701, Amsterdam, The Netherlands, October 2008.

[6] L. Liang, C. H. Liang, X. W. Zhao, and Z. J. Su, "A novel broadband EBG using multi-period mushroom-like structure," in *Proceedings of the International Conference on Microwave and Millimeter Wave Technology (ICMMT '08)*, pp. 1609–1612, Nanjing, China, April 2008.

[7] R. Coccioli, F. R. Yang, K. P. Ma, and T. Itoh, "Aperture-coupled patch antenna on UC-PBG substrate," *IEEE Transactions on Microwave Theory and Techniques*, vol. 47, no. 11, pp. 2123–2130, 1999.

[8] M. Rahman and M. A. Stuchly, "Circularly polarised patch antenna with periodic structure," *IEE Proceedings: Microwaves, Antennas and Propagation*, vol. 149, no. 3, pp. 141–146, 2002.

[9] M. F. Abedin, M. Z. Azad, and M. Ali, "Wideband smaller unit-cell planar EBG structure and their application," *IEEE Transactions on Antennas and Propagation*, vol. 56, no. 3, pp. 903–908, 2008.

[10] W. J. Krzysztofik, "Modified Sierpinski fractal monopole for ISM-bands handset applications," *IEEE Transactions on Antennas and Propagation*, vol. 57, no. 3, pp. 606–615, 2009.

[11] M. Y. Fan, R. Hu, Q. Hao, X. X. Zhang, and Z. H. Feng, "New method for 2D-EBG structures research," *Hongwai Yu Haomibo Xuebao/Journal of Infrared and Millimeter Waves*, vol. 22, no. 2, pp. 127–131, 2003.

[12] C. Puente, J. Romeu, and A. Cardama, "Fractal-shaped antennas," in *Frontiers in Electromagnetics*, D. H. Werner and R. Mittra, Eds., pp. 48–93, IEEE Press, Piscataway, NJ, USA, 2000.

[13] C. Puente-Baliarda, J. Romeu, R. Pous, and A. Cardama, "On the behavior of the sierpinski multiband fractal antenna," *IEEE Transactions on Antennas and Propagation*, vol. 46, no. 4, pp. 517–524, 1998.

[14] H. O. Peitgen, H. Jurgens, and D. Saupe, *Chaos and Fractals, New Frontiers in Science*, Springer, New York, 1992.

[15] N. S. Holter, A. Lakhtakia, V. K. Varadan, V. V. Varadan, and R. Messier, "On a new class of planar fractals: the Pascal-Sierpinski gaskets," *Journal of Physics A*, vol. 19, no. 9, article 047, pp. 1753–1759, 1986.

Permissions

The contributors of this book come from diverse backgrounds, making this book a truly international effort. This book will bring forth new frontiers with its revolutionizing research information and detailed analysis of the nascent developments around the world.

We would like to thank all the contributing authors for lending their expertise to make the book truly unique. They have played a crucial role in the development of this book. Without their invaluable contributions this book wouldn't have been possible. They have made vital efforts to compile up to date information on the varied aspects of this subject to make this book a valuable addition to the collection of many professionals and students.

This book was conceptualized with the vision of imparting up-to-date information and advanced data in this field. To ensure the same, a matchless editorial board was set up. Every individual on the board went through rigorous rounds of assessment to prove their worth. After which they invested a large part of their time researching and compiling the most relevant data for our readers. Conferences and sessions were held from time to time between the editorial board and the contributing authors to present the data in the most comprehensible form. The editorial team has worked tirelessly to provide valuable and valid information to help people across the globe.

Every chapter published in this book has been scrutinized by our experts. Their significance has been extensively debated. The topics covered herein carry significant findings which will fuel the growth of the discipline. They may even be implemented as practical applications or may be referred to as a beginning point for another development. Chapters in this book were first published by Hindawi Publishing Corporation; hereby published with permission under the Creative Commons Attribution License or equivalent.

The editorial board has been involved in producing this book since its inception. They have spent rigorous hours researching and exploring the diverse topics which have resulted in the successful publishing of this book. They have passed on their knowledge of decades through this book. To expedite this challenging task, the publisher supported the team at every step. A small team of assistant editors was also appointed to further simplify the editing procedure and attain best results for the readers.

Our editorial team has been hand-picked from every corner of the world. Their multi-ethnicity adds dynamic inputs to the discussions which result in innovative outcomes. These outcomes are then further discussed with the researchers and contributors who give their valuable feedback and opinion regarding the same. The feedback is then collaborated with the researches and they are edited in a comprehensive manner to aid the understanding of the subject.

Apart from the editorial board, the designing team has also invested a significant amount of their time in understanding the subject and creating the most relevant covers. They scrutinized every image to scout for the most suitable representation of the subject and create an appropriate cover for the book.

The publishing team has been involved in this book since its early stages. They were actively engaged in every process, be it collecting the data, connecting with the contributors or procuring relevant information. The team has been an ardent support to the editorial, designing and production team. Their endless efforts to recruit the best for this project, has resulted in the accomplishment of this book. They are a veteran in the field of academics and their pool of knowledge is as vast as their experience in printing. Their expertise and guidance has proved useful at every step. Their uncompromising quality standards have made this book an exceptional effort. Their encouragement from time to time has been an inspiration for everyone.

The publisher and the editorial board hope that this book will prove to be a valuable piece of knowledge for researchers, students, practitioners and scholars across the globe.

List of Contributors

Moises Ferber, Sanaa Zangui, Christian Vollaire and Laurent Krahenbuhl
Laboratoire Ampere, CNRS UMR5005, Ecole Centrale de Lyon, Universite de Lyon, 69130 Ecully, France

Carlos Sartori
Laboratorio de Eletromagnetismo Aplicado, Universidade de Sao Paulo, 05508900 Sao Paulo, SP, Brazil

Ronan Perrussel
LAPLACE, CNRS UMR5213, INP-ENSEEIHT, Universite de Toulouse, 31071 Toulouse, France

Aixin Chen, Weiwei Jiang and Jiaheng Wang
School of Electronic and Information Engineering, Beihang University, Beijing 100191, China

Zhizhang Chen
Department of Electrical and Computer Engineering, Dalhousie University, Halifax, NS, Canada

Kentaro Nishimori
Faculty of Engineering, Niigata University, Niigata-shi, Niigata 950-2181, Japan

Koshiro Kitao and Tetsuro Imai
NTT DOCOMO, INC., 3-5 Hikari-no-oka, Yokosuka-shi, Kanagawa 239-8536, Japan

Hoi-Shun Lui
Department of Signals and Systems, Chalmers University of Technology, 412 96 Gothenburg, Sweden
School of Information Technology and Electrical Engineering, The University of Queensland, Brisbane, QLD 4072, Australia

Hon Tat Hui
Department of Electrical and Computer Engineering, National University of Singapore, Singapore

Shang-Te Yang and Hao Ling
Department of Electrical and Computer Engineering, The University of Texas at Austin, Austin, TX 78712, USA

Gonzalo Exposito-Dominguez, Jose Manuel Fernandez-Gonzalez and Manuel Sierra-Castaner
Radiation Group, Department of Signals, Systems and Radio Communications, University of Madrid, 28040 Madrid, Spain

Pablo Padilla
Department of Signal Theory, Telematics and Communications, University of Granda, Granda 18071, Spain

Yawgeng A. Chau and Karl Y.-T. Huang
Department of Communications Engineering, Yuan Ze University, 135 Yuan-Tung Road, Chung-Li 32003, Taiwan

Gui Gao, Gongtao Shi, Huanxin Zou and Shilin Zhou
School of Electronics Science and Engineering, National University of Defense Technology, Changsha, Hunan 410073, China

Joong-Han Yoon
Department of Electronic Engineering, Silla University, San 1-1 Gwaebeop-Dong, Sasang-Gu, Busan 617-736, Republic of Korea

Young-Chul Rhee
Department of Information and Communication Engineering, Kyungnam University, 449Wolyong-Dong, Masan, Kyungnam 631-701, Republic of Korea

Woo-Su Kim
School of Electrical and Computer Engineering, Georgia Institute of Technology, North Avenue, Atlanta, GA 30332, USA

Yuhuang Ye, Jiade Yuan and Kaixiong Su
College of Physics and Information Engineering, Fuzhou University, Fuzhou, Fujian 350108, China

Abdelhamid Tayebi, Josefa Gomez, Ivan Gonzalez and Felipe Catedra
Electromagnetic Computing Group, Computer Science Department, University of Alcala, Escuela Politecnica, Alcala de Henares, 28871 Madrid, Spain

Carolina Tienda and Jose A. Encinar
Department of Electromagnetism and Circuit Theory, Technical University of Madrid, 30 Complutense Avenue, 28040 Madrid, Spain

Manuel Arrebola
Department of Electrical Engineering, Group of Signal Theory and Communications, University of Oviedo, Academic Campus, 33203 Gijon (Asturias), Spain

Jong-Eon Park, Ki Young Kim and Jae-Won Song
School of Electrical Engineering and Computer Science, Kyungpook National University, Daegu 702-701, Republic of Korea

Thomas Kaufmann
School of Electrical and Electronic Engineering, The University of Adelaide, Adelaide, SA 5005, Australia
The Auto-ID Laboratory, The University of Adelaide, Adelaide, SA 5005, Australia

Christophe Fumeaux
School of Electrical and Electronic Engineering, The University of Adelaide, Adelaide, SA 5005, Australia

Damith C. Ranasinghe and Ming Zhou
The Auto-ID Laboratory, The University of Adelaide, Adelaide, SA 5005, Australia

Hou Zhang, Hai-Yang Xu, Bo Tian and Xian-Feng Zeng
Radar Engineering Department, Missile Institute of Air Force Engineering University, Shaanxi Province Sanyuan 713800, China

Maifuz Ali
Information & Electronics Research Institute, Department of Electrical Engineering, Korea Advanced Institute of Science and Technology (KAIST), Daejon 305-701, Republic of Korea

Seong-Ook Park
Department of Electrical Engineering, Korea Advanced Institute of Science and Technology (KAIST),
Daejon 305-701, Republic of Korea

Carlos Delgado, Josefa Gomez, Abdelhamid Tayebi, Ivan Gonzalez and Felipe Catedra
Computer Science Department, University of Alcala, 28871 Madrid, Spain

Peter Knott, Thomas Bertuch, Helmut Wilden, Olaf Peters, Andreas R. Brenner and Ingo Walterscheid
Fraunhofer FHR, Fraunhofer Street 20, 53343 Wachtberg, Germany

Mei-yan Zheng and Ke-song Chen
School of Electronic Engineering, University of Electronic Science and Technology of China, Chengdu 611731, China

Hong-gang Wu
The Second Research Institute of Civil Aviation of China, Chengdu 610041, China

Xian-pan Liu
School of Astronautics and Aeronautics, University of Electronic Science and Technology of China, Chengdu 611731, China

Ching-Chih Hung, Chia-Mei Peng and I-Fong Chen
Department of Electronic Engineering, Institute of Computer and Communication Engineering, Jinwen University of Science and Technology, No. 99 An-Chung Road, Hsintien City 23154, Taipei, Taiwan

Miyuki Hirose, Hironobu Yamamoto and Takehiko Kobayashi
Wireless Systems Laboratory, Tokyo Denki University, 5 Senju-Asahi-Cho, Adachi-Ku, Tokyo 120-8551, Japan

Gianluca Gennarelli and Giovanni Riccio
D.I.E.I.I., University of Salerno, Via Ponte Don Melillo, Salerno, 84084 Fisciano, Italy

Do-Gu Kang and Youngje Sung
Department of Electronic Engineering, Kyonggi University Yeongtong-gu, Suwon-si, Gyeonggi-do, Republic of Korea

Jing Jing Yang, Ming Huang, Fu Chun Mao, Guang Hui Cai and Yao Zhong Lan
School of Information Science and Engineering, Yunnan University, Kunming 650091, China

Tamer Aboufoul, Akram Alomainy and Clive Parini
School of Electronic Engineering and Computer Science, Queen Mary University of London, Mile End Road, London E1 4NS, UK

Huynh Nguyen Bao Phuong
School of Electronics and Telecommunications, Hanoi University of Science and Technology, Hanoi 10000, Vietnam

Dao Ngoc Chien
Department of High Technology, Ministry of Science and Technology, Hanoi 10000, Vietnam

Tran Minh Tuan
National Institute of Information and Communications Strategy, Ministry of Information and Communications, Hanoi 10000, Vietnam